THE SPATIAL IMPACT OF TECHNOLOGICAL CHANGE

THE SPATIAL IMPACT OF TECHNOLOGICAL CHANGE

EDITED BY JOHN F. BROTCHIE,
PETER HALL & PETER W. NEWTON

CROOM HELM
London • New York • Sydney

Croom Helm Ltd, Provident House, Burrell Row,
Beckenham, Kent, BR3 1AT

Croom Helm Australia, 44-50 Waterloo Road,
North Ryde, 2113, New South Wales

Published in the USA by
Croom Helm
in association with Methuen, Inc.
29 West 35th Street
New York, NY 10001

British Library Cataloguing in Publication Data

The spatial impact of technological change.
Industry 2. Technological innovations
3. Geography, Economic
I. Brotchie, John F II. Hall, Peter, *1932*-
III. Newton, Peter W
338′.06 HC79.T4

ISBN 0-7099-5006-3

Laser set by Dead Set Publishing and Information Services Pty Ltd,
Melbourne, Australia
Printed and bound in Great Britain
by Billing & Sons Limited, Worcester.

CONTENTS

FIGURES

TABLES

ACKNOWLEDGEMENTS

This book is produced as a second report on an international study of technological change and spatial impacts convened by the International Council for Building Research Studies and Documentation (CIB) as Working Commission 72 (CIB W72). The book is based on an international workshop held as part of the study in Melbourne, in 1985. The first report was published as *The Future of Urban Form: The Impact of New Technology* (Brotchie *et al.* (eds), Croom Helm, London and Nichols, New York) in 1985.

*

The editors are most grateful to Dr. Gyula Sebestyén, Secretary General of CIB, for inviting the study; Commonwealth Scientific Industrial Research Organization, Australia; the Ministry of Industry, Technology and Resources, Victoria; and the Ministry of Planning and Environment, Victoria, for financial support; to Professor Walter Isard and the World University for sponsoring it as the initial activity of their Australasian office; to the contributors for their enthusiastic participation in discussion and manuscript preparation; to Rhyll Reed for typing the manuscript; Jill Connor for compiling the bibliography; and Veronika Gazner for drafting the illustrations.

PREFACE

The developed world is presently involved in a new industrial revolution—the information technology revolution—which is the most profound, most rapid and most widespread transformation to date. It is changing our processes of production, and the goods and services produced; facilitating new goods and services; changing the nature of work and its organisation; and substituting information and knowledge for labour, materials, energy and land as the prime factors of production. It is blurring the previous distinctions between manufacturing and service jobs, between employment, self-employment and informal activity, between work, education and leisure activities, and between male and female work roles.

Global networks and global markets operating continuously throughout the day and night are being established. Industry is marshalling at a transnational level to exploit these changes and new products and services based on this technology are emerging. In fact information technology is pervading virtually all forms of human endeavour: work, education and leisure, communication, production, distribution and marketing, and the time scheduling of these. It is changing the scale and content of information networks, the interdependence of organisations, and how as well as where we live, work, shop, learn, communicate and play.

The dynamics of this change are providing information never before available on economic systems and their sensitivity to change—which should provide new understanding of processes of human endeavour and their interactions. Information technology is providing the means of acquiring this information and of analysing and displaying it to an extent never before possible.

The first stage of this information revolution, now well advanced, consists of the automation of predefined data processing and of associated physical tasks. The second phase, just beginning, consists of the automation of knowledge processing and of intellectual tasks. The first phase is already affecting the precision, quantity, even diversity and types of goods and services produced. The second will affect quality in more profound ways and will cause even more comprehensive shifts in work, in leisure activity, and in comparative advantage between nations, between industries and between people.

THE STRUCTURE OF THE BOOK

Though written by over thirty different authors, the chapters in this book all fit within a common structure and share common themes. In Part 1, the focus is on the emerging information economy which is already characteristic of all the advanced industrial nations. In Part 2, the focus is on the specific role of the new information technologies within that economy. Part 3, the longest section, looks at the geographical patterns of production and consumption of the new technologies which are emerging worldwide. Part 4 turns to the role that government policy is playing, and might play, in promoting the development of the new technologies and the transition to the information economy. A concluding section turns to broader speculations and generalisations on the nature of society in the information age.

Part 1: The New Information Economy

The developed world is presently involved in a new industrial revolution. Fundamental transitions are taking place in relation to the types of economic activity being undertaken, by whom, and in what locations. In the context of the formal (market) economy, terms such as post-industrial, service, post-service and several others have been employed (not without some reservation) as descriptors of the new face of urban economic development.

In Chapter 1, Hall indicates that, in a statistical-descriptive sense, the existence of post-industrial economies have long been substantiated—as reflected in significant shifts in employment from goods-producing to service-providing, from manufacturing and goods-handling to information-handling. The economies of advanced urban societies are now firmly centred upon information activity (hence the term *information economy*). For many, the adjective 'information' is more precise than service or post-industrial for describing the fundamental dynamic element that will shape the economy and society of the future. The chapters on high-technology industry all indicate that the new economic system will be based on knowledge-intensive technologies.

The spatial implications of a shift from an industrial to an information economy are canvassed by Hall. Future growth prospects for existing world cities, national centres, provincial cities and smaller towns are variously assigned according to their financial and human (knowledge) capital linkages with the headquarters of multinational companies in advanced technology and advanced services industries, as opposed to the once-important manufacturing-mining operations. Growth is occurring in places quite different from the old manufacturing centres: larger international cities, as well as some more specialised, smaller service centres. Much of it is shifting from the cores of the major world cities to smaller cities nearby. But the old, bigger industrial cities lose out.

Distinctions are not only being sharpened spatially, but also socially. Encel (Chapter 2) observes that growth-oriented, technological capitalism exploits the readiness of many people to accept inequality as the price of rising aggregate living standards. Despite rises in real income per head of population in advanced economies, the distribution of wealth has become more unequal over the past quarter century. An increased level of unemployment in the *formal economy* is testament to a tendency for new technology to displace more labour than it creates. Who, then, are the people living on the fringes of the institutionalised world and what are their long term prospects?

In the remaining chapters of Part 1 emphasis shifts to the two oft-forgotten sectors of a national economy, namely the *informal* (or community) *economy* and the *household economy*. Pym's essay takes up a theme from Encel, that the information society may create nations of people for whom the informal or community economy—based on idealism and on mutuality—may form the only workable antidote. For Pym the informal economy constitutes the self-employed and the more independent folk working from and around where they live. In contrast to the networks of the knowledge workers and their institutions which are facilitated by advances in information and communication technologies (technologies seen as enabling economic *control* on a *global* scale), the range of work undertaken in the community (informal) economy is defined at the scale of local, interpersonal communication—generally without documentation. Typical of informal enterprise are: household maintenance and renovation, local health and care, and small-scale contracting in a host of services requiring good knowledge of local requirements yet limited capital.

Resilience of a national economy is also located among its households. According to Ironmonger almost two-thirds of total intellectual capital (a central ingredient in post-industrial societies) in the US is in the household economy, with one-quarter in business. The percentages change to 44 and 32 respectively with the inclusion of tangible capital; so taken overall, the strength of the household economy as a producer and consumer of goods and services is considerable. The information revolution is facilitating more home-based activities, introducing more services and technology to perform them in the home. It is also increasing leisure time to utilise these services and to facilitate a wider range of life styles. Demographic and social factors are adding to the number of households created. The net result is that the household is increasing in importance as a social and economic unit. Furthermore, the potential for new technologies and products to radically alter the organisation and structure of households (which products are consumed, how time is spent, etc.) is considerable; to illustrate, Ironmonger tracks the impact of new technologies on households over the past century.

Part 2: New Technology and the Information Economy

The essays in Part 2 consider the impact of new technologies on the *formal economy*. Deken argues that the information and communication revolution is not simply an additional technological factor introduced into existing economic and social structures; rather, it constitutes a new *medium* with profound implications for social, economic and spatial organisation. The ultimate personal impact of computing is seen as breaking the tyranny of personal space—the mismatch between the freedom and capacity of the human *mind* and the absolute physical constraints of the *body*. Deken proceeds to argue that the significant role of the computer medium over the next few decades will be, firstly, to encapsulate all of the synergistic elements of *presence* (sense, interaction and effect—only partially accommodated at present by telephone and television) and, secondly, to enable presence to become transportable and negotiable. The result: individuals and organisations will be capable of direct and effective *telepresence* far beyond their physical capabilities.

Harnessing these new technological opportunities facilitates industrial control by electronic devices. The location of the 'control panel' can be independent of the location of the factory. The ability to *separate economic functions* in the productive process becomes a central theme in several of the essays in this volume; for Deken this facilitates the re-establishment of the home as centre of economic activity and social life. The chapter by Moss builds on its predecessor. New communications technologies are seen as *space-extending*, allowing individuals and firms to function within geographically larger sets of boundaries, permitting exchange of information and ideas without interpersonal contact, and permitting what were once spatially separate economic activities to become highly integrated.

Evidence assembled by Moss (and later by Langdale and Gordon and Kimball) suggests that the new communications technologies are strengthening selected principal world cities (that is, those with global headquarter functions across key industrial groupings and possessing CBDs capable of providing the advanced services such as finance, law, marketing, management, consultancy, etc., necessary for the co-ordination and control of widely dispersed enterprises) as well as those urban centres, within the same country or overseas, with which they have close economic relations. The headquarters of independent firms that once thrived in smaller cities are now increasingly subsidiaries of large multinational enterprises, the result being a weakening of the traditional autonomy of the smaller cities. Further, the emerging web of new information infrastructure such as fibre-optic cable systems indicates that the strength of key centres with existing comparative economic (and information infrastructure) advantage will, in general, be reinforced.

Langdale, analysing the takeup of electronic information services, finds that there is a decided hierarchy of information centres dominated by the largest cities—a process reflecting the increased concentration of business corporations and by the control that they exert from the cities in which they are headquartered. Tucker, following this, focuses on the phenomenon of traded services. The global networks discussed above provide the highways for a new international traffic in information services. This traffic is rapidly increasing and is already approaching in value the international trade in goods along the conventional trade routes. The barriers to trade in goods (e.g. via regulation) influence the trade in services and it remains to be seen what constraints might be applied to the development of these new global service industries. Examples are provided, however, which indicate that new information technologies are likely to overcome these regulatory effects, by making services freely available electronically. Another emerging trend seems, according to Tucker, to be one where consumers are becoming less dependent on local production capacity in formal market arrangements (although perhaps more so in relation to the informal or community economy; see Pym).

The changes induced by information technology are pervasive and affect the retail sector in several ways—through the services associated with retailing, and through automation of production, distribution and ordering of the goods involved. The interaction between new technology and labour and the effects on information technology diffusion and labour participation rates are considered in particular by Fisher. He analyses the impact of new information technology on retailing in Australia, a country with strong labour unions and close regulation of trading hours. He argues that new technology will benefit larger chains, operating in suburban centres, at the expense of the smaller street-corner shop. The very tightness of the labour market will speed the introduction of innovatory technology, but this, he thinks, will in turn just increase the pressure for deregulation, as has long been true in the United States and is just about to occur in Britain.

The rate and manner in which information technology diffuses throughout the various sectors of industry will depend on a range of economic, organisational, political and behavioural factors that interact in complex ways. For example, Roessner indicates that the cost equation associated with office automation is not clearly defined. In many service industries (as opposed to manufacturing) automating an existing task, which may or may not confer cost savings, is less important than *enabling new tasks*—expanding the range of services which can be offered. (The introduction of automated teller machines was a move that initially did not prove profitable for banks, but spread rapidly in an environment where competition is based on *product quality*, rather than product or service price.) In Fisher's essay it is also clear that cost savings in retailing are attained not only via introduction of new technology, but also through labour practices (e.g. casual, part-time, female, youth employment). By the end of the 1990s most industrial sectors, according to Roessner, will have experienced the shock of and resistance to the initial exposure to information technology: the next decade will see office work and occupations restructured to take full advantage of continuously improving technology.

Part 3: New Technology and Space

Part 3 provides a cross-national set of studies on high-technology development and related public policy. Hall opens this section with a chapter comparing the progress and the locational patterns of high-technology manufacturing industry in Britain and the United States. Contrary to popular belief in Britain, high tech there—in sharp contrast to the American story—has actually been losing jobs, though in both countries the Silicon Valley myth is at least partly true: high tech is highly concentrated in a few core areas (California, Massachusetts; the crescent around west London, centred on the Western Corridor), though within them it is tending to diffuse somewhat. The key to these patterns, Hall concludes, is conscious government policy which has influenced the distribution of crucial research contracts, overwhelmingly defence based. Paradoxically, research in itself does not appear a significant high-tech generator: it must be the right sort, perhaps at the right time.

Gordon and Kimball's chapter underlines the point that in Silicon Valley, State intervention for military purposes was critical at first; this was followed by demand from large transnational corporations. Such a formula, they argue, is not reproducible in other places. Their conclusion is a bleak one for regions and cities not now on the high-tech terrain. Batty, surveying the British scene in the following chapter, suggests a related aspect: an innovative, information-technology oriented culture may develop in regions of high-tech concentration, thus fortifying their initial advantage and denying other areas the chance to enter the club. In a related fashion, but at the scale of the individual enterprise (which range from single plant-single site firms to multinational corporations), Taylor's chapter employs a core-periphery perspective to explore the development and transfer of new technology, using case studies from the UK and the Pacific region.

Three papers from Western Europe seem to reach a considerable degree of agreement. Funck and Kowalski argue that the need is to identify the spatial innovation potential of regions and cities. Big cities in centrally-located regions, in their finding, are well located and even older industrial areas have some good potential features. Giese stresses that an innovation-oriented regional policy demands positive policy intervention in the form of tax concessions, government contracts and technology-transfer organisations; the problem is that at present in the Federal Republic of Germany (and doubtless elsewhere) research and development policy is not integrated with regional policy. Nijkamp and Mouwen make similar points: the need is to create knowledge centres, or spatial concentrations of scientific and technological communication, information and research centres, serving both public and private sectors. In general, big cities are still best placed to perform this role, though their innovation potential may be diminishing; but in a small country like the Netherlands, regional boundaries provide no barrier to the transfer of information.

For Western observers, a particular interest thus attaches to Japan's policy of creating a network of such centres. Toda shows that the nineteen 'technopolies', established by the Ministry of International Trade and Industry, combine industry, academic research and residential areas, and are deliberately located so as to provide a counterweight to the big cities. They aim not merely to introduce new high-tech strains, but also to foster the development of advanced technology in the area's existing industries. But, of course, it is too early to judge their success.

The final group of chapters in this section deals with the Australian experience. The macro-level analysis undertaken by Newton and O'Connor utilises an equivalent methodology to that of Hall and his associates. The results indicate that Australia is performing poorly—less than half of 1 per cent of its workforce could be classified as high-tech employees working in high-technology industry. For every high-tech employee in high-technology industry within Australia, there are twenty in the UK and 170 in the USA (significantly different from the employed persons' ratios of 1:5:20). It would seem that Australian high-tech activity is primarily downstream or user activity, rather than producer activity; that is, high technology is being used in production—it is part of the process, not the product. Location of high-tech industry in Australia is characterised by concentration—in areas where the necessary networks of information and access are found. At the national level, concentration is centred on the two most populous eastern States; at State level almost 90 per cent of activity is centred on the two capital cities, Sydney and Melbourne; and at metropolitan level, most high-tech industries are located in the high amenity middle-class suburbs of these principal cities. Searle and Hutchinson, stressing the same points, show that Sydney's leading position in high-technology industries is in general—biotechnology excepted—quite highly concentrated in a series of industrial clusters. Overall, the picture to emerge in relation to high-technology industry is that there is no ubiquity, no model high-tech development capable of being mirrored worldwide, nationwide or citywide. Instead there are dominant nations at global level, core urban centres at sub-national level and favoured sectors of cities at intra-metropolitan level.

Three routes to explanation of the developmental pattern of high-technology industry are identified in the chapter by Gordon and Kimball:

1. traditional location theory and its emphasis upon *inter-firm material linkages* within a framework of spatial constraint (the distribution of factors and markets);
2. reformulated location theory and its focus upon a specification of the production requirements of high-tech industry within a geographical framework (i.e. the *areal attributes* of high-technology industry). The research reported by Hall and Newton and O'Connor adopt this theoretical path to explanation; and
3. world economic systems-structuralist perspectives and their emphasis on organisational changes situated within the system of production. A focus upon the *historical changes* in the *global organisational capacity of capitalist enterprise* is central to their study of the post-war micro-electronics industry in California.

In fact, a synthesis of all three perspectives is required for a satisfactory understanding of high-technology development.

Part 4: Government Intervention and High Tech Industry Policy

Currently, in most advanced economies, governments are actively pursuing policies and programs designed to encourage technological innovation. The arguments for government intervention, as outlined by Ronayne, are rooted in the observations of Schumpeter who placed innovation at the centre of his theory of economic development; as well as the arguments of the market-failure theorists who contend that, if left to itself, the free market will tend to under-invest in research. The pros and cons of this thesis are examined by Ronayne as well as Macdonald; viz, whether government intervention stifles rather than promotes development; whether many high-tech policies seem to encourage firms simply to become itinerant (e.g. factors of production allow footlooseness; government incentives positively reinforce this capability). The consensus seems to be that since governments will continue to intervene, how best can this be undertaken, in the present case, in relation to high-tech industry? Several chapters serve to evaluate the attempts of governments in a number of different countries to address this issue.

A common conclusion that seems to emerge is one where conventional industrial policy and theory appears not to be appropriate for high-technology industry. Macdonald indicates that high-technology industry's main requirement is *information*—land, labour and capital issues are secondary. (Other authors also point to the knowledge intensive nature of this industry, in relation not only to technical aspects, but also to commercial and managerial facets.) High-tech industry can only thrive where vast quantities of appropriate information are readily available; and even here it is the information network which is important, and these are difficult to achieve.

For many existing policies, need for access to information has been translated by the bureaucracy into the need for physical proximity to a knowledge base—hence the predilection for technology parks. The deficiencies inherent in such policies are identified in several earlier chapters. Nijkamp, for example, reveals that in the Netherlands the pattern of innovative activity did not correlate well with the distribution of knowledge centres. This suggested that the *entrepreneurs' information net extends beyond regional boundaries* (Funck and Kowalski point to the role of new telecommunication and information technologies in overcoming friction of distance, while Taylor highlights the gatekeeper role of multinational corporations in technology transfer and technology acquisition) and/or that the knowledge centres (especially those in the central, core regions) proved unsuited to the particular high-tech industrial activities being undertaken locally.

Although seemingly self-evident, it is clear that an understanding of what constitutes high-technology industry is necessary *before* public policy at any level is likely to be effective. The work reported by Elzinga represents an attempt by government to identify the spectrum of emerging technologies and select those capable of being winners within that particular country's economic framework. Related issues of a policy-sensitive nature, frequently overlooked in targeting public programs, concern *types of high-tech factory*, of which Toda earlier identifies three (R & D—prototype; production to order/on demand—small lots; mass production); and *types of R & D activity*, of which Toda identifies two (frontier type—original R & D; transfer and adapting type). The industrial permutations and combinations possible among these two sets of factors, leaving aside the sectoral aspects (e.g. whether to set higher priorities for agriculture-biotechnology versus manufacturing-robotics, etc.) confront governments with the need to make decisions as to how much and what type of support to give to assist new (sunrise) industry to achieve mastery of a particular high-tech innovation as opposed to assistance to existing (traditional) industry in applying high-tech products and/or processes in an endeavour to restructure and increase productivity. The locational implications of the two alternative policy supports are potentially quite different. For example, if high-tech activity in a nation or region is primarily downstream or user activity, then the location of high-tech activity is determined by the locations of those industries using it, and no policy is relevant, except to encourage more of it where it adds to the productivity and competitive advantage of the industries concerned.

Much attention has been given to the supply side in relation to high-tech industry. The demand side of the equation is also significant. As Ronayne indicates, demand considerations are major determinants of inventive effort—with the proviso that some inventions will not be possible until the *knowledge* upon which they are based is uncovered. Hence the importance of R & D, the major source of knowledge, and information networks and information-technology media appropriate for accessing/disseminating this knowledge to potential users. Herein lies the challenge/problem for contemporary urban planners concerned with issues of equity in particular, namely the uneven concentrations of populations with capability of social and economic innovation (the information and knowledge workers). Newton and O'Connor found that the suburban areas to emerge as preferred high-tech environments mostly lay across the large middle-class belt of suburbia, generally pulling up short of the working-class industrial areas. In this selectivity, high-technology industry is further strengthening a pattern of well-off and less well-off areas within the city. Batty's chapter provides a further insight here: the uneven penetration of microcomputers within Britain clearly highlights those regions that could be classed as information rich; and it is *these information-rich environments which are proving to be the twentieth century hearths for high-technology industry.*

Wegener turns to the role of urban planning in an information economy and develops the model of a participatory democracy based on instant feedback through universally available information, a model, he stresses, that depends on somewhat optimistic assumptions about diffused technology and broad-based democracy. Its potential for good and for ill and the reasons for its limited success to date as a public-sector planning tool are discussed. Relatedly, Harris argues that we do not know what will happen when information technology is widely available; it could create a highly computer-literate society, but it can never in itself achieve optimal choices, as was once hoped.

Part 5: The Dimensions of Change

This final section deliberately moves out into the realm of speculation. Opening it, Meier gives his vision of a totally new social structure: robots perform many of the mechanical functions once provided by blue-collar workers, leaving most of the human population divided into family-servers and community-servers, but with higher strata serving increasingly wide client publics up to the level of the global and even the universal. Gappert gives an idealised view of post-industrial America, developing via three alternative scenarios: one based on competition, a second on quality of life and a third dependent on a non-economic combination of the two.

The final chapter by Brotchie, Hall and Newton suggests that the transition to an information society involves increasing interdependence between production, information technology and transport at the regional and global levels, increasing emphasis on knowledge and intelligence as factors in this system, increasing informality of human organisation within it, increasing integration of global and regional network nodes and their markets, increasing concentration of corporate decisions and their support services in these nodes, increasing dispersal of production, and a continuing transition from an energy-based society to an information-based society.

Not all readers are likely to agree with these visions. But they may serve as challenges to think about alternatives.

John Brotchie
Peter Hall
Peter Newton

PART 1.
THE NEW INFORMATION ECONOMY

Chapter 1

THE GEOGRAPHY OF THE POST-INDUSTRIAL ECONOMY

P. Hall

Soon after the term was coined by Daniel Bell (1973), the post-industrial society turned from a sociological insight into a media cliché. But that does not necessarily make it invalid. It was also savaged by a new generation of sociological and economic critics, who argued that Bell's concept of the future was superficially optimistic, culture-bound and ideology-bound, and hopelessly uncritical. Now that the dust has settled, and after a decade and a half, it seems useful, in the context of the present volume to re-evaluate the thesis.

In a strict statistical-descriptive sense, Bell's hypothesis has long since been true. As A. G. B. Fisher and Colin Clark forecast in 1935 and 1940 respectively (Fisher, 1935; Clark, 1940), only a minority of workers in advanced countries are now engaged in making goods; the majority are in service industries. Table 1.1 shows the extent of this shift over recent decades for four of the leading advanced industrial countries. This process is most advanced in the United States where workers in goods production made up only 35 per cent of the labourforce in the early 1960s and 28 per cent in the early 1980s. In the United Kingdom workers in goods production have declined from over 46 per cent of the total to less than 34 per cent in only twenty years. In the Federal Republic of Germany the fall has been least marked, from just under 49 per cent to exactly 42 per cent; in Japan, alone of the four, there has been a slight increase—albeit a fall over the most recent ten-year period. By the early 1980s, service workers in all four nations made up a majority of the labourforce: over 68 per cent in the United States, nearly 64 per cent in the United Kingdom, 56 per cent in Japan and over 52 per cent in Germany.

Two kinds of reason are quoted in the literature for the rapid growth of service employment at the expense of goods-producing employment. The conventional explanation is that, at least until recently, it was possible to

Table 1.1: Industrial Employment, Selected Countries, 1963-1983

	United States					United Kingdom				
	1963 %	1973 %	1983 %	83/63 %/yr	83/73 %/yr	1963 %	1973 %	1983 %	83/63 %/yr	83/73 %/yr
1. Agr, Hunt, For, Fish	7.1	4.2	3.5	-1.5	-0.1	4.4	2.9	2.7	-2.6	-1.4
2. Mining & Quarrying	0.8	0.8	0.9	2.6	3.7	2.8	1.5	1.4	-3.6	-1.3
3. Manufrg	26.6	24.8	19.8	0.5	-0.5	34.8	32.4	24.6	-1.9	-3.2
4. Elec, Gas, Water	1.3	1.1	1.2	1.6	2.5	1.7	1.4	1.4	-1.0	-0.3
5. Constrn	6.3	6.5	6.1	1.8	1.0	7.2	7.2	6.1	-0.9	-2.0
6. Who & Ret Trade, Rests & Hotel	-	21.1	22.3	-	2.3	17.2	18.1	20.3	0.7	0.6
7. Transp, Storage & Cmn	-	5.8	5.3	-	0.7	7.1	6.4	6.1	-0.9	-1.1
8. Finance, Insrce, Real Est, Busn Svs	-	7.3	9.7	-	4.6	4.0	6.3	8.7	3.8	2.8
9. Community, Soc, Pers Svs	-	28.5	31.3	-	2.7	20.9	23.8	28.8	1.4	1.4
TOTAL	100.0	100.0	100.0	2.0	1.7	100.0	100.0	100.0	-0.2	-0.5
1. Primary	7.1	4.2	3.5	-1.5	-0.1	4.4	2.9	2.7	-2.6	-1.4
2. Industry	35.1	33.2	28.0	0.9	0.0	46.4	42.4	33.6	-1.8	-1.8
3. Services	57.8	62.6	68.5	2.9	2.6	49.2	54.6	63.7	1.1	1.0

Source: OECD (1985)

obtain productivity gains in manufacturing through the substitution of capital for labour in a way that was not possible for the services sector. This thesis proceeds to argue that since the mid-1970s the economic recession has made employers ever-more cost conscious, causing them to pursue ruthless rationalisation of manufacturing processes (process innovation) in mature, established industries (Massey and Meegan, 1982; Massey, 1984). This in itself has led to the phenomenon of de-industrialisation (Blackaby, 1979; Bluestone and Harrison, 1982). Here, however, an important distinction emerges between the United Kingdom and the United States: in the latter, this process has been partly compensated for by the growth of high-technology industries; in the former, these industries have actually contributed a small part to the decline (Hall, this volume; Markusen *et al.*, 1986; Breheny *et al.*, 1986).

The alternative explanation is more novel: it is that the declining share

Germany					Japan				
1963 %	1973 %	1983 %	83/63 %/yr	83/73 %/yr	1963 %	1973 %	1983 %	83/63 %/yr	83/73 %/yr
11.9	7.3	5.6	-	-3.3	26.0	13.4	9.3	-4.0	-2.8
1.8	1.5	1.3	-	-2.1	0.7	0.2	0.2	-5.8	-2.6
37.7	36.7	32.5	-	-1.9	24.1	27.4	24.5	1.2	-0.3
1.0	0.8	1.0	-	1.7	0.5	0.6	0.6	1.8	0.6
8.2	8.5	7.2	-	-2.3	6.3	8.9	9.4	3.2	1.5
14.6	14.8	15.0	-	-0.5	-	20.6	22.9	-	1.9
5.8	6.1	6.2	-	-0.6	5.3	6.4	6.1	1.8	0.4
3.3	4.9	6.4	-	2.0	-	3.0	6.4	-	-
15.7	19.4	24.8	-	1.8	-	19.1	20.3	-	-
100.0	100.0	100.0	-	-0.7	100.0	100.0	100.0	1.1	0.9
11.9	7.3	5.6	-	-3.3	26.0	13.4	9.3	-4.0	-2.8
48.7	47.5	42.0	-	-1.9	31.7	37.2	34.8	1.6	0.2
39.3	45.2	52.4	-	0.8	42.3	49.4	56.0	2.5	2.1

of goods-producing employment is due to an organisational shift within modern capitalist industry, whereby—with increasing concentration and worldwide organisation—more and more of the total employment in goods production is not in direct production but in administrative, marketing and sales functions. Therefore, properly, these functions should be treated as part of goods-production employment. But, because many of them are contracted to specialist organisations (advertising agencies, legal and accountancy offices, truck companies, airlines, etc.), they appear under services according to the statistical conventions (Gershuny, 1978; Gershuny and Miles, 1983; Pahl, 1984).

These explanations are in fact not mutually exclusive; it is quite possible that both are true. Capital substitution has been greater in direct goods-production than in ancillary functions. More and more of these functions have been contracted and so appear as service employment.

However, the social and spatial impact of the two processes has been quite profound. First, as is generally known, the result has been a massive shift of employment away from blue-collar occupations towards white-collar ones. In Britain during the 1970s alone, the fastest-growing occupations were all white-collar or at least pink-collar ones: professional and management, literary, artistic and sport, security services, clerical and personal services (Beacham, 1984, p. 7). Conversely, the declining jobs were the blue-collar ones: processing and processing services, transport and storage, construction and mining. Secondly, and associatedly, the growing jobs were disproportionately those with high rate of female participation while the declining jobs were disproportionately male ones: personal service, clerical, and the welfare, education and health professions. It is not surprising therefore that in Britain, female employment rose by 40 per cent over the thirty years 1951-81, from 6.5 million to 9.1 million, mainly through the addition of married women into the workforce. Conversely, the proportion of adult males who are economically active has been declining—and among them, in every age group a higher percentage are unemployed than among the same group of women (Beacham, pp. 7-9).

There has, of course, been a vast feminist literature on this subject. Much of it suggests that the phenomenon arises from the need of capitalist firms to try to maintain their flagging profits during times of crisis. They do so by shedding organised, highly-paid men and recruiting unorganised, poorly-paid women. But this ignores the obvious fact that the main source of the change has been the structural shifts described above, coupled with sex stereotyping of occupations that has been evident at least since the end of the nineteenth century. In fact the British evidence shows that though women are under-represented in the very top white-collar jobs—the professional and intermediate categories—they are actually over-represented when the large category of skilled, non-manual workers is added. The men are, in fact, disproportionately over-represented in the bottom categories of the semi-skilled and unskilled. It is among those categories, concentrated in the goods-producing industries, that the most savage reductions of employment have been recently recorded. Women have been much more successful than men in occupying the middle-range niches in the new service economy—even if they may still earn less than men for equivalent work, their greater skills help compensate for this.

The shift in the economy also has a profound spatial component. Traditional smokestack industry has remained heavily concentrated in those regions and cities where it originally located when young: the port, coalfield and transportation cities of the nineteenth century. Within them, further, it has often stayed in the inner-urban locations established by the founding fathers, over-crowded and inefficient though these might be. As is well known, since 1950 this picture has been profoundly transformed, most of all in Britain, where the phenomenon of de-industrialisation has been so stark. Large tracts of the inner cities have simply been denuded of industry and employment. Associated goods-handling services like ports and railways, interestingly, have been similarly affected. The result is a huge concentration of unemployment in these areas, some of it consisting of older, skilled workers whose jobs have been lost through rationalisation or through simple competitive failure on the part of their former employees (Massey and Meegan, 1982); another large part consists of unskilled, unqualified, young people. Among these, as among other age groups, male unemployment rates are higher than female rates.

This decline in smokestack manufacturing has been compensated for—but then only in part, and not always—by employment growth in high-technology manufacturing (Hall, this volume). Even in the United States it is clear that the new employment in high-tech manufacturing falls well short of what would be needed to compensate for the loss in older industrial jobs (Castells, 1984, p.81; Markusen *et al.*, 1986). As has been evident throughout this chapter, over-all the shift is from manufacturing to services. There is also strong evidence that over-all, both high-tech industry and advanced services share the same labour structure, with a minority of highly-paid professionals and managers and a majority of low-paid processers—chiefly female (Castells, 1984, p.90; Teitz, 1984). In the United Kingdom the same processes are occurring , but the losses in traditional employment are so much the greater and the gains in new employment so much the weaker.

There is a close—and so far not completely understood—relationship between the growth of high-tech manufacturing and the growth of certain associated services. Reviewing these parallel trends, Castells suggests that for the United States we are witnessing 'a dramatic process of inter-sectoral differentiation. Some industries literally die while others explode.' (Castells, 1984, p.75). This is also a spatial differentiation: the declining sectors are disproportionately localised in the old manufacturing cities, while high-tech, together with energy resources and some advanced services, provide the basis for the growth of the new sunbelt cities; but at the same time, a few high-level cities manage to attract these services both in frostbelt and snowbelt (ibid.). A related pattern is evident for Britain but within the framework of a much poorer over-all economic performance (Hall, this volume). There, London continues to decline in manufacturing and goods-handling services while expanding modestly in information services; a crescent to the north and west of the capital proves to be the main magnet to indigenous and international high technology and to decentralising business services.

The key to the patterns of growth in both high-tech and advanced services, indeed, may be concentrations of innovatory information in the form of Research and Development. In both the United States and in Great Britain, research has shown that R & D is geographically concentrated—

and that at least some of it is associated with the growth of the new urban complexes (Malecki, 1980, 1981; Buswell and Lewis, 1970; Howells, 1984). Recent research in the United States has cast doubt on this association, at least in a strict statistical sense, but there is still an open question for further work here (Hall, this volume).

THE NEW SERVICE JOBS

The new service jobs are located in various places. Some are in the very cores of the larger national and regional centres (New York, Los Angeles, San Francisco, London, etc.), which form the centres of the new corporate power, in a variety of both producer and consumer services. Others are in what Castells (1985) calls high-tech nests, some on the fringes of major metropolitan areas (Los Angeles, San Francisco, London), some in more isolated, smaller cities associated with defence industries. Some are in recreational and resort cities, or in educational centres (Stanback, 1985). But very few are in the inner cities, or in the old manufacturing cities generally. In Britain, some major provincial cities seem to have done particularly badly in retaining service employment.

In order to better understand these shifts, researchers have increasingly found the need to classify the service occupations. There are many alternative ways of classifying and dividing them, including producer and consumer services, goods-handling and information-handling services, and public and private services (Singelmann, 1978; Stanback and Knight, 1970; Stanback, 1979; Stanback *et al.*, 1981; Stanback and Noyelle, 1982; Noyelle and Stanback, 1984). Table 1.2 shows how the main service industries fall into these various taxonomies. Clearly, there is room for a great deal of argument about them; many services do not fall clearly into one or another category, and some fall into one category in some countries and other categories in others (as, for instance, transportation and communication, which are now mainly private in Britain and the United States but are still mainly public in most European countries). Nevertheless, it can be argued that the table does usefully demonstrate the multi-dimensional character of the service industries. The private-public division, clearly, depends in large measure on the politico-economic system of the country in question; the other two classifications spring, however, from the intrinsic nature of the services themselves. The goods-information division refers to the nature of the end product, while the producer-consumer division refers to the role within the economic system as recorded in an input-output table.

The classification does prove to have important implications for an understanding of the nature of service work, the possibilities of substituting capital for labour, and the location of the activity. In this regard, the

Table 1.2: The Service Industries: Alternative Classifications

1. *Producer Services*: Transportation, Communication, Business Services, FIRE (Finance, Insurance and Real Estate), Corporate Services, Government.

   Consumer Services: Public Utilities, Wholesale and Retail Trade, Personal Services, Non-profit Services.

2. *Goods-Handling Services*: Transportation, Public Utilities, Wholesale and Retail Trade, Personal Services.

   Information-Handling Services: Communication, Business Services, FIRE, Corporate Services, Non-profit Services, Government.

3. *Private Services:* Transportation*, Communication*, Wholesale and Retail Trade, Business Services, FIRE, Corporate Services, Personal Services.

   Public Services: Public Utilities, Government, Non-profit Services.

* Position depends on the politico-economic system of the country.

goods/information-handling division is probably the most important. As Castells has emphasised, the fundamental change that is now occurring in the economies of the advanced world is a shift from an industrial to an information economy (Castells, 1985, p.11). This does not necessarily mean a decline in industrial output, though—due to rising productivity—it means a slowing or even a reversal of growth in manufacturing employment (Castells, 1984, Table 16) and, accompanying it, a similar trend in the goods-handling services. Many of the goods-handling services have traditionally been male employing and have, like manufacturing, been subject to extreme capital displacement and relocation in recent years. One of them, personal service, is, however, quite different: it is traditionally female (hotel and restaurant services, hairdressing and similar activities) and by its nature is not subject to capital substitution; it must also locate close to consumer demands. Information-handling services in contrast have not been greatly subject to capital substitution, and have been primarily responsible for the big jump in female employment over the past thirty years; but there is intense speculation about the impact of the new information technologies. Producer services have traditionally clustered in major business centres but are now tending to deconcentrate locally in search of lower rents and salary bills; consumer services are partly residentiary in character, but increasingly cater for the demands of a shifting population (tourism, business travel, etc.).

The most intensive work on these changes has been undertaken by Noyelle and Stanback in the United States. Table 1.3, based on their work,

Table 1.3: United States: Employment Change Rates, Selected Industry Groups, by Type of Metropolitan Economy, 134 Largest SMSAs, 1976-1983

Type of SMSA	No. of Cases	Annual Rate of Employment Change 1976-1983						
		Total	(Rank)	Mfrg	Who Retl	FIRE	Othr Svcs	Govt
National Nodal	4	1.4	(6)	-0.9	1.5	3.4	4.1	-0.2
Regional Nodal	19	2.6	(3)	0.7	2.8	3.6	4.9	1.2
Subregional Nodal	16	2.2	(5)	0.1	2.3	2.8	4.8	1.6
Functional Nodal	24	1.0	(9)	-1.5	1.9	2.6	4.1	0.4
Govt.-Education	14	2.5	(4)	1.2	3.2	4.4	5.0	1.2
Education-Mfrg.	5	1.4	(7)	-1.3	2.0	2.3	4.3	1.3
Manufacturing	22	0.6	(10)	-1.6	1.6	2.9	3.8	0.4
Industrial-Military	12	3.1	(2)	2.0	3.9	4.1	5.9	1.4
Mining Industrial	12	1.1	(8)	-1.3	2.3	3.5	4.3	1.2
Residential-Resort	11	5.0	(1)	4.4	4.9	6.7	6.6	2.4

Source: Stanback (1985)

highlights some of their most important conclusions. It takes a sample of some of the larger service industries and records their recent rates of growth in different types of metropolitan areas. Nodal centres are the major service cities, hierarchically ordered (national, regional, subregional), with concentrations of distributive and producer services, and often non-profit and consumer services also. Functional nodal centres are basically manufacturing cities with corporate headquarter functions, and thus have a more restricted service base. Government-education are mainly state capitals, university cities, or both. Education-manufacturing centres are predominantly old industrial cities that have become university centres. Production centres are manufacturing cities that lack headquarters or research functions. Industrial-military centres have large concentrations of government and military personnel. Finally, resort-residential centres include some places on the edges of the big national nodal centres, as well as some fairly recent resort and retirement centres.

Stanback's analysis shows that the fastest-growing metropolitan areas have been those favoured by the post-World War II patterns of agglomeration: resort-residential areas providing consumer services; industrial-military centres, fuelled by re-armament; and government-educational centres related to higher-level activities in these fields (State governments, major universities). The slow-growing areas are those that developed in the previous industrial era: manufacturing centres, functional-nodal (industrial headquarters) areas and mining-industrial areas. But viewed more closely, Table 1.3 demonstrates that the national-nodal, functional-nodal and education-manufacturing centres have recorded heavy losses in manufacturing jobs; their over-all gains arise from compensating gains in FIRE and other business services (Goddard and Smith, 1975). They have been relatively successful in changing their economic structures, while the pure manufacturing (and mining-industrial) centres have shown little capacity to make the transformation to the service economy (Stanback, 1985, pp.127-8).

There is no corresponding analysis as yet for Britain—though there are good partial studies of office location (Daniels, 1975, 1977, 1978, 1983a,b, 1984; Daniels and Holly, 1983; Evans, 1973; Goddard, 1975; Goddard and Morris, 1975; Marquand, 1979; Marshall, 1982; Westaway, 1974a,b). The interesting point of comparison is that Britain's is a very small national space-economy, which should properly be seen as only a part of a wider European and worldwide economy (Thrift, 1985). Viewed in this light, London is undoubtedly a true world city with controlling and organising functions that extend globally. Like New York, it has seen a massive decline in manufacturing jobs; unlike New York, it has failed to compensate through growth in service employment though the decline in the latter has been only marginal (Table 1.4). But the major provincial British cities, ten or so in number, may each represent the apex of regional systems that are now technologically archaic and in rapid decline. With the improvement of both personal movement (through motorways, high-speed rail and air services) and telecommunications, they occupy a niche that is no longer necessary. Additionally, they were never pure regional-nodal cities in Stanback's sense, but depended also on manufacturing or goods-handling services for their livelihoods, functions that are now disappearing, with serious implications for their remaining business-service functions. Thus both Liverpool and Manchester, the twin regional centres of north-west England, have recorded heavy losses in business services (Cheshire *et al.*, 1984; Tym and Partners, 1981). It is the cities at the next level of the urban hierarchy, as identified by Smith (1967)—the medium-sized county towns, especially in southern England—that seem to have been the main recipients of decentralised service growth.

Table 1.4: Greater London and New York City: Employment Changes, 1977/8-1983

	Greater London Employment Change				New York City Employment Change			
	1978	1983	1978-83		1977	1983	1977-83	
	000s		Abs	%	000s		Abs	%
Manufacturing	769	594	-175	-8.6	540	432	-108	-20.0
Construction	176	144	-32	-18.2	66	87	+21	+31.8
Services	2728	2621	-107	-3.9	2568	2824	+256	+10.0
Goods - Handling	NA	NA	NA	NA	880	847	-33	-3.7
Info-Handling	NA	NA	NA	NA	1688	1977	+289	+17.1
Total Non-Agr	3633	3359	-314	-8.6	3175	3334	+159	+3.1

Source: SERPLAN (1984), Table 3.6; Castells (1984), Table 17

One other feature of the Stanback analysis, which emerges strongly from Table 1.3, is the strength of the finance, insurance and real estate and other services (chiefly business and non-profit services) sectors. They grew rapidly everywhere, but especially in the residential-resort, government-education and industrial-military centres. Of course, there is an element of circularity in this: these sectors may have grown rapidly there simply because these places were growing. But additionally, there may be a tendency to outward deconcentration of such services from the larger (national and regional) nodal centres to such smaller places, in search perhaps of lower rents and lower labour costs. Here the work of Baran (1985) is significant. She has analysed in detail the changing patterns of work in the American insurance industry, a major segment of the FIRE group. She shows that during the 1970s the industry made a dramatic shift from male to female labour and that a part of this process was the search for cheaper, flexible labour that would adapt to the new pattern of computerisation. Routine clerical jobs are being eliminated; new professional jobs are being created, though new technology is tending somewhat to deskill them. The result is a big expansion of fairly high-skill jobs at fairly low rates of pay, filled by well-educated, white, suburban housewives who are willing to accept them in order to combine career and family responsibilities. Logically, the industry has moved out to this labour pool, leaving the army of low-paid, routine, clerical workers (many of them inner-city minority females) in danger of losing their jobs (Baran, 1985).

It seems clear that exactly the same processes are at work in Britain. Just as American FIRE offices have relocated from New York to places like Stamford (Connecticut), and from San Francisco to Walnut Creek (California), so the British equivalents have gone from the city and West End of London to places like Reading and Basingstoke. There is an extraordinary parallel between the reconstruction of downtown Stamford, 40 miles from New York, and the reconstruction of central Reading, 40 miles from London. In Britain there is also evidence that firms are moving even further afield into country towns 100 and more miles from London. But the pattern is clear: the industry is seeking middle-class labour and is rejecting the inner city and minority areas.

SUMMING UP

The evidence can be summarised thus. Traditionally, FIRE and business services were located in national or regional nodal centres; government services and high-level educational and health services were concentrated in American State capitals and English county towns. The former have tended to deconcentrate somewhat from the larger (national) centres toward suburban locations, especially to suburban towns on the fringe of the major metropolitan area. Governmental-eductional centres have tended to show strong growth in FIRE and business services. New specialised service centres (government-military and recreation-retirement) have also shown dynamism in both countries. New technology has had an important impact here, by creating a demand for well-educated (but relatively low-paid) female labour. The older industrial cities—including the larger among them, which were also regional or functional nodes—tend to have been unattractive precisely because they lacked this quality of labour, and because the collapse of their manufacturing base reduced the demand for business services.

For the future, Baran doubts that white-collar jobs will be displaced by new technologies, even in such a labour-intensive, paper-processing operation as insurance; too many office jobs are unique, unstructured, or dependent on personal communication. Rather, the likelihood is a further growth of semi-skilled, predominantly female jobs drawing on a middle-class labourforce (Baran, 1985, p.168). From a European perspective it might be added that cultural and linguistic barriers will also impose severe limits to the process of spatial decentralisation: business service offices will not find it as easy as factories to move their operations to Third World locations, or even to lower-cost locations in other European countries.

There is yet another striking contrast between the two economies: the remarkable development in the United States of a low-tech, low-wage, urban economy associated with a new burst of immigration from the developing world. When an estimate is made for the huge volume of illegal immigration, it is likely that the absolute numbers of immigrants in the United States in the early 1980s were close to, or even above, the record set in the first decade of the twentieth century. Disproportionately these new immigrants come from Latin America and the Far East; disproportionately they remain in a few large cities of entry such as New York, Miami, Houston and Los Angeles. The result has been the development of a separate Third World kind of economy in the hearts of these cities, as documented by Soja *et al.* (1983) for Los Angeles, which exists side by side with advanced services—and in Los Angeles, high-tech manufacturing—in a segmented, ghetto-ised economy. Because of stricter immigration restrictions there is no equivalent phenomenon in the United Kingdom, though perhaps the Bengali sweatshops of London's Whitechapel provide a miniature parallel.

We should beware, then, of the simplistic notion that the new high-tech and service economy will be a new golden age. For those countries and those regions that can share in it, it may not be all that bad either. It will provide some good jobs, very many more middle-level jobs in which people (chiefly women) will be employed rather below their level of competence, and (in some places) many low-paid entry-level jobs in manufacturing and consumer services. The critical question is whether it can offer any real prospects for progressive upgrading of skills and prospects, or whether the logic will lead inexorably to deskilling. On that, the only reasonable verdict at present must be an open one.

REFERENCES

Baran, B. (1985)'Office Automation and Women's Work: The Technological Transformation of the Insurance Industry', in M. Castells (ed.) *High Technology, Space, and Society*, Urban Affairs Annual Reviews, vol. 28, Sage, Beverly Hills.

Beacham, R. (1984)'Economic Activity: Britain's Workforce 1971-81', *Population Trends*, 37, 6-14.

Bell, D. (1973) *The Coming of Post-Industrial Society: A Venture in Social Forecasting*, Basic Books, New York.

Blackaby, F. (ed.) (1979) *De-Industrialisation*. Economic Policy Papers, 2, Heinemann, London (for the National Institute of Economic and Social Research).

Bluestone, B. and Harrison, B. (1982) *The Deindustrialization of America: Plant Closures, Community Abandonment, and the Dismantling of*

Basic Industry, Basic Books, New York.
Breheny, M., Hall, P., Hart, D. and McQuaid, R. (1986) *Western Sunrise: The Genesis and Growth of Britain's High-Tech Corridor*, George Allen and Unwin, London.
Buswell, R. J. and Lewis, E. W. (1970)'The Geographical Distribution of Industrial Research Activity in the United Kingdom', *Regional Studies*, 4, 297-306.
Castells, M. (1984) *Towards the Informational City? High Technology, Economic Change and Spatial Structure: Some Exploratory Hypotheses*, Working Paper no. 430, Institute of Urban and Regional Development, University of California, Berkeley.
——— (ed.) (1985) *High Technology, Space, and Society*, Urban Affairs Annual Reviews, vol. 28, Sage, Beverly Hills.
Cheshire, P. C., Hay, D. and Carbonaro, G. (1984)'Regional Policy and Urban Decline: The Community's Role in Tackling Urban Decline', Draft Final Report, Joint Centre for Land Development Studies, Reading (mimeo).
Clark, C. (1940) *The Conditions of Economic Progress*, Macmillan, London.
Daniels, P. W. (1975) *Office Location*, Bell, London.
——— (1977)'Office Location in the British Conurbations: Trends and Strategies', *Urban Studies*, 14, 261-74.
——— (ed.) (1978) *Spatial Patterns of Office Growth and Location*, Wiley, Chichester.
——— (1983a)'Business Service Offices in British Cities: Location and Control', *Environment and Planning A*, 15, 1101-20.
——— (1983b)'Service Industries: Supporting Role or Centre Stage?', *Area*, 15, 301-9.
——— (1984)'Business Service Offices in Provincial Cities: Sources of Input and Destinations of Output', *Tijdschrift voor Economische en Sociale Geografie*, 75, 123-39.
——— and Holly, B. P. (1983)'Office Location in Transition: Observations on Research in Great Britain and North America', *Environment and Planning A*, 15, 1298.
Evans, A. W. (1973)'The Location of the Headquarters of Industrial Companies', *Urban Studies*, 10, 387-96.
Fisher, A. G. B. (1935) *The Clash of Progress and Security*, Macmillan, London.
Gershuny, J. (1978) *After Industrial Society: The Emerging Self-Service Economy*, Macmillan, London.
——— and Miles, I. (1983) *The New Service Economy: The Transformation of Employment in Industrial Societies*, Frances Pinter, London.
Goddard, J. B. (1975) *Office Location in Urban and Regional Development*, Oxford University Press, Oxford.

——— and Morris, D. (1975) *The Communications Factor in Office Decentralization*, Pergamon Press, Oxford.

——— and Smith, I. J. (1975)'Changes in Corporate Control in the British Urban System 1972-1977', *Environment and Planning A*, 10, 1073-84.

Howells, J. R. L. (1984)'The Location of Research and Development: Some Observations and Evidence from Britain', *Regional Studies*, 18, 13-29.

Malecki, E. (1980)'Corporate Organization of R & D and the Location of Technological Activities', *Regional Studies*, 14, 219-34.

——— (1981)'Government-Funded R & D: Some Regional Economic Implications', *Professional Geographer*, 33, 72-82.

Markusen, A., Hall, P. and Glasmeier, A. (1986) *High Tech America: The What, How, Where and Why of the Sunrise Industries*, George Allen and Unwin, London.

Marquand, J. (1979) *The Service Sector and Regional Policy in the United Kingdom*, Centre for Environmental Studies, London.

Marshall, J. N. (1982)'Linkages Between Manufacturing Industry and Business Services', *Environment and Planning A*, 14, 1523-40.

Massey, D. (1984) *Spatial Divisions of Labour: Social Structures and the Geography of Production*, Macmillan, London.

——— and Meegan, R. (1982) *The Anatomy of Job Loss*, Methuen, London.

Noyelle, T. J. and Stanback, T. M. (1984) *The Economic Transformation of American Cities*, Rowman and Allenheld, Totowa, New Jersey.

OECD (1985) *Labour Force Statistics 1963-1983*, Department of Economics and Statistics, OECD, Paris.

Pahl, R. (1984) *Divisions of Labour*, Blackwell, Oxford.

Standing Conference on London and South East Regional Planning (SERPLAN) (1984) *Regional Trends in the South East: The South East Regional Monitor 1983-84*, SERPLAN, London.

Singelmann, J. (1978) *From Agriculture to Services: The Transformation of Industrial Employment*, Sage Library of Social Research, vol. 69, Sage, Beverly Hills.

Smith, R. D. P. (1967)'The Changing Urban Hierarchy', *Regional Studies*, 2, 1-19.

Soja, E., Morales, R. and Wolff, G. (1983)'Urban Restructuring: An Analysis of Social and Spatial Changes in Los Angeles', *Economic Geography*, 59, 195-230.

Stanback, T. M. (1979) *Understanding the Service Economy: Employment, Productivity, Location*, Johns Hopkins University Press, Baltimore.

——— (1985)'The Changing Fortunes of Metropolitan Economies', in M. Castells (ed.) *High Technology, Space, and Society*, Urban Affairs Annual Reviews, vol. 28, Sage, Beverly Hills.

———, Beare, P. J., Noyelle, T. J. and Karsach, R. A. (1981) *Services: The New Economy*, Allenheld, Osmun, Totowa, New Jersey.
——— and Knight, R. V. (1970) *The Metropolitan Economy: The Process of Employment Expansion*, Columbia University Press, New York.
——— and Noyelle, T. J. (1982) *Cities in Transition: Changing Job Structures in Atlanta, Denver, Buffalo, Phoenix, Columbus (Ohio), Nashville, Charlotte*, Allanheld, Osmun, Totowa, New Jersey.
Teitz, M. (1984)'The California Economy: Changing Structure and Policy Responses', in J. J. Kirlin and D. R. Winkler (eds) *California Policy Changes, 1984*, School of Public Administration, University of Southern California, Los Angeles.
Thrift, N. (1985)'Research Policy and Review 1. Taking the Rest of the World Seriously? The State of British Urban and Regional Research in a Time of Economic Crisis', *Environment and Planning A*, 17, 7-24.
Tym, Roger and Partners (1981) *Capital of the North: The Business Service Sector in Inner Manchester/Salford*, Report to the Manchester/Salford Inner City Partnership, Roger Tym and Partners, London.
Westaway, J. (1974a)'The Spatial Hierarchy of Business Organizations and its Implications for the British Urban System', *Regional Studies*, 8, 145-55.
——— (1974b)'Contact Potential and the Occupational Structure of the British Urban System 1961-66: An Empirical Study', *Regional Studies*, 8, 57-73.

Chapter 2

PERSPECTIVES ON POST-INDUSTRIAL SOCIETY: AN AUSTRALIAN VIEWPOINT

S. Encel

The concept of post-industrial society is at once an explanation of the past and a prediction about the future. As a sociological metaphor it imparts historical and philosophical profundity to our concern with the relationship between technological innovation and social change. In a relatively short chapter it is only possible to explore a few aspects of this relationship; nor is it possible to comment on more than a few contributions to the vast body of discussion already generated by the term since it came into widespread use twenty years ago. I have therefore been selective both as to topics and to authors.

Reviewing this vast literature, a recent commentator notes that the theory of post-industrial society is essentially a minor development of the concept of industrial society (Badham, 1984). This assumes that 'the development of industry and its impact on society are the central features of modern states. Social thought then becomes concerned with the social requirements of industrial development, the social structures that either facilitate or hinder the efficient pursuit of industry, and the impact of industrial development upon society.'

Generally speaking, theories of industrial society incorporate six elements:

1. A basic transition from an agrarian to an industrial economy;
2. a stress on the essentially progressive character of this transition;
3. the view that class conflicts, which predominate in the earlier stages of the transition, gradually become of minor importance;
4. the development of liberal democracy as the political counterpart of the industrial economy;

5. the view that traditional societies will become modernised along the same lines as existing industrial societies; and
6. a belief in the ultimate convergence of currently diverse industrial societies.

Post-industrial society, on the other hand, appears to have five distinctive features:

1. a further transition from the industrial to the service economy;
2. substantial growth of white-collar, professional and technical occupations;
3. the rise of information technology as the dominant branch of technological activity;
4. the key role of scientific and theoretical knowledge; and
5. growth of concern with leisure and the quality of life. (Badham, 1984).

These five characteristics do not represent a decisive break with the past, and most of them were foreseen by nineteenth century writers like St-Simon, Weber and Durkheim. This is not wholly true as regards information technology, whose spectacular development was not predictable until the advent of electronics and quantum mechanics. Hence, the concept of the information society has emerged as a subset of the broader theory of post-industrial society. I shall return to this point later.

The term post-industrial society is associated particularly with the name of Daniel Bell, and much of the critical literature is directed at his ideas. In Bell's work, scientific discovery, technological innovation, industrial development and rationalisation become the moving forces in generating social mutations instead of the class struggle postulated by Marx as the engine of social transformation.

In Bell's words, the 'axial principle' of post-industrial society is the 'centrality of theoretical knowledge' as the source of innovation and of policy formulation. A dominant role will be played by professional and technical experts, armed with the new intellectual technology. The new scientific and technical elites will displace existing powerful groups and society will be run on the basis of rational decision-making. The roots of post-industrial society, according to Bell, lie in the 'inexorable influence of science on productive methods. The scientific estate—its ethos and its organisation—is the monad that contains within itself the image of the future society.' (Bell, 1973).

Bell's conception of post-industrial society rests partly on a theory of history and partly on two main sources of data, one dealing with the workforce, the other with education. His early papers on the subject were clearly influenced by the pioneering work of Fritz Machlup on the growth of the information industry (Machlup, 1962). In some respects, Bell's work is less significant than that of Clark Kerr, who was one of the first writers to emphasise the broader socio-economic implications of Machlup's analysis. 'What the railroads did for the second half of the last century and the automobile for the first half of this century may be done for the second half of this century by the knowledge industry; that is, to serve as the focal point for national growth.' (Kerr, 1963). Kerr gave this argument an important locational twist when he foresaw the development of industrial complexes linked with institutions of higher education and research. At the time he wrote, this was already well developed along Route 128 in the Greater Boston area, to be followed by the growth of Silicon Valley in Santa Clara County, California. Since then, of course, the idea has become part of conventional planning wisdom. In this vision of the post-industrial society, knowledge has become the prime resource and the professionally-trained expert has replaced the entrepreneur or the financier. Consider Bell's own words:

> If the dominant figures of the past 100 years have been the entrepreneur, the businessman, and the industrial executive, the 'new men' are the scientists, the mathematicians, the economists, and the engineers of the new computer technology. And the dominant institutions of the new society—in the sense that they will provide the most creative challenges and enlist the richest talents —will be the intellectual institutions. The leadership of the new society will rest, not with businessmen or corporations as we know them (for a good deal of production will have been routinised) but with the research corporation, the industrial laboratories, the experimental stations, and the universities. (Bell, 1976).

Those are fine words, and when they were written they may have sounded plausible. Few people would now endorse that kind of optimism, even if they take a generally cheerful view of the future. As one of Bell's American critics has remarked sardonically, the post-industrial society 'was a period of 2 or 3 years in the mid-sixties when GNP, social policy programmes, social research and universities were flourishing. Things have certainly changed' (Miller, 1975).

PESSIMISTIC VISIONS

The writings of Kerr, Bell and others in a similar vein reflect a basic optimism about the social role of science and technology. To others, the concept of post-industrial society is either misleading or downright sinister.

Touraine argues that post-industrial society is not very different from industrial society. Class relationships remain central to capitalism and industrialism. What has changed, because of the impact of science and technology, is the quality of these relationships, particularly because of the gulf created between the information rich and the information poor (Touraine, 1971). Writers like Braverman and Gorz, concerned especially with the nature of work, reject the optimistic arguments about the premium placed on skills and knowledge in the post-industrial society. According to Braverman, the supposed increase in skill levels due to technological change and the expansion of education is no more than an illusion. In reality, there is a massive process of deskilling which sharpens distinctions within the working classes (Braverman, 1974). Gorz argues that the important point remains the control of the means of production, and that the impact of technology depends on decisions which are either class based, as in the West, or politically based, as in the USSR, which Gorz (1976) regards as a State-capitalist society.

The description of the Soviet Union as a system of state capitalism is relevant to the emphasis on science and technology which has marked official doctrine since the 21st Congress of the Soviet Communist Party in 1961. At the meeting, the late Nikita Khrushchev defined the scientific-technological revolution (STR) as the 'key growth factor of social productive forces'. The 24th Congress, in 1971, declared that the STR was 'the main lever for building the material and technical basis of communism'. Before the Congress, D. M. Gvishiani, First Deputy-Chairman of the USSR State Committee for Science and Technology, wrote in the journal *Science and Life* that the STR involved the mechanisation and automation of both the physical and mental activities of man, which will bring about a profound change in all sectors of production and will affect all aspects of social life (Gvishiani, 1972).

These predictions bear some striking similarities to the ideas of Kerr, Bell and other post-industrial utopians. Two recent Soviet commentators, Modrzhinskaya and Kosolapov, are good examples. Progress, they argue, is to be measured by the development of productive forces—a theme which harks back to the work of Trotsky, Preobrazhensky, Bukharin and other Bolshevik theorists of industrialisation in the 1920s. Bukharin once described four stages in the development of socialism—ideological, political, economic and technological. Once the revolution had been accomplished (the political stage) and a socialist economy established, productive forces would take charge without the earlier constraints. Thus, Modrzhinskaya writes that the pace of scientific and technological development has gained unprecedented strength, and it is difficult to imagine the speed with which man's mastery over nature will increase in the future. Kosolapov foresees that knowledge will create the scientific and technological basis for producing universal prosperity, and that social transformations will be generated by the irresistible momentum of technological change. Progress will be limited not by scientific and technological possibilities, but by labour, financial resources, and the social structure (Freeman and Jahoda, 1978).

The Soviet Government has attempted to translate these ideas into practice by setting up industrial—scientific complexes along the lines described by Kerr. The first of these, Akademgorodok (Academic City) was established at Novosibirsk, near Lake Baikal, in 1962; another centre was inaugurated near Vladivostok ten years later. The Soviet authorities have also tried to bring research and technological development closer together by locating laboratories and factories on adjoining sites. The Institute of Cybernetics, administered by the Ukrainian Academy of Sciences, is designed to promote development in the computer industry by this means (Rigby and Miller, 1976).

This Soviet orthodoxy stands in sharp contrast not only to the neo-Marxist critique but to the even more radical scepticism of other Left-wing critics, especially those who write from an anarchist viewpoint. Thus, Ivan Illich defines the present stage of capitalism as hyper-industrial, marked by biological degradation, radical monopoly, over-programming, polarisation, obsolescence and frustration. For him, the post-industrial age has yet to come; the character which he gives it is almost a mirror image of the picture painted by Bell and Kerr. In Illich's post-industrial society there will be a new balance, in which the industrial mode of production will disappear, and with it the specialised professions generated by the industrial system (Illich, 1973). The evangelical quality of Illich's preaching is not to everyone's taste, but his message is essentially similar to that of a much more balanced and reflective exponent of the anarchist position, E. F. Schumacher. Schumacher notes that modern technology has deprived man of the kind of creative work which he enjoys most and replaced it with fragmented tasks which he does not enjoy at all. Unfortunately, he argues, 'the type of work which modern technology is most successful in reducing or even eliminating is skilful, productive work of human hands' (Schumacher, 1973).

PESSIMISM AND OPTIMISM IN AUSTRALIA

The economic slump which has affected the entire capitalist world since the early 1970s induced considerable thought about the future in Australia—most of it pessimistic in tone, and most of that on the Left-wing side of politics. Business spokesmen generally take an up-beat view of the future, if only because it is difficult to stay in business without being confident about future prospects. This is what Cotgrove calls the cornucopian view, and he found its supporters among engineers, consultants, directors, managers, marketing and sales people, and industrial scientists. By contrast, the pessimistic, catastrophist view was espoused by research scientists, academics, doctors, teachers, clergy, social workers, writers and actors (Cotgrove, 1982).

A typical business view is presented by a management consultant who foresees the development of a 'more entrepreneurial and innovative Australia' through the investment of venture capital in high-technology industries (Christie, 1984). A few years ago, we were also encouraged to take an expansive view of the future by the archpriest of technological optimism, Herman Kahn. Essentially, Kahn saw Australia profiting from the post-industrial society, provided only that the trade unions were kept in line and that the spirit of entrepreneurship was encouraged. The real danger was in the kind of complacent hedonism reflected in bumper stickers declaring that 'I'd rather be sailing' (Kahn and Pepper, 1980). This view is not dissimilar from the emphasis on economic libertarianism expressed in the collection of essays by a group of conservative economists, published contemporaneously with Kahn's work (Kasper *et al.*, 1980).

A very different conception of the future may be found in the work of the biologist Charles Birch, whose concerns are similar to those of other environmentalists like Barry Commoner and Paul Ehrlich. Birch would like to see a 'sustainable Australian society in a sustainable world'. This would require an appropriate strategy for the use and management of non-renewable resources like nickel, iron ore, uranium and coal, of which Australia holds some of the world's largest reserves. To achieve this demands a resource policy that takes into account the needs of the world now and in the future, as well as our own needs. In particular, the export of minerals should be aimed at the welfare of the underdeveloped countries of Oceania and South-East Asia, which should also be helped by policies that encourage them to sell their own goods to the rest of the world.

As a biologist, Birch is concerned predominantly with food, population, resources, environment and conservation. His picture of Australia turns essentially on two propositions: excessive pre-occupation with economic growth, and failure to accept responsibility for the problems of neighbouring areas of the world (Birch, 1975).These issues are also of special importance to other writers concerned with the environment, whether natural or social. Thus, Sandercock maintains that capitalism—industrial or post-industrial—is responsible for uncontrolled urban growth and environmental disruption. 'The structure of responsibility for ecological destruction', she writes, 'is bound deeply into the social stratification of our form of society... Capital is the critical factor in determining development, organisation, methods and amounts of production, in fact the structure of industrial society.' The central issue of the future will be the capability of society to deal with the environment, but this may well exceed the obdurate limits of the democratic system, and lead to an authoritarian society (Sandercock, 1975).

Sandercock's writings are strongly influenced by the work of Hugh Stretton, whose studies of urban life represent a major contribution to Australian social thought. Whereas Bell sees theoretical knowledge as the axial principle of post-industrial society, the struggle for equality occupies the corresponding place in Stretton's intellectual scheme. Growth-oriented, technological capitalism exploits the readiness of many people to accept inequality as the price of rising living standards. Despite rises in real income per head, the distribution of wealth, ownership and privilege has become more unequal since the 1950s. The social and political issues of the future, he maintains, will be centred on the relations between growth, conservation and equality (Stretton, 1976).

The pessimism of these writers reflects a widespread public mood which is manifested in a number of opinion surveys. Although the findings of national opinion surveys on complex issues should always be treated with caution, more detailed analyses also indicate an underlying pessimism, which appears to arise partly from economic instability and partly from international tension. A carefully structured survey by Crompton (1982) makes this clear. He surveyed a group of 120 professionals drawn from a variety of backgrounds, and asked them to indicate the most important problems that were likely to confront Australia and the world at large during the remainder of the century. Energy shortages and the prospect of technological changes to provide alternative energy sources far outstripped any other aspect of technology. Alternative energy sources were seen as the most important technological change by 49 per cent of the respondents, compared with 12 per cent for communications technology.

Crompton also asked his respondents to select probable scenarios of the future, following the approach advocated by Miles, who argues that most forecasts have distinct economic and political dimensions (Miles, 1978). On this basis, Crompton constructed four scenarios:

1. Aggressive competition—high economic growth, minimalist government, uneven distribution of wealth.
2. Planned economy for growth—high economic growth, central planning, egalitarian social policy.
3. Consumer society—low growth, decentralised economy, egalitarian social policy.
4. Socio-economic disarray—low growth, inegalitarian society with authoritarian government.

The survey-respondents were asked two questions about those four alternatives—first, which future they would *prefer*, and which they thought *most likely*. The results show an interesting reversal. The clearest statement relates to preferences. In this case, 67 per cent of the group favoured the second scenario, planned economy for growth. Clearly, no one preferred the fourth scenario, and of the other two, 19 per cent favoured aggressive competition with 8 per cent in favour of the consumer society. When asked to predict which future was likely, the respondents divided much more evenly; 44 per cent saw aggressive competition as most likely, and 39 per cent predicted socio-economic disarray. In other words, there is no clear majority for the probable continuance of the *status quo*, and a very large minority who foresee things getting worse (Crompton, 1982).

SOCIAL AND INTELLECTUAL IMPLICATIONS

Because the affluent, capitalist countries share a similar pattern of socio-economic and techno-economic structures and processes, the problems posed by post-industrial society are also broadly similar in each case. Australia differs from the other countries in this group because of its size, geographic location, dependence on rural and extractive industries, and some distinctive political institutions. In this section, I shall not make any particular attempt to separate what is specifically Australian from what is generic to the subject as a whole.

It may be observed, first of all, that the concept of post-industrial society carries some profoundly inegalitarian implications. The theory harks back to the ideas of St-Simon, who looked forward to a world run by scientists, engineers and entrepreneurs. In its modern form, this theory has been described by Goldthorpe as technocratic historicism. It treats the capitalist-industrial norm as one towards which other societies will converge, and the end point will be a post-industrial society dominated by scientific and technological elites (Goldthorpe, 1971). Kerr, for his part, foresees that the technostructure, which he identifies with the 'educational and scientific estate', will supersede the old-fashioned businessman, and that the structure of the post-industrial society will be determined by the nexus between technology, education, occupational structure, and organisation (Kerr, 1969). Bell presses the argument further and negates the views of John Rawls on social justice. He maintains that Rawls's emphasis on equal treatment as the basis of justice and fairness will lead to mediocrity, whereas the post-industrial society demands a strong emphasis on meritocracy (Bell, 1973). In his companion volume, *The Cultural Contradictions of Capitalism*, Bell launches a strong attack on the American educational system, which he sees as the embodiment of mediocrity (Bell, 1976).

The stress on information and knowledge as the central phenomena of the post-industrial society leads, naturally, to a special concern with education, and with higher education in particular. This is clearly evident in the writings of Kerr, Bell and their numerous *epigoni*. It was the knowledge industry, Kerr declared, which had transformed the social role of higher education, and he suggested the terms 'multiversity' and 'ideopolis' to epitomise these new relationships (Kerr, 1963). Bell, for his part, relates the growth of occupational specialisation to the expansion of higher education. The growth of R & D, in particular, was responsible for the rapid development of post-graduate studies (Bell, 1973). These views, in their turn, are clearly the product of economic analyses associated with the names of writers like Solow, Denison and Schultz. Edding, reviewing levels of enrolment at different stages of the educational process since the beginning of the century, suggested the operation of an iron law of educational expenditure which makes it grow faster than national output, because student enrolments and teacher-pupil ratios have positive income elasticity (Edding, 1962). A more general picture of this growth is given by Poignant, who demonstrates that the present balance between the various levels of education and training is very much a post-1945 phenomenon (Poignant, 1973). As a historical footnote, it may be observed that Russian economists in the 1890s were already arguing that educational expenditure was related to productivity (Anderson and Bowman, 1966), a point of view also put forward by Adam Smith in *The Wealth of Nations* and endorsed a century later by Alfred Marshall in *The Principles of Economics*. In the Soviet Union, the orthodox view from the beginning has been that educational expenditure is an essential component of capital expenditure (Strumilin, 1962).

It now appears, however, that Edding's iron law may also operate in reverse. With the decline of growth rates (made even sharper in real terms because of inflation) there is a widespread tendency to cut spending on education even faster than the general downward trend of government expenditure, and the higher levels are feeling the greatest pinch. In Britain, for example, the Thatcher Government has imposed a number of restrictive measures, including the closure of departments, encouragement of early retirement among academic staff, reduction of student grants, increases in fees, and cuts in research funds. In France, the permissive policies which followed the student disturbances of May 1968 have been sharply reversed in the 1980s, and during 1985 the Government announced measures to make higher education more useful and socially relevant. In Australia, there have been repeated efforts since 1975 to redefine the role of higher education in terms of social usefulness and the development of excellence, which seems to be closely related to industrial profitability. Thus, the Williams report on education, training and employment (1978) stressed the need for skills required in the job market (*Vestes*, 1979).

POLITICS AND POLICY

As Kumar notes, the post-industrial theorists made the complacent assumption that 'more money spent on education and research, and larger numbers involved in them, *ipso facto* means a better-educated, more self-conscious and more "knowledgeable" society.' (Kumar, 1978).These complacent assumptions were, to a considerable extent, influenced by the political climate of the 1960s. Dedijer has collected an anthology of pronouncements about the crucial role of science in promoting national strength and prosperity (Dedijer, 1966). Perhaps the best known, and certainly one of the most florid, was Harold Wilson's famous metaphor about the changes in British society which would be brought about by the'white heat of the scientific revolution' (Wilson, 1963). In this climate, the Australian Labor Party adopted a science-policy platform for the first time in 1965, which declared roundly that'science is the fountainhead of human progress' (Encel, 1981).

All governments have subsequently retreated to a much more sceptical position since the heyday of this overblown optimism. The Whitlam Government, in 1975, made a gesture towards science policy by establishing the Australian Science and Technology Council (ASTEC) but gave it very little in the way of resources and ignored its advice on a couple of crucial questions almost from the time it was created (Encel, 1981). ASTEC was very nearly abolished by the Fraser Government; although it survived, its influence has been limited mainly to advice on decisions about the funding and organisation of specific projects.

An examination of the role of science and technology in advanced industrial countries illustrates one of the major fallacies of the theory of post-industrial society, which is to ignore the enormous importance of the military-industrial complex in the economies of these countries. The military budget has been the main source of support for the many areas of scientific research and advanced technology. The aerospace industry, which spends the highest proportionate amount on R & D, would virtually cease to exist if it was not underwritten by governments; similarly, many of the major advances in electronics, solid-state physics, and nuclear engineering have depended on the military budget. Applied mathematics and computing have also been major beneficiaries, and most research on artificial intelligence is supported by military agencies. In the USA, nearly two-thirds of R & D activity relies on military funding; in Britain the proportion is close to one-half.

By contrast, areas of R & D which do not have military support are endangered by cut-backs in government funding, or are under increasing pressure to demonstrate their relevance (that is, the possibility of profitable exploitation). In Britain, for example, an official report in 1985 recommended the reduction of national expenditure on research in high-energy physics, including a cut in Britain's contribution to the European Organisation for Nuclear Research (CERN) in which British scientists and engineers have played a leading role since its establishment in 1954 (*High Energy Physics*, 1985). CERN provides the only major laboratory for British high-energy physicists, since there are no such facilities in Britain itself.

In Australia, the effects of governmental cutbacks have been felt particularly by Commonwealth Scientific and Industrial Research Organisation (CSIRO), which accounts for about 25 per cent of all R & D expenditure in Australia. In 1977, a major inquiry into CSIRO's activities criticised its 'remoteness from the marketplace' and recommended that its work should be directed more closely to the needs of manufacturing industry (CSIRO, 1977). A further inquiry was conducted by ASTEC in 1985, which laid even greater stress on the need for CSIRO to be responsive to the needs of industry and to orient its organisational structure and research priorities accordingly (ASTEC, 1985).

The proposition that we are living in a post-industrial or post-service society in which information technology is a central resource has been ardently promoted in Australia by Barry Jones, now Minister for Science in the Federal Government. Jones's well-known book, *Sleepers, Wake!*, published not long before the Australian Labor Party came to power in 1983, provides a valuable analysis of technological and occupational changes since World War II, and argues the case for vigorous government action to encourage technological innovation in socially desirable directions (Jones, 1982).

While Jones is too easily seduced by the metaphors of post-industrialism, he is acutely conscious of the fallacies inherent in the complacent optimism of writers like Kerr and Bell. In particular, he stresses that one of the prime consequences of information technology is its adverse impact on work. Although the new technology also creates new jobs, its net effect is destructive. Apart from reducing employment opportunities, information technology has a deleterious effect on the quality of working life, as it polarises work roles between the highly skilled information technologist on the one hand and the machine operator on the other. Rapid technological change, he notes, can have the effect of intensifying the worst aspects of a competitive society.

Jones's book contains a number of concrete suggestions, a few of which have actually been put into practice. Apart from endeavouring to make CSIRO more relevant, the Federal Government has introduced a range of fiscal incentives to encourage innovation in private industry, where the level of expenditure on R & D has fallen to its lowest level for thirty years. These include an incentive scheme for the establishment of management investment companies (MICs), whose role is to provide venture capital for technologically innovative enterprises and, most recently, a 150 per cent tax write-off for expenditure on industrial R & D—a proposal originally made in the 1960s and rejected in favour of a grants scheme, which had the effect of subsidising large companies whose R & D activities were already well established (Encel, 1971). It remains to be seen whether tax incentives will work better.

One of the interesting consequences of this rapid increase of governmental interest in industrial innovation has been the shift of political action from Canberra to the State-Government level. Since 1975, most States have established ministries whose title includes the word 'technology'. In Western Australia, for example, the *Technology Development Act 1983* was passed, setting up a Science, Industry and Technology Council, a Technology Development Authority, and a Technology Directorate. A Department of Computing and Information Technology was established in 1984, with the goal of promoting the effective use of information technology in both public and private sectors. Following the Route 128 model, a Technology Park was inaugurated in 1984, close to the campuses of Murdoch University and the West Australian Institute of Technology.

Corporatism

These moves at the State level represent an extension of a familiar pattern in Australian history, whereby governments have played a leading role in stimulating industrial employment and capital accumulation through direct investment or fiscal incentives to private capital. Interdependence between government and industry has become particularly close in the post-industrial era, and some writers have suggested that this amounts to the emergence of a neo-corporatist form of relationship between the state and society. In a paper published a few years ago, I suggested six aspects of contemporary capitalism to which the growth of corporatist tendencies can be attributed (Encel, 1979). These are as follows:

1. Techno-industrial changes, especially the shift from labour-intensive methods of production, which create massive problems of investment , retraining and industrial competitiveness.
2. The dominance of multinational corporations, whose interests are served by a consensual style of politics and centralised industrial relations. At the same time, this very dominance generates attempts to protect locally owned industry from multinational competition.
3. Chronic instability of the workforce.
4. A high level of welfare expenditure, which gives state activity a crucial function in maintaining standards of living.
5. A continuing high level of unemployment, to which governments are under pressure to respond with job-creation policies.
6. A continuing high level of inflation and budget deficits, which also evoke continuous attempts by government to contain them.

Corporatism means, among other things, an attempt to integrate employers and trade unions into a tripartite framework of co-operation. According to Crouch, there has been a general tendency in Europe towards the reduction of the role of political parties by promoting deals with producer groups, and manoeuvring unions into acting as instruments for the mobilisation and integration of the workers (Crouch and Pizzorno, 1978). Panitch, a Canadian writer, notes the attempts of the former Canadian Prime Minister, Pierre Trudeau, to propagate a corporatist ideology of co-operation between the government, the employers, and the union movement. Corporatism, he argues, is an ideology of consensus imposed on a situation of class conflict, which works most successfully in countries with a strong social-democratic tradition, like Scandinavia and Australia (Panitch, 1977; Castles, 1978).Viewed in this perspective, we can interpret the policies of the Australian Labor Government which took office in 1983 as an experiment in neo-corporatism. While still president of the Australian Council of Trade Unions (ACTU) the Prime Minister, Mr R. J. Hawke, regularly used the rhetoric of partnership between government, industry and the unions as his preferred strategy for dealing with economic and social problems. The most spectacular use of a corporatist strategy was undoubtedly the economic summit conference convened by Mr Hawke a few months after becoming Prime Minister. This device involved a direct appeal to interested parties and culminated in a wages-prices Accord which became the Government's major instrument of economic policy (Kemp, 1983; Beilharz and Watts, 1983). The Accord was remarkably successful in achieving industrial peace, wage restraint and price stability, despite an unfavourable external trade situation. Among other things, the Accord provided for another corporatist strategy of encouraging workers' participation in industrial management, which Mr Hawke had vigorously supported as president of the ACTU. Corporatist theory also emphasises the growth of the administrative functions of the state and the blurring of traditional distinctions between public and private sectors. This has taken place in a number of countries, although with important institutional variations reflecting different political systems. In the United States, the proliferation of think tanks has created a new, hybrid structure of policy-making which forms a kind of external bureaucracy. In Britain, the acronymic term 'quango' became popular a few years ago to describe a typically British form of public-private hybridisation. In Australia, the most interesting development has been the establishment of a Senior Executive Service at the top level of the Australian Public Service, whose object it was to create a stratum of 'generalists' who would be deployed on a more flexible basis. In particular, the Senior Executive Service would be open to recruitment from the private sector; a process of interchange would also take place by which managerial staff were seconded from the public to the private sector and vice versa. The new legislation (enacted in 1984) also provides for the employment of ministerial consultants on short-term contracts (up to three years). Emphasis is to be given to the development of general management and policy skills (in public service jargon, 'helicopter vision').

THE POST-INDUSTRIAL SOCIETY AND THE THIRD WORLD

As noted at the beginning of this chapter, the concepts of industrial and post-industrial society embody the view that traditional societies will become modernised along the same lines as existing industrial societies. This involves a theory of history enunciated by the Enlightenment philosophers of the eighteenth century, espoused in a particularly compelling form by Marx, and existing in several variants at the present time. In its post-industrial form, as we have seen, it lays particular stress on the determining role of technology as an instrument of modernisation. Theories of this type have exercised a profound influence on international relations, which can be seen in the very use of categories like modernisation, development and underdevelopment.

Modernisation has indeed taken place in the traditional societies of the Third World—another concept resting on debatable assumptions. But, like other social changes, it has evoked strong reactions and violent conflicts. Instead of the implementation of development plans by modernising elites, many Third World countries are governed by military dictatorships whose prime interest is to stay in power and enhance their privileges. Attempts at industrialisation have worsened the economic situation of many developing countries and generated the north-south problem, a new metaphor which underlines the fact that world poverty is growing rather than decreasing.

Most spectacular of all is the emergence of Islamic fundamentalism. Whereas the emergence of military dictatorships and other corrupt, repressive regimes underlines the incidental difficulties of modernisation, fundamentalist Islam poses an ideological challenge of the most basic kind to the very idea of modernisation itself. Theories based on the assumption that the scope of rational thought will steadily increase have no answer to such phenomena as deliberate martyrdom on the part of Shi-ite terrorists, emulating their predecessors, the original 'assassins' of the eleventh and twelfth centuries. A recent analysis of the Iranian revolution of 1979 and its galvanic effects on Islamic society in the Middle East argues that the violence and fanaticism of militant fundamentalism can only be understood as a deliberate rejection of the values of Western culture, which is seen as a force destructive of Islamic values and independence (Wright, 1985). In more general terms, this confrontation provides strong evidence for the view that social change is a manifold process in which technology is one element, and not always the decisive one; it suggests, further, that technological progress is likely to contribute to an even deeper gulf between the affluent West and the countries of Africa and Asia.

CONCLUSION

Although the metaphor of the post-industrial society is still popular, it has lost a considerable amount of its original appeal as a vision of the future. Since it came into vogue twenty years ago, we have been reminded with increasing force that technological change and the social changes which accompany it have enormous potential for disruption, conflict and inequality. As Jones observes, technology can be used to promote greater economic equity, more freedom of choice, and a more participative society. Conversely, it can also be used to intensify the worst aspects of a competitive society, to widen the gap between rich and poor, to make democratic goals irrelevant, and to lead to a technocratic authoritarianism (Jones, 1982). In addition, technology harnessed to military purposes has the potential to make the world unlivable.Partly because of the ambiguities and inadequacies inherent in the concept of post-industrial society, and partly because of the spectacular expansion of information technology, there has been a shift of attention to the information economy and the information society. This is clear in the work of the Japanese writer Masuda, who attracted attention with his *Plan for an Information Society*, first published in 1981 (Masuda, 1981). In a more recent book, Masuda rejects post-industrial as an excessively vague term in favour of the information society, which will be characterised by the fact that 'the production of information values and not material values will be the driving force' (Masuda, 1985). This society, which he also calls Computopia, will move away from centralised government and class hierarchies to a system of participatory democracy, horizontally functional, maintaining social order by autonomous and complementary functions of a voluntary civil society.

This change of labels does not necessarily indicate greater clarity or profounder insight into the relationships between science, technology, and social change. If anything, the move from post-industrial society to information society may be seen, in theoretical terms, as a retreat from the crudities of technocratic historicism to the even greater crudity of technological determinism. Marien, in a sceptical comment on Computopia, notes that it is but one of many labels generated by the advance of information technology, none of which has lasted very long. He identifies eight areas in which the growth of information technology is likely to have ambiguous or conflicting consequences: work, commerce, health, entertainment, education, politics, intergroup relations, and family life (Marien, 1985).

The outcomes in these and other areas will be decided not simply by the technology, but by economic, social, political and cultural factors—the very opposite of technological determinism. We can identify certain situations in which it is clearly important and constructive to promote the use of information technology and to plan accordingly—as in the improvement of labour productivity, the improvement of weather forecasting, educational technology, and so on. But there is no way in which we can use metaphors like post-industrial society or information society as a basis for predicting the future in some general sense.

REFERENCES

Anderson, C. A. and Bowman, M. J. (eds) (1966) *Education and Economic Development*, Frank Cass, London.

Australian Science and Technology Council (ASTEC) (1985) *New Directions for CSIRO*, AGPS, Canberra.

Badham, R. (1984) 'The Sociology of Industrial and Post-Industrial Societies', *Current Sociology*, 32(1), 1-70.

Beilharz, P. and Watts, R. (1983) 'The Discovery of Corporatism', *Australian Society*, 2(10), 27-30.

Bell, D. (1973) *The Coming of Post-Industrial Society*, Basic Books, New York.

——— (1976) *The Cultural Contradictions of Capitalism*, Basic Books, New York.

Birch, L. C. (1975) *Confronting the Future*, Pelican, Ringwood, Victoria.

Braverman, H. (1974) *Labor and Monopoly Capital*, Monthly Review Press, New York.

Castles, F. G. (1978) *The Social Democratic Image of Society*, Routledge & Kegan Paul, London.

Christie, N. (1984) 'Venture Capital—A Window of Opportunity' in National Information Technology Committee, *Impact of Information Technology 1984*, AGPS, Canberra.

CSIRO (1977) *Report of the Independent Inquiry on CSIRO*, AGPS, Canberra.

Cotgrove, S. (1982) *Catastrophe or Cornucopia?*, Wiley, London.

Crompton, R. W. (1982) *Australia—What Future?* TSP Paper No. 3, University of New South Wales, Sydney.

Crouch, C. and Pizzorno, A. (eds) (1978) *The Resurgence of Class Conflict in Western Europe Since 1968* (2 vols), Macmillan, London.

Dedijer, S. (1966) 'Research Policy: From Romance to Reality', in M. Goldsmith and A. Mackay (eds) *The Science of Science*, Pelican, Harmondsworth.

Edding, F. (1962) in I. Svennilson, F. Edding and H. L. Elvin, *Targets for Education in Europe in 1970*, OECD, Paris.

Encel, S. (1971) 'The Support of Science', *Minerva*, 9(3), 327-41.

——— (1979) 'The Post-Industrial Society and the Corporate State', *ANZ Journal of Sociology*, 15(2), 37-44.

——— (1981) 'Pushing the Barrow Uphill', in S. Encel, P. Wilenski and B. Schaffer (eds) *Decisions*, Longman Cheshire, Melbourne.

Freeman, C. and Jahoda, M. (eds) (1978) *World Futures—The Great Debate*, Martin Robertson, London.

Goldthorpe, J. H. (1971) 'Theories of Industrial Society', *European Journal of Sociology*, 12(2), 263-88.

Gorz, A. (ed.) (1976) *The Division of Labour*, Harvester Press, Hassocks.

Gvishiani, D. M. (1972) 'Soviet Science Policy', *Science Policy Reviews*, 1, 22-31.

High Energy Particle Physics in the UK (1985) Report of a Committee Chaired by Sir John Kendrew, HMSO, London.

Illich, I. (1973) *Tools for Conviviality*, Calder and Boyars, London.

Jones, B. (1982) *Sleepers, Wake! Technology and the Future of Work*, Oxford University Press, Melbourne.

Kahn, H. and Pepper, T. (1980) *Will She Be Right? The Future of Australia*, University of Queensland Press, St Lucia.

Kasper, W., Blandy, R., Freebairn, J., Hocking, D. and O'Neill, R. (1980) *Australia at the Crossroads*, Harcourt Brace Jovanovich, Sydney.

Kemp, D. (1983) 'The National Economic Summit', *Economic Record*, 166(59), 209-19.

Kerr, C. (1963) *The Uses of the University*, Harvard University Press, Cambridge, Mass.

——— (1969) *Marshall, Marx and Modern Times*, Cambridge University Press, Cambridge, England.

Kumar, K. (1978) *Prophecy and Progress*, Pelican, Harmondsworth.

Machlup, F. (1962) *The Production and Distribution of Knowledge in the United States*, Princeton University Press, Princeton, N.J.

Marien, M. (1985) 'Some Questions for the Information Society', in T. Forester (ed.) *The Information Technology Revolution*, Blackwell, London.

Masuda, Y. (1981) *The Information Society As Post-Industrial Society*, World Future Society, Bethesda, Md.

——— (1985) 'Computopia', in T. Forester (ed.) *The Information Technology Revolution*, Blackwell, London.

Miles, I. (1978) 'World-Views and Scenarios', in C. Freeman and M. Jahoda (eds) *World Futures*, Martin Robertson, London.

Miller, S. M. (1975) 'Notes on Neo-Capitalism', *Theory and Society*, 2(1), 10-35.
Panitch, L. (1977) *Social Democracy and Industrial Militancy*, Cambridge University Press, Cambridge, England.
Poignant, R. (1973) *Education in the Industrialised Countries*, Martinus Nijhoff, The Hague.
Rigby, T. H. and Miller, R. F. (1976) *Political and Administrative Aspects of the Scientific-Technological Revolution in the USSR*, Occasional Paper No. 11, Department of Political Science, Australian National University, Canberra.
Sandercock, L. (1975) *Cities for Sale*, Melbourne University Press, Melbourne.
Schumacher, E. F. (1973) *Small is Beautiful*, Blond and Briggs, London.
Stretton, H. (1976) *Capitalism, Socialism and the Environment*, Cambridge University Press, Cambridge, England.
Strumilin, S. (1962) 'The Economics of Education in the USSR', *International Social Science Journal*, 14(4), 633-46.
Touraine, A. (1971) *The Post-Industrial State*, Random House, New York.
Vestes (1979) 'Symposium on the Williams Report', 22(2), 3-33.
Wilson, H. (1963) Preface to *Labour and the Scientific Revolution*, British Labour Party, London.
Wright, R. (1985) *Sacred Rage: the Wrath of Militant Islam*, Simon and Schuster, New York.

Chapter 3

THE INFORMAL SECTOR: ON THE PROSPECTS FOR DOING THINGS TOGETHER

D. Pym

The informal sector is a politically inspired misnomer for a re-emerging oral tradition which grows out of and is marginal to the institutions of industrial society. Its essential characteristics derive from the reintegration of social and economic life and as such it would be more appropriately described as the community economy.

The community economy still lacks an institutional basis and is therefore invisible to the literate mind and even to those who sustain it, the self-employed and more independent folk working from and around where they live. It does not exist as employment does for the employee but is composed of self-sustaining networks of people whose primary references are local and geographically limited. Consequently, knowledge workers and those who gain their living directly from institutions number among its least effective members. However, the impact of more flexible employment contracts and unemployment affect the community economy as these lead people to attend to the socio-economic possibilities around where they live.

In the short term, the crisis of Western societies is inextricably linked to the imprisonment of its thinkers and policy-makers by their own employment and the artefacts (writing, printing, computers, etc.) through which we attempt to define reality. The educated, literate mind cannot begin to grasp the potentials of the informal sector for reuniting economic and social life nor the rapidity with which community life might be revitalised because its most essential activities are invisible and covert and therefore lack legitimacy. To dismiss this under-institutionalised and apparently vacuous world as insignificant in the evolution of society is to advertise our trained incapacity to recognise what is real.

The scope for employment and its institutions in fashioning meaningful work, adding to the creation of wealth through people and enabling social order, is declining. By contrast the potentials of the informal sector, or what I prefer to call the community economy, are vast and will continue to grow as people spend more of their time in and around the household. Such prospects will improve vastly when we are able to break those collusive arrangements which currently prevent people from acknowledging the trivialising, demeaning and fraudulent aspects of modern employment (Pym, 1976, 1983).

Progress in this direction is helped by our acknowledgement that although the community economy is succoured by the visual literary order, its roots reach into a different tradition, to a re-emerging oral tradition. It grows from the crisis of Western societies and is aided by a shift in historical awareness from eye to ear, to a new balance in the sensorium which is already driving back the old visual literary order (McLuhan, 1964; Ong, 1982). This phenomenon is nurtured in part by the electronic feast but by a different course to that which we associate with the so-called Information Society.

The difficulties of knowledge workers in recognising and coming to terms with a tradition which rejects the basis of their power and influence is appropriately illustrated in the use of the concept 'spatial' throughout the chapters of this book.

Space is widely assumed to have only visual meanings, a perception which gives authority to distance, physical and social, and the abstraction of the visual. However, as we get physically closer to the subject of our attentions so a purely visual representation of reality gives way somewhat to an awareness of what we feel, hear, smell and taste. Ivan Illich uses the expression 'convivial' to describe living in this context and he wishes to advance those tools and techniques which enhance conviviality. I prefer the word 'intimacy' because it demands more of us; more obviously depends on physical proximity between people and between people and machines and nature. It reflects, too, on the limitations of informing devices in these respects.

The community economy is founded on the oral tradition, household, human resourcefulness, a maintenance ethic and the use of electronics to extend human communication. It contrasts with the information economy which represents a technology-driven continuation of the literary tradition, employment, consumption and the enlarged authority of the institutions of industrial society whose legitimacy we must now seriously begin to doubt (Habermas, 1979).

DEFINING TERRITORY AND TERMS

The title of my chapter is both ambitious and daunting and its proper consideration beyond my capabilities but it is essentially about basics, people doing things together, and how we might begin to value these things. I trust the subject is familiar but nobody can be sure in this world of contrived complexity where the simple and natural, like talking to one's fellows, have become more difficult. (This is no idle assertion. At the institution which employs me, the London Business School, the most popular courses in the behavioural area involve teaching people interpersonal skills.) So I am not proposing some radical new perspective but a way of thinking upon and valuing our affairs which elevates the everyday. Therein lies the way out of our institutional morass.The phrase 'the informal sector' is not mine but I am more than happy to explore its meaning and origins. I think we know to what it refers: it has derivative overtones; it is less than the formal; it grows out of the frailty of assumptions which bestow on employment a monopoly over the creation of wealth; it implies that we have been driven, in our search for new employment, into acknowledging the existence of economic activities outside employment; it signals the demise of universal employment. It is, indeed, unknown, threatening, dangerous territory. A simple analogy might just help the reader to grasp the difficulties in contemplating the subject despite its essential nature. In terms of the characteristics of an established community economy, consider the problems faced by General Westmoreland in estimating the strength of the Viet Cong and the likely nature and direction of their attack shortly before the 1968 Tet Offensive.

The informal sector encompasses the economic pursuits of people operating in and from the household and within the community. Their self-generating and small-business activity, individually and collectively based, gains from and operates around the institutional world. Those who draw attention to the parasitic elements of the informal sector are typically ignorant of its existence now and before men and women went out from their homes to factory, office and laboratory to fulfil their servile duties as employees. For propaganda purposes, I prefer to describe the informal sector as the community economy, the community being the place where people's economic and social lives can be reunited through doing things together.

LIMITATIONS IN THE TRADITIONAL PERSPECTIVE: THE FORMAL ECONOMY

We must begin by exploring some dimensions of our difficulty in recognising both the crisis we have made for ourselves and how it might be tackled. This necessitates some reflections on the formal economy from a viewpoint sympathetic to our subject. The perceptions and thinking of liberal-educated man are catastrophically constrained by his status as employee and the artefacts through which he perceives and tries to relate to his environment. In the prophetic words of William Blake (Keynes, 1977) what we perceive, we have become. So we value organisations while discounting organisation. We attribute the process of informing to devices which locate authority in exchanges outside people, that is, they deform or 'outform', perhaps? We confuse informing and communicating and deny the essential intersubjective nature of human communication and its fundamental role in good order.

In identifying organisation as a physical entity represented by abstractions—annual reports, charters, balance sheets, budgets, plans and the like—we locate authority outside people and their actions. This view is reinforced by what ownership has come to mean to us. As employees we do not own our skills, products, time and space. In Marx's time this ownership was frequently in the hands of other people but often other people known to the employees. Today ownership, where it exists, is typically either a euphemism for debt to money lenders (e.g. home ownership) or in the hands of some abstraction—the state, stockholders or multinational corporation.

Without the advent of writing, of print and now the computer, the extent of authority vested in organisations could not have come about. The founding fathers of this tradition were more wary about its likely detrimental effects than we have proved. Socrates believed that our artefacts were without value. The purpose of his questioning was to discover the good in oneself. He valued experience, particularly manual work, in the development of wisdom (Hackforth, 1933). In the *Dialogues* (Jowett, 1895, p. 284) Plato has Socrates respond to the view of the inventor of writing that it would improve memory, wit and wisdom (human attributes most valued by the ancients) thus:

> Inventors of an art, cannot judge its uses. On the contrary, writing will lead to a loss of memory as people come to trust and depend on external written characters. Such people (the servants of modern organisations?) will be hearers of many things and have learned nothing, appear omniscient and know nothing and they will be tiresome company having the show of wisdom without the reality.

A more appropriate definition of organisation and one not technically determined could be expressed as 'people doing things together'. This apparently simplistic representation ignores the time and space of context and the critical influence of what it is that people do in their relationships around a shared task. Nevertheless, it ensures that the locus of authority stays with the participants. To be fair to the prospects for the revival of community, we must continue with such definitions of terms, a none-too-easy responsibility.

Even an apparent sympathiser with the informal economy like Ray Pahl (1984), when faced with the choice between representing his subject in its own terms and pandering to institutionalised academic sociology, chose the corporate researcher's glass and saw nothing of consequence, though at least he rendered my argument a service. Pahl relied for the bulk of his evidence on an interview-questionnaire study conducted by another party and on historical documents. His concern was to render the invisibility of his subject visible and the covert, overt. His findings showed that informal sector economic activity in the Isle of Sheppey was at an all-time low. He expected people in a one-off interview to reveal personal details about their less legitimate economic activities to a stranger, and this in a society in which the private is, in many respects, sacrosanct. But let us suppose Pahl's observations are correct. Then we must also recognise the subject of his concern and his methodologies as a remarkable concession to his employment and to a conceptually empty methodological empiricism which offers little or no hope to those whose prospects and opportunities within employment are, for all intents and purposes, non-existent. Pahl thinks he has captured the truth when he has merely taken the alienated condition as normal. This, one might conclude, represents an advertisement for his own condition. He makes it clear that questions of morality and hope represent no serious concern to the person as researcher. Nevertheless we can be sure that somewhere among Pahl's sample are to be found people living with dignity on the fringes of the institutional world.

Just how biased the rational empirical mind can be is evident in the covert assumption involved in the cry for data. For within this perspective, information, that is not expressible as an abstraction of the visual sense, lacks legitimacy. To repeat myself, we assume that information is reducible to the affairs of one sense, abstracted. This is a remarkably impoverished view of ourselves. Little wonder so many of us are short on dignity and self-respect. And it is good for business to propose extending this impoverishment. The manufacturers of electronic-information devices tell us we need still more visual, literary and numerical data, faster. It clarifies the situation. With all this additional data people will know more and their judgement will be enhanced. Fantastic, but also insane, for confusion is now everybody's experience.

Employment as a condition helps too. When the employee complains upon the lack of information in communication, as most do in large corporations, he or she is placing the responsibility for knowing on some other party. The word 'inform' looks more and more like a politically and technically determined misnomer.

Unfortunately the inadequacies of orthodoxy don't stop there. Walter Ong (1982) of the University of St Louis is now my authority. In all one-way informing devices, he observes, the message sender must fictionalise the message receiver and visa versa. This means the state of body and mind of the other party is not known and must be invented. Such invention is not random but is based more or less on previous experience or on what some external authority tells us. As the experience fades into the past so our presuppositions easily become cast iron stereotypes. Such stereotypes can be modified, but if the actor is to remain in control, some direct contact between himself and his subject is necessary for this to happen. Informers, that is those whose employment give them authority in this respect, are not generally disposed to this tack. Indeed we are likely to regard our informing devices as substitutes for human contact. That too is one basis on which most informing devices are sold. So our increasing dependence on informing devices creates views about ourselves and the world around us which compound the feelings of isolation and powerlessness informing techniques visit upon us, making us the easy prey of the institutional world.

The information explosion belongs less to the future than to a rearguard action in defence of the industrial order, its institutions and the linear sequential way of handling experience which underlies that order. This then is electronics in the service of the industrial estate—big employers, central authority, bureaucracy, abstract education and rampant consumerism. The information society, as we currently reckon it, is a dangerous nonsense because it fictionalises people not as gods but as one-dimensional, one-sense, less-than-human beings. It advances, too, those very cultural artefacts which preserve the most dehumanising aspects of industrialisation, namely the erosion of meaning through its divorce from the realities of the everyday, the deification of the abstracted and the acquiescence of socially grounded notions of freedom to an abstract totalitarianism which the formal, public, institutional domain is beginning to represent to more and more citizens. In terms of social dislocation, this fifth column, the real enemy of good order, has greater impact than any revolutionary political movement.

HUMAN COMMUNICATION, ORGANISATION AND COMMUNITY

In his survey of the direction which organisation theory might take, Pfeffer (1982) identified three strands which he associated with organisation as process; namely attention to social networks, physical environment and demographic aspects of population. Pfeffer offers no context of these developments but we can assume that for him these would express themselves within the formal system, within organisations. They are also among the more important elements of a viable community economy.

A Community Predisposition

The picture below is uneven, romanticised and banal because such a description reflects the way many of those who live a community life might represent it if they were able and willing. My purpose is not to present the facts, those grim abstracted details, but to explore the possibilities, the fantasies and realities of those who already struggle to earn their crust on the fringe of the institutional world. In short, my task is one of legitimation.

The pre-occupation of people in the community economy is with getting by and they do this in no small measure by living their task(s). Often these people are sustained by idealism, an idealism that may be rooted in, or fly in the face of, reality. A strong belief in, for example, the sacredness of the task, independence, ownership, the appropriate use of tools, more 'natural' ways of living and any of a range of religious and political ideologies, provides the driving force among people whose energy, enthusiasm and human compassion frequently contrast with the lethargy and quiet despair of those who feel themselves prisoners of the formal system (which includes most of the unemployed). In the view of the knowing observer, people living within the community may endure relative material deprivation and seldom take holidays (but they live with dignity), which contrasts with the familiar public assessment of them as fiddlers and tax dodgers.

The word 'community' carries with it a hundred meanings and many of the misconceptions could be attributed to the authority of the formal system. As a meaningful, if vague, notion in our terms, it springs less from the activities of people in and around local government, business, parish council, school, church or voluntary organisation than from those fluid, invisible, more private relations between people engaged in a mass of commonplace, taken-for-granted, mundane, everyday activities and routines in which we all engage to varying degrees. Its most profound feature is its invisibility (relations could not represent anything else to the literate eye) which of course constitutes no problem for the insider. For this community depends on organisation rather than organisations, and an acceptance of intimacy as the basis of social order. The human brain fed by the whole sensorium, not the computer, is its memory story.

Though we are not all employed we are all party to a miriad of activities, some pursued in isolation, others jointly which, employment notwithstanding, make up the greater proportion of the wealth created by the direct efforts of people. The community economy begins in the household where some two-fifths of all capital is invested and where we spend more than half our waking lives (in contrast with the less than 10 per cent in employment). In Britain the focal human unit in such activities is the person rather than the family, the key figures being self-employed or those renowned for their independence in their employment. This is not a place for those who advertise themselves as unemployed, even though some of its members may be on the dole. Most likely, the wag will tell you, he's unemployable. Here social and economic activities are inextricably interlaced and yet bound by time and space. Territory has meaning for people who will forego financial and other opportunities when these lose their social meanings, whatever the justifications given for doing so.

Organisation without clear, visible structure calls for a strong maintenance ethic. Time spent with others maintains the organisation, ensures connections and coherence, and builds commitments. So time spent yarning in the pub, club, cafe, household kitchen, street corner or on the telephone may be lost work time but they provide the place for those exchanges which create the social glue that we in the institutional world try to achieve by written contracts, procedures, regulations and a million pieces of paper. In a world which elevates context, the hero is a local not a cosmopolitan. Stories which establish and embellish the standing of local characters and exemplars carry more important social messages in reflecting on their ingenuity, resourcefulness, capacity to make things happen, and keep machines or relations going or success in beating the system. The fixer or bricoleur is our mythical private hero who contrasts with the public achiever or entrepreneur (Pym, 1985). By contrast, anti-social behaviour, identifiable by its association with responding complaint or hostile gesture, represents failures in reciprocity, basic incivilities, greed and egocentrism. Such are the themes of stories based on an undercurrent of complaint. The importance of reciprocity in informal social networks can also be reflected in the intrusions of the sensitive, outsider. The doing of good turns for another which come spontaneously to less materially advantaged folk is likely to represent a matter of continuing awkwardness and concern to the more privileged trying to break into this exciting if unsettling scene.

Still on questions of maintenance, the ordering of the day through personal or shared routines and small rites presents few problems, but these are easily ignored in a sustaining community. Being around and performing our daily rituals are the taken-for-granted and ignored essentials of community. A community based on itinerants and people pre-occupied with the novel is a hazardous affair, as most holiday towns so aptly demonstrate. Rites and ceremonies of greater substance are necessary too and, in my view, remain weak or absent. Street parties, feasts, and celebrations in regular form are needed to complement births, marriages and deaths. However, in the United Kingdom there are now signs of a resurgence of live entertainment and seasonal rites in pub life.

In personal matters the demand for exercising judgement and choice far outrun experiences in employment. These may also be the raw ingredients in the development of dignity and self-respect. Minds must be continually made up over jobs, priorities, matters of association, the use of tools and moral questions; second nature to the sophisticate but disturbing for the novice. 'Live and let live' constitutes a familiar value in everyday dealings. In tighter circumstances justice takes priority over the law. The oral mafia, as Marshall McLuhan (1969) observes, can always master a literate, mechanistic structure by simple by-pass. Social debt, repayments in kind, is preferred to financial debt not the least because relative material deprivation (contrary to formal mythology) is a common experience. Neighbours, whether loved or hated, offer better prospects for co-operation than their abstractions—bank and building society. As Hugh Stretton (1976) tells us, a high proportion of Australian homes built before 1940 were erected by the original occupants and their neighbours. These were homes for living in rather than speculating upon. Money lenders have now done people out of homes, neighbours out of jobs and would convert the community to a virtual wasteland.

Security or its absence remains a pre-occupation of the self-employed. Keeping the jobs coming in and the money too is a big issue because it is seldom discussed. To this end saving takes the form of uncollected debts and the banking of good turns. Those who find themselves carrying a pile of unsettled debts can assume they have been assessed as reliable and safe payers by their colleagues and so able to act as a bank for a rainy day. Matters of this kind frequently remain unrecorded and appear to be treated with amazing casualness. Neither interest nor inflation exist in this world of controlled expectation so every man can think he gains. The authority of the institutional world is reduced to manageable proportions by representing its entirety in one or two personal contacts. In this simplistic, personalised representation, our man *is* the council, the bank or the school.

Rapid psychological obsolescence in the consumer society makes yesterday's still-new machines and tools available and relatively inexpensive. Once people know what you're after, calls and suggestions roll in. Equipping at low cost, provided the recipient has the maintenance skills, is the insider's privilege in a milieu in which scavenging and scrounging are the norm and the smart local—mobile, observant, quick to do himself or a neighbour a good turn—is ever-ready to grasp an opportunity. Priorities and knowing what is needed work-wise is greatly simplified by the centrality of task. In contrast with the trivialisation and denial of task in employment, task in community matters, is never far from one's thoughts, is often sacred. Task modifies all other predispositions, no doubt to the horror of the theory-obsessed academic, for people must live and survive doing with others those things which sustain them materially, socially and spiritually.

Optimising the Prospects

Inevitably there are circumstances which favour the evolution of a viable community economy. It is easier to get going if you are self-employed, in the centre of a large city, country town or a rural area and where there are not excesses of unemployed people or highly homogenous populations, as in seaside retirement or single-industry towns, and many suburbs. In general, opportunity increases with population density. Unlike in the formal economy, women who are not imprisoned by the home are likely to find themselves more favourably placed than men. Recent figures in the UK show the new self-employed to be predominantly women. People with manual skills and basic social skills will find the going easier than professionals and executives. There are signs, not yet in the statistics, that young people without employment are staying on at home longer, a situation which does not have to be disadvantageous provided home space is available to them and parents encourage their initiatives, economic and otherwise, on a local basis.

The very nature of some work offers better prospects than others. Small builders and repairers of household effects establish their socio-economic networks and clients by virtue of having to be out and about on the job. By contrast arts-and-crafts folk are isolated by the nature of their work and frequently by personal disposition; many see themselves as refugees from the rat race. The 30,000 or so arts and craftsmen and women in the UK have low incomes (average less than £4,000 per year) and widely report their isolation and loneliness. Yet it is not difficult to recognise some of the ways in which this vicious circle might be broken. People in such a predicament might move to live nearer others similarly occupied, muscle in on the networks of builders and repairers, go to more country shows and town festivals, use agents and retailers, spend more time selling, or establish contact with small farmers and local businesses who could do with the craftsman's labour in exchange for workshop space and so get the craftsman into networks that would sustain him.

Work in the Community Economy

We are inevitably moving towards the kind of work that is most available in the community field. To recap: wealth-creating activity in this sphere is characterised by the integration of social and economic concerns, a strong maintenance bias, dependence on human-scale tools and techniques and the importance of context. People belong to it by virtue of their territorial allegiances both in terms of association and markets. This excludes the self-employed professionals working from home in splendid material isolation, for they are financially, socially and spiritually part of the institutional world unless or until they turn their attentions and skills to the immediate environment.

Household Maintenance remains one of the largest if least exploited opportunities for paid work outside employment. Surveys of household effects and possessions reveal an ever-increasing mass of machinery, furnishings and artefacts which consumerism would have us discard long before their useful life has been realised. A decline in the production-consumption ethic, the emergence of the ecological perspective, limitations of available finances will enhance the prospects of self-employed people who can undercut the corporations' after-sales services and provide more personalised attention. Furthermore, reductions in sales will have less effect on employment as time goes by, because the growth in household maintenance will substitute machine based production with work for people.

Building and Renovation represent a major growth area in self-employment over the past decade. Raw-material costs, shortages of building land, building restrictions and high labour-costs are providing a field day for small builders, joiners, plasterers, plumbers, electricians and the like who gain by the shift in building practices from green-field developments to renovations and redemption. The high cost of energy has also led many householders to alter and establish heating and insulation arrangements, which represents still more opportunity for the small operator. The rising interest in history, conservation and ecological issues will sustain these trends. Opportunities in community production would appear to be patchy. This category includes not only small local manufacturers who have gone out of existence but also those setting up to produce quality consumer goods—clothing, furniture, specialists in one off machinery requirements of the formal economy and the food and flowers from market gardening operations. The whole range of arts and crafts might be included here; experience in these activities would represent a good measure of the opportunities. Although long hours and small financial rewards define the lot of those in cottage industries, the situation is improving.

Local Health and Care will continue to offer opportunities as the state finds itself unable to carry the mounting cost of industrialised and depersonalised health and welfare. In spite of opposition from the medical profession and health services, a whole range of alternative medical services (herbalists, hydrotherapists, dieticians, psychologists, osteopaths, 'natural' healers, etc.) are emerging whose roots are in community health. The pre-occupation with health through diet similarly helps the small holder and organic grower and exposes the political undercurrents in the word 'inefficient' used to dismiss the small farmer's efforts over several decades. Local household groups are able to assist, too, with childcare, support for the handicapped and aged, leisure and sporting facilities, the provision of meals at home and the like.

Small Scale Contracting in a host of services needs most of all a good knowledge of local requirements and limited capital to get going. Again the problem is less one of opportunities than a way of exploring and representing the world in which we find ourselves. Neither education nor youth training have much to offer in this respect, which brings us back to the three prerequisites for the adventurer in the community economy, namely independence, resourcefulness and the ability to get on with others on a permanent basis.

Government Attitudes and Action That Would Help

Although my purpose has been to emphasise the economic elements of community organisation, the underlying purpose is social, that of community reconstruction. An appropriate government response would be to concern itself more with the practice of democracy and less with its mechanisms. This necessitates a shift from the current pre-occupations with legislation and control towards the role of enabler and protector of good order.

The regeneration of inner-city areas, an emphasis in building policy on renovation and redemption, efforts to conserve and protect history and environment and the like, would all aid reconstruction, as would liberal taxation policies towards the self-employed and those who work from home.

Within this frame of reference government would concern itself with freeing rather than fragmenting and limiting the use of time and space. It might encourage employers and employees to seek more flexible employment contracts, for example, rather than go for legislation on shorter working hours, consideration might be given to work sharing and broadening the basis of corporate ownership. Such legislation would reduce the efficiency of the formal economy while sustaining its excessive authority. By the same token, many of what Richardson (1982) describes as industries of tomorrow—information technology, defence, financial and investment services, tourism, high fashion—are in reality the industries of latter day industrial society. By contrast, the development of two-way communications—through videophones, two-way cable, cellular radio, satellite transmission—could maintain the locus of authority with people and enhance the community economy.

There is one major mechanism that would boost the community economy (and the whole idea of a dual economy, Pym, 1980) and make it a viable political force in the short term. This is the substitution of the dole or unemployment relief by a social wage or basic income to be paid to every citizen of voting age. This idea is not new. It is an old plank in Social Credit policy. Tax thresholds or supplementary benefits already determine a minimum standard of living. This idea is already being explored by representatives of all political parties and a variety of interested bodies through the Basic Incomes Research Group in London. Such a policy is important for the following reasons: it ensures a minimum income for all citizens; frees the labour market; denies the monopoly of employing institutions over the creation of wealth; generally contributes to the scope for true democracy; and recognises the large proportion of wealth already generated by machines in the formal sector—a heritage in which we should *all* share.

CONCLUSION

In summary, we must represent the community economy as a return to fundamentals rather than as an alternative, as a second Renaissance, and as an opportunity to elevate people and their everyday affairs. The need to do so is more serious than we literate folk can know, for the crises of our time strongly suggest that the industrial establishment may have lost the capacity to renew itself.

ACKNOWLEDGEMENTS

My thanks are due to the Management Department, University of Western Australia, and to Dr L. V. Entrekin and Dr M. Goldstein whose support gave me the time and space to write this document.

REFERENCES

Habermas, J. (1979) *Communication and The Evolution of Society*, Heinemann, London.

Hackforth, R. (1933) 'Great Thinkers: Socrates', *British Journal of Philosophy*, 8, 259-72.

Keynes, J. (1977) *The Writings of William Blake (1757-1827)*, Nonesuch Press, London.

Jowett, B. (1895) *The Dialogues of Plato*, Volume 1, Clarendon Press, Oxford.

McLuhan, M. (1964) *Understanding Media: The Extensions of Man*, McGraw-Hill, New York.

——— (1969) *Counterblast*, Rapp-Whiting, London.

Ong, W. (1982) *Orality and Literacy*, Methuen, London.

Pahl, R. (1984) *Divisions of Labour*, Blackwell, Oxford.

Pfeffer, J. (1982) *Organisations and Organisation Theory*, Pitman, Boston.

Pym, D. (1976) 'The Demise of Management and The Ritual of Employment', *Human Relations*, 28(8), 675-98.

——— (1980) 'Towards the Dual Economy and Emancipation from Employment', *Futures*, 13(3), 223-37.

——— (1983) 'Emancipation and Organisation', in N. Nicholson and T. Wall (eds) *Theory and Practice of Organizational Psychology*, Academic Press, London.

——— (1985) 'Heroes and Anti-Heroes—The Bricoleur in Community Reconstruction', *Futures*, 17(1), 68-76.

Richardson, C. (1982) 'Urban Revitalization: Appropriate for To-day', *Work and Society*, Report No. 5, HMSO, London.

Stretton, H. (1976) *Capitalism, Socialism and Environment*, Cambridge University Press, Cambridge, England.

Chapter 4

THE IMPACT OF TECHNOLOGY ON THE HOUSEHOLD ECONOMY

D. Ironmonger

Individually, households are small units; collectively they constitute a large economy. On a world scale, households may produce more than 50 per cent of all valuable output. Even in advanced countries with well-developed market sectors, the household produces about one-third of all output and uses about half of all productive time. The impact of technological change on household production is thus potentially as significant as the impact of technological change on market production. Moreover, the purchases by the household economy of energy, materials, services and equipment from the market economy provide the avenue for many major technological diffusion processes.

This chapter begins by reviewing some of the estimates of the value of production, time input and capital use in the household. It then explores some of the major processes of technology diffusion that have taken place in the flow of commodities from the market to the household and concludes that these processes have provided the major mechanism for the achievement of economic growth, as judged by the evidence for the United Kingdom in the 1920s and 1930s.

Household Production in Great Britain

Colin Clark's (1958) calculation for Great Britain showed that in 1871 the value of work done in the home was greater than that done away from the home. Since then, although a substantial amount of work has moved out of the household, much has remained; some work has moved back to the home. A good example is laundry which, after having a decade or so outside the home in large commercial steam laundries, moved back to the home with the widespread use of mechanical washing machines. Dry cleaning and some laundry such as curtains and drapes are still largely commercial activities in the market economy, but do-it-yourself laundrettes, though technically in the market economy, require large inputs of non-market time from householders. Clark's study calculated that in Britain unrecorded household production declined from over 100 per cent of recorded net national product in 1871 to 43 per cent in 1956 as shown in Figure 4.1. It is likely, however, that Clark's estimates are underestimates of the total value of household production because his method of estimation involved some very rough assumptions, a common problem for national accounting pioneers.

Figure 4.1 Great Britain: 1871 to 1956. Unrecorded Household Production (Per Cent of Recorded National Product). **Source:** Clark (1958).

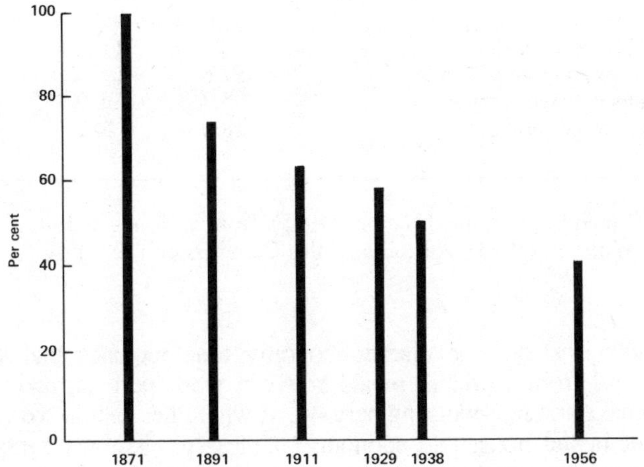

Time-Use Studies

In the 1960s and 1970s a number of time-use studies were made, one international (Szalai, 1972) and the others mainly in the United States. These show that in the 1970s people in the industrial countries spent about as much time in non-market productive work as they spent in market work. An illustration of the results from these surveys is shown in Table 4.1 which gives the distribution of productive time for housewives, women in paid work and market-employed men in the United States in 1965, in Britain in 1972 and in Australia in 1974. The household tasks included in the household economy include cooking, cleaning, housework, childcare and shopping but do not include personal care, self-education

Table 4.1: Productive Time: Hours Per Week

	Household Economy	Market Economy	Total
United States: Forty-five Cities, 1965			
Housewives	55.3	0.6	55.9
Women in market employment	28.1	40.4	68.5
Men in market employment	12.8	48.0	60.8
Britain: London and Home Counties, 1972			
Women not in paid work	45.5	-	45.5
Women in part-time paid work	35.3	26.3	61.8
Women in full-time paid work	23.1	40.2	63.3
Men	9.9	49.5	59.4
Australia: Melbourne, 1974			
Women, non-wage earning	57.6	-	57.6
Women, wage earning	35.7	30.2	65.9
Men, wage earning	18.6	44.2	62.8

Source: United States, United Nations (1975); Britain, Young and Willmott (1973); Australia, Cities Commission (1975)

and voluntary work. The market-economy time includes that spent in travel to and from work, in meals between work periods, and in other regular breaks and non-working periods. At what rates should we value the time spent in the household economy? Unless productivity ratios differ markedly, approximately equal time inputs should produce approximately equal outputs of goods and services. Thus, if we value commodities produced by the household at comparable rates to those produced by the market, household output in the United States, Britain and Australia in recent years would be about the same order of magnitude as market output in those countries.

Modern Estimates of the Value of Household GDP

Several estimates of the value of household GDP have been made by using the data from time-use studies. Hawrylyshyn (1976) presents a survey of some of these. Morgan, Sirageldin and Baerwaldt (1966) estimate that the inclusion of unpaid housework in 1964 would have increased United States GNP by 38 per cent, Sirageldin (1969) estimates the value of unpaid output for the American family in 1964 at approximately 50 per cent of its disposable income, and Leuthold (1981) estimates that for the American family in 1975 home production income was approximately $8,400 or roughly 52 per cent of family disposable income.

Gronau (1980) finds that the value of home production associated with the work at home of United States wives in 1973 exceeded 60 per cent of the family's money income before taxes and 70 per cent of the family's income after taxes. These estimates were based on the *Michigan Study of Income Dynamics* using the 1974 panel data for the 1973 activities of white, married couples. According to Gronau's estimates, the average value of home production of a United States household in 1973 was over $7,500. It was close to $6,500 when the family did not have pre-school children and reached almost $10,000 when it had. These estimates ignore the value of home production due to husbands and other members of the family.

The estimates need closer scrutiny, not only to examine the basis for the valuation of labour inputs, but also to see whether a contribution from physical capital has been included. Again, nearly all the studies of time-use omit productive inputs from young people. Furthermore, although the time studies record the work we do for ourselves each day on personal-care activities such as bathing, shaving, dressing, undressing and going to the toilet, the valuations of household production exclude this time. Time inputs in personal care and from young people may each add another 10 per cent to household productive time.

To resolve the issue of whether household production in modern industrial countries is one-third, one-half, or more of all economic activity would require closer examination of the data, the assumptions and the hidden value-judgements incorporated in their compilation. In earlier times for these countries a much higher proportion of economic production was carried out in the household or on adjoining household land. This would also be true for most countries of the developing world today. In spite of its impressive figures, the market economy was and is less important than we have been led to believe.

Robert Eisner and colleagues at Northwestern University (Eisner, 1978, 1980a, 1980b; Eisner and Nebhut, 1981; Eisner *et al.*, 1982) have made various studies which have produced augmented GNP estimates for the United States. These studies produce a series of accounts which cover productive activities within the household and which are known as Total Income Social Accounts (TISA). In the 1982 study they report: 'The business sector in 1976 produced only 48 per cent of TISA GNP, households produced 38 per cent. The remainder comprised government production, 10 per cent and non-profit and government enterprise output, 3 per cent.' In these estimates unpaid household labour services are valued at the wages of paid domestic employees and estimates are included of the contribution of household capital to gross product.

Capital Used in the Household

The Eisner estimates show for the United States in 1976 a tangible capital value of $3,052 billion in the business sector (44 per cent of total tangible capital in the economy) and of $2,185 billion (32 per cent) in the household sector. The remaining tangible capital comprised $256 billion (4 per cent) in government enterprises, $1,293 billion (19 per cent) in general government and $150 billion (2 per cent) in the non-profit sector.

Tangible or physical capital excludes intangible or intellectual capital—the R & D investment by businesses and human capital investment by households. The inclusion of estimates for the value of intangible capital completely turns around our picture of the relative importance of the business and household sectors (see Table 4.2). On this more inclusive basis, the TISA estimates show that some 63 per cent of the total capital of the United States is in the household economy, only 25 per cent is in the business sector, 9 per cent in general government, 2 per cent in government enterprises and 1 per cent in the non-profit sector.

Although human capital is, of course, hired out to the business and government sectors, it is owned by and employed in the household sector. The household sector gains a rather greater importance than usual when we realise that almost two-thirds of the capital of the United States is used in that sector, though not exclusively.

Table 4.2: Total Capital: United States of America, 1976

Sector	Tangible	$ Billion Intangible	Total	%
Household	2185	6473	8658	63
Business	3052	329	3381	25
Government enterprise	256	-	256	2
General government	1293	-	1293	9
Non-profit	150	-	150	1
TOTAL	6936	6802	13738	100

Source: Eisner, Simons, Pieper and Bender (1982, p. 167)

NEW TECHNOLOGY IN THE HOUSEHOLD: SOME DIFFUSION PROCESSES IN THE UNITED KINGDOM, 1940-1960

Changes in the technology of household production can operate through two avenues. The first is through changes in the characteristics of the commodities transformed by the household into final consumption services. Thus the introduction of new forms of food and drink such as coffee, tea, margarine, canned vegetables and ice-cream enabled households to improve the variety and quality of household meals.

The second avenue is through changes in the characteristics of the equipment used by the household to effect these transformations. Thus sewing machines and clothes washers enabled improvements in household production and maintenance of clothing; electric stoves and refrigerators improved the household meal-production process through the saving of time and effort in cooking and through the prevention of spoilage and the maintenance of food freshness.

Table 4.3: New Household Technology: Timetable of Introduction, United Kingdom, 1840-1960

	Materials and Services	Equipment
1840s	Tea; railways	
1850s	Oranges and lemons	
1860s	Dry cleaning	Sewing machines
1870s	Margarine	Bicycles; perambulators
1880s	Telegrams; eating chocolate; tramways	Pneumatic tyres; electroplate
1890s	Canned and preserved foods; newspapers and periodicals; bananas	Telephones; motor cycles; reliable bicycles
1900s	Proprietary medicines; soft drinks; electricity	Plywood furniture; electric lights
1910s	Cinema; milk chocolate; cigarettes; packaged tobaccos	Photographic goods; furniture suites; safety razors
1920s	Wrapped bread, pasteurised milk; grapefruit; ice-cream	Motor-cars; electric heaters; electric cookers; radios; shorter skirts; zip fasteners
1930s	Bottled milk; sliced bread	Rayon and artificial silk; crease-resistant fabrics; refrigerators
1940s		
1950s	Detergents; colour film	Television; nylon and terylene fabrics; motor scooters; long-playing records; magnetic recording; ball-point pens; polythene

Both avenues—improvements in the materials and improvements in the equipment of household production—have been the subject of technological change. Table 4.3 (above) illustrates the changes in both areas of household production in Britain during the years 1840 to 1960. This table is adapted from Ironmonger (1972, p.133) and indicates for each new technology the decade during which it was introduced into the household. This introduction period has been taken as the time when there was a sharp take-off in the rate of consumption of the commodity. This generally occurs at a slightly later period than the first sales of the commodity. For example, although the first shipment of bananas came to Britain in 1878, the take-off in the consumption of bananas does not appear to have occurred until the 1890s. A counter-example is radio, where the take-off occurred immediately upon introduction in 1923.

Duration of Technology Diffusion

Although some changes in household technology were completed in the comparatively short time of about twenty years, many changes in household technology seem to have taken fifty or sixty years to complete and some are still in progress after seventy or eighty years. For a number of the commodities listed in the timetable of introduction to the household, statistics were obtained which enabled approximate estimates of the duration of the diffusion process. The statistics were expressed in terms of annual consumption per head of population.

Not all new commodities continued to grow without pause. Some declined, as did coffee in the period from 1850-1910, and rail passenger journeys after 1908. Nor are the rates of growth very even from year to year or from commodity to commodity. Tea seems to have been one of the slowest-growing commodities, whereas radio was among the fastest.

Table 4.4 shows the approximate duration of the diffusion processes for six commodities with apparently completed diffusion processes and two commodities with incomplete diffusion processes. Those with completed diffusion processes have tended to have shorter durations as time goes on, decreasing from ninety years to twenty years. However, there are two commodities with incomplete processes that, it would appear, will eventually have durations of seventy and eighty years. Although there is evidence that lengthy diffusion processes are still proceeding, two hypotheses are suggested. These are:

1. The rate per decade at which new technologies are introduced into, and accepted by, the household has increased.
2. The speed of absorption of new technologies is faster now than was the case a century ago.

These hypotheses reinforce each other and are supported by the expectation that there is a greater current awareness on the part of market suppliers of the opportunities for exploiting new products, and on the other hand a greater degree of communication between households and between suppliers and households today than formerly. Households on high incomes should also be more prepared to hazard the risk involved in buying an unknown new product than households on low incomes. Thus, insofar as incomes have increased, it would be expected that there would be an increase in the amount of experimental purchases and the quicker discovery of suitable, new, household technology.

Table 4.4: Diffusion of New Household Technology: United Kingdom

New Technology	Take-off Date	Completion Date	Duration Years
Tea	1840s	1920s	90
Railway journeys	1840s	1900s	70
Oranges and lemons	1850s	1930s	70
Bananas	1890s	1930s	50
Radio	1923	1950s	40
Television	1950	1970	20
Cars	1920s	1990s?	70?
Electricity	1930s	2000s?	80?

Source: Ironmonger (1972)

To test these hypotheses thoroughly would involve making a complete catalogue of the new products introduced into the household market and estimating the rates of growth in the purchase of these commodities. As the bias of the quantity of available information on commodities is towards recent periods, there would be a bias towards finding more new commodities in the most recent past than earlier. Even if this bias could be overcome in some way, the fact that the rate at which new technologies are introduced and accepted into the household does not mean that their importance necessarily has increased to the same degree. One major new technology may be as important as ten minor ones. The gigantic task of making this catalogue of changes in household technology has not been undertaken. However, it is clear from an analysis of the data for total expenditures by households in Great Britain that the diffusion of new household technology in the period 1920-1938 had an impact on the growth of household expenditure which was greater than the combined impact of all other factors, including income, prices and population growth (Ironmonger, 1972, p.164).

It is also clear from the statistics that for many new commodities in the United Kingdom, rates of growth of consumption per head, often well over 10 per cent per annum, took place for continuous periods of two or more decades. By any standard this must be regarded as phenomenal. The best example is that of electricity which, after six decades of growth, much of which was at the rate of more than 15 per cent per annum, still grew in the 1960s at a rate of 7 per cent per annum. Parallel phenomena can be observed in most other countries. For example, Kuznets (1942) catalogues some of the changes in household expenditure in the United States and points out the growing importance of household expenditure on new commodities. With recently introduced information-technology products such as colour television, video-casette recorders and home computers, even faster rates of growth are apparent as real prices are decreasing and real performance capabilities are increasing.

CONCLUSION

The diffusion of new commodities to the household, either as new types of capital equipment or as new forms of food, materials, energy and services, has provided the basis for the very large increase in household production of the last 100 years or so. Although households use small-scale technology, the total investment in and output from household technology is larger than that of any single sector of the market economy. Consequently there is an enormous potential for gain in human welfare by improving the technology of the household. Where these innovations will take place is difficult to predict. As Gershuny suggested almost a decade ago, it is possible that the same forces that produced the home washing-machine will in the future produce the home hospital-machine or the home university (Gershuny, 1978, p. 91).

In the future, as we move towards an information society, it is likely that the diffusion of new technologies for the household will involve a greater component of information processing equipment and materials than in the past. Households, of course, have already made a large investment in sound and image-processing machinery—radios, television sets, record and tape players and recorders, and, more recently, video recorders. Although this equipment is used mostly for entertainment, it can be used for information storage and retrieval and thus for diagnostic and educational purposes. It may not be such a big step for each household to have its own in-house medical centre and university.

REFERENCES

Cities Commission (1975) *Australian's Use of Time: A Contribution of Social Planning to Urban Development and Land Use Design*, AGPS, Canberra.

Clark, C. (1958) 'The Economics of Housework', *Bulletin of the Oxford University Institute of Statistics*, 20, 205-11.

Eisner, R. (1978) 'Total Incomes in the United States, 1959 and 1969', *Review of Income and Wealth*, 24, 41-70.

——— (1980a) 'Capital Gains and Income: Real Changes in the Value of Capital in the United States, 1946-1977', in D. Usher (ed.) *The Measurement of Capital, Studies in Income and Wealth*, Volume 45, University of Chicago Press, Chicago (for the National Bureau of Economic Research).

——— (1980b) 'Total Income, Total Investment and Growth', *American Economic Review*, 70(2), 225-31.

——— and Nebhut, D. H. (1981) 'An Extended Measure of Government Product: Preliminary Results for the United States, 1946-76', *Review of Income and Wealth*, 27, 33-64.

———, Simons, E. R., Pieper, P. J. and Bender, S. (1982) 'Total Incomes in the United States, 1946-1976: A Summary Report', *Review of Income and Wealth*, 28, 33-174.

Gershuny, J. (1978) *After Industrial Society? The Emerging Self-Service Economy*, Macmillan Press, London.

Gronau, R. (1980) 'Home Production—A Forgotten Industry', *Review of Economics and Statistics*, 62, 408-16.

Hawrylyshyn, O. (1976) 'The Value of Household Services: A Survey of Empirical Estimates', *Review of Income and Wealth*, 22, 101-31.

Ironmonger, D. S. (1972) *New Commodities and Consumer Behaviour*, Cambridge University Press, Cambridge, England.

Kuznets, S. (1942) *Uses of National Income in Peace and War*, Occasional Paper No. 6, National Bureau of Economic Research, New York.

Leuthold, J. H. (1981) 'Taxation and the Value on Non-market Time', *Social Science Research*, 10, 267-81.

Morgan, J. N., Sirageldin, I. A. and Baerwaldt, N. (1966) *Productive Americans: A Study of How Individuals Contribute to Economic Progress*, Survey Research Center Monograph 43, Institute for Social Research, University of Michigan, Ann Arbor.

Sirageldin, I. A. H. (1969) *Non-Market Components of National Income*, Survey Research Center, Institute for Social Research, University of Michigan, Ann Arbor.

Szalai, A. (ed.) (1972) *The Use of Time: Daily Activities of Urban and Suburban Populations in Twelve Countries*, Mouton, Paris.

United Nations (1975) *Towards a System of Social and Demographic Statistics*, Department of Economic and Social Affairs Studies in Methods Series F No. 18, United Nations, New York.

Young, M. and Willmott, P. (1973) *The Symmetrical Family: A Study of Work and Leisure in the London Region*, Routledge & Kegan Paul, London.

PART 2.
NEW TECHNOLOGY AND THE INFORMATION ECONOMY

Chapter 5

PERSONAL COMPUTING AND PERSONAL ARCHITECTURE

J. Deken

Contemporary innovations in information and communication technology mark the beginning of a worldwide economic and cultural revolution. As clearly as ancient epochs such as the Stone Age and the Iron Age are defined by the media which shaped them, we can now recognise our own position at the dawn of the Computer Age. And since our human use of information and our human use of space are closely intertwined, the spatial impacts of the unfolding information and communication revolution will be profound.

Far more fundamentally than other, extrinsic technological advances such as mass production of automobiles, modern information technology will reshape the lives of individuals. A widely-computerised environment will change not just ways of living, but ways of thinking. For historical comparison, one can only imagine the simultaneous invention of the printing press and the engine. Even more difficult, it must be realised that this new 'press' can print out engines by the thousands, and that physical force and work, as well as sensation and perception, can now be governed entirely by language. The computer revolution represents no less than that radical innovation (Deken, 1981).

From the perspective of individuals, technology has already crossed a watershed of proliferation—microcircuit fabrication has enabled computers and other sophisticated logic devices to emerge from the realms of government, large organisations and businesses to become the tools, servants and companions of ordinary people. Artificial logic is beginning now to expand beyond traditional information processing and communication devices, and will become thoroughly interwoven with all advanced technology. Logic engines and language-responsive devices will gradually replace all of the physically-linked media of conventional machinery and processes. Historic constraints on physical location and movement of individuals will consequently weaken and disappear, allowing a far wider range of personal choice. Enfranchised (literate) individuals will become the architects of their own personal space rather than simply inhabiting highly constrained locations for living and working.

In assessing the personal spatial impact of the computer revolution though, the certainty of change is mingled with the uncertainty inherent in any radical upheaval—small variations in initial conditions can produce vastly different eventual outcomes. In addition, the revolution developing from new information processing and communication technology is hardly monolithic; in its present volatile beginnings it must advance on many disjointed frontiers: device technology and hardware-systems design (including robot sensors and actuators); hardware production; software and human-interface design; applications software production; distribution and maintenance; network design and implementation; user training and public education.

Within individual geographic and political regions, progress is being made on each of these frontiers at vastly different speeds. When viewed globally, enormous variability in technological innovation and in its assimilation appears from one country to the next. A failure to develop even a single aspect can produce a critical bottleneck. In the light of such variability, instability and interdependence, early assessments of the impact of computing technology can hardly extend to detailed global predictions or precise local timetables of events; it is more productive now to focus on the fundamental underlying forces of change.

In developed countries, for example, the technological capability to fabricate computers is far ahead of methods for producing and maintaining software or for training organisations to use new information systems productively. Rapid innovations in technology, rather than contributing to the spread of effective computer use, now often discourage individuals and organisations, as their hard-won new equipment and training rapidly become obsolete. Another key stumbling block holding back the effect of computer innovations is that, unlike electrical power which flows freely in developed countries to drive physical devices, the information power of data and programs does not yet circulate in any widely and easily accessible form. An unstable positive feedback (vicious circle) presently retards both the development of software and the installation of computer networks: software developers rely on a large installed base of computers to justify development costs; potential computer buyers remain on the sidelines because suitable software is not available. The utility of connection to a computer network depends directly (and circularly) on the number of other users and the amount of information connected to the network.

Hardware standards (even if these standards are only suboptimal, *de facto* consequences of competition weeding out all but the largest manufacturers) are now emerging. With standardised features providing a wide base of target machines as a market, incentives for software development are increasing. Data communications protocols and inexpensive dedicated processors designed as bridges between incompatible systems have also begun to stimulate a substantial spread of computer networks. In favourable circumstances, these network 'power lines' for information have already become established, linking together office and factory workers, sales forces, or researchers. Their reach may span single buildings or entire nations. But, as with any positive feedback system, the evolution of software and the installation of computer networks can be held back extensively by negative events or expand explosively over a short period of time.

The character of life in developed countries is thoroughly shaped by the fact that electrical power, as well as the devices which use this power—whether electric motors, air-conditioners or television receivers—are universally available. Information networks, the corresponding large scale distribution systems for information, are essential to the spread of logical devices. Like the historic effect of electrification, the ultimate impact of intensive worldwide information networking will be profound.

THE COMPUTER AS MEDIUM

On the surface, the driving force propelling humankind into the Computer Age is the proliferation of cheap and powerful (electronic) logical devices. We now usually call such logic machines computers, from the numerical applications of their origin. These familiar electronic computers and microprocessors will continue to proliferate, to improve in speed and capacity as well as in the sophistication of their uses. It may even seem in the mid-eighties that we know what computers are if not yet exactly how to deal with them. That is hardly the case. Present-day computers are only primitive precursors of future logical devices. Whether rough hand-tools such as pocket calculators, cumbersome and limited 'home' computers, or lumbering mainframe dinosaurs the size of rooms, today's computers are all threatened with obsolescence by rapidly emerging new generations of hardware.

By the end of this century, in fact, the vintage 1980s concept of computing itself will seem severely limited. Logic devices based on optics will emerge to complement and challenge their electronic predecessors. As both electronic and optical-logic circuits are further miniaturised, complete logical systems in small physical packages will become vastly more powerful and ever more widespread. Even now, the sheer number of logic processors and the amount of information accessible to them increases dramatically with each passing year.

More importantly, as computer technology reaches farther in breadth it also increases in depth. Artificially processed information now goes beyond numbers and words to abstract symbols of all sorts, and beyond symbols to direct sensation and action (robots) in the real world. By the end of this century 'computer' will certainly be far too weak a word to describe the myriad roles of logical devices, ranging far from the conventional manipulation of numbers and text to span intelligent information extraction and synthesis, natural language understanding and perceptive, active robotics.

To obtain a reliable grasp of the impact of computing technology it is necessary to look beyond the present inventory of computing machines and their commonest numerical and text-processing uses. Rather than focus on the multiplicity of computer devices, it is important to discern the underlying unity of their role. What is the role of artificial logic and communication technology? On close examination, it becomes clear that the information and communications revolution is not simply an additional factor introduced into our existing economic and social structures. Rather, the computer represents a new *medium*, of unprecedented power and versatility (Deken, 1983). By working in that medium, humankind will ultimately shape an entirely new framework for productive economies and societies.

The word 'medium' itself has three major meanings, each meaning representing a significant role of modern logical processors (computers).

Communications Medium

In common usage, we most often speak of media in reference to print, film and broadcast communication channels. In this sense, as a medium for transmission, the computer is rapidly replacing and transforming all other media. The logically coded information transmitted by electronic and optic devices is inherently transportable and negotiable. Transportability means that information of all sorts—words, numbers, pictures, programs or the sensations of remote receptors—can be transmitted, processed, stored and retransmitted indefinitely, and at lightning speed. Negotiability means that information, in computer form, can go directly to work as it is received. The computer is an active medium, able to translate information automatically between a form appropriate for the sender (human or machine) and any other form, appropriate for the recipient. Immediately upon receipt at its destination, negotiable data will directly drive and control not just ordinary computer operations but physical processes, from machine tools to lasers and robots. In its role as a communications medium, the computer forms a wide bridge to join any willing pair of individual minds or to join mind and machine, whether the intervening distance spans cities, nations or the reaches of space.

Modelling Medium

Like the artist's colours, the sculptor's clay or the engineer's wind tunnel, the computer also represents a medium in which concepts can be translated into sensible, compelling reality. An aircraft can be modelled in the computer, electronically flown with colour graphics on the interactive video display, and redesigned a thousand times before it is physically manufactured. Modern integrated design and manufacturing systems (CAD/CAM) show the power of the computer as a modelling medium. In a CAD/CAM system, the design developed within the computer medium exists not just as a scale model for planning, but finally functions to control the production tools which create a finished device. Like a mould for casting metal or a form for shaping concrete, the computer model in CAD/CAM becomes a template which holds an entire design; once raw material is added, the system will produce a physical object.

The possibilities for building in the computer modelling medium span the spectrum from engineering to science and the arts (Deken, 1983). Moulding a musical composition with the computer instead of pen and ink, a composer can hear effects instantly as the score is written, even designing and auditioning new instruments to add to the ensemble. And unlike the words of the poet or the paint of the artist, the computer medium is infinitely malleable, restorable and co-operative. A computer painting can be redone in different tones or from a different point of view upon command without destroying the original. Using not only written words but notions and gestures as well, modern computing languages and interface systems allow computer models to be marshalled more like a skilled staff of apprentices or a talented troupe of actors than like the unresponsive heaps of pigments or letter forms now common to art and literature. As modelling media to support creative and accurate designs, the versatility of artificial-logic machines is unprecedented and as yet unfathomed. Every realm of human thought and artefact will ultimately be reshaped.

A Nutrient Medium

The computer, as it becomes woven into the personal environment of individuals, also begins to function as an intellectual nutrient medium, a life-support system for cultivating ideas. In the supportive environment which a powerful, graphic and responsive personal computing milieu provides, new ideas and new approaches can flourish like biological life forms in a sustaining ecology (Bolt, 1984). Seymour Papert, a pioneer in teaching even young children to launch powerful ideas within this nutrient medium, speaks of his computer worlds as 'environments for learning' and 'incubators for knowledge' (Papert, 1980). It is widely recognised by computer scientists that once tools are built within a supportive, interactive, computer environment, the tools' effects multiply and propagate exponentially through all subsequent efforts. As a nutrient medium, the computer is fertile. We are initiating in it an intellectual recombinant genetics.

THE VANISHING TYRANNY OF PERSONAL SPACE

The human mind, in its problem solving, creativity and imagination, is unhindered by the circumstances of time and place which constrain the human body. We may easily plan the future of a city or a highway system, or think of a friend on another continent, but physically we cannot exist except at a single time and location. Countless tasks, no matter how routine or boring, simply do not get done unless human presence and attention are devoted to them. This mismatch between the freedom and capacity of the human mind and the absolute physical constraints of the body is the tyranny of personal space. The ultimate, personal impact of computing will be to decisively break that tyranny. On a personal level, computing technology will not only broaden the horizons of individual minds, but also free those minds to act and interact powerfully beyond their inherent bodily limitations.

In emphasising here the historic tyranny of personal space, I do not mean to assert that constraints of time and place are universally oppressive. It is difficult to imagine people at all without the essential cultural and social viewpoint which their physical context provides. Nonetheless, the removal of physical constraints on individuals is ordinarily welcomed as civilisation advances. In primitive cultures, foragers and hunters spent most of their waking hours compelled to arrange a coincidence of their personal time and space with that of their next meal. Even in relatively advanced societies, workers whose productivity depends on physical labour are obviously forced to remain in direct physical contact with their digging, chopping, harvesting and so on. With time and civilisation though, food production has progressed from hunting to agribusiness and biotechnology. Power production has developed from muscle to combustion, electricity and nuclear reactions. Where technological advances have provided other options, few individuals today prefer to live by hunting or hard labour, or to unnecessarily accept the tyranny over their personal space which their remote or recent ancestors experienced.

Despite ten millennia of technology, however, significant unavoidable constraints of time and place retain their force over people. Individuals still must largely live near their livelihood, if not near their immediate food sources. Forced migrations to follow work occur worldwide, and include the daily migration of commuting. The travelling salesman still travels, but now at the peril of jet lag. Mega-aggregations of people, generated by technically outdated economies of scale, persist despite the inconvenience, hardship, or even danger they pose to individuals.

Sense + Interaction + Effect = Presence

Telephony and television are familiar examples of electronic media which have already demonstrably enlarged the personal space of individuals in developed societies, directly expanding their sensory awareness. At a speed which will circle the globe nearly instantaneously or travel to the moon and back in a fraction of a second, electronic and optical signals serve to retrieve sights and sounds. Through telephone networks, direct personal interactions are routinely maintained even across international distances. A single television news program can immerse its viewers in the sights and sounds of all the continents of the world.

Telephone and television, if only in the limited realm of sight and sound, provide a partial surrogate for being there. No matter how physically remote, the horror of a war or famine or the spectacle of sports vividly leaps to the colour video screen. Advice, instructions and agreements can be subtly and effectively conveyed by electronic conversation. But the complete equation by which individuals will ultimately absolish the tyranny of personal space is: Sense + Interaction + Effect = Presence.

The synergy on the left side of the above equation is at present only weakly suggested by telephone and television. The television image of a faraway situation is confined to two dimensions rather than three. The interaction of telephony is immediate and direct, but limited to sound. No worldwide system of effect-at-a-distance (tele-operation) exists at all. The significant role of the computer medium over the next few decades will be to encapsulate all of the fundamental synergistic components of presence—sense, interaction and effect—so that presence itself becomes as transportable and negotiable as images, voice and conventional data are today. Sensing interactively and acting through the computer medium, individuals will be capable of direct and effective telepresence far beyond their physical capabilities. The tyranny of personal space will be removed.

Sensation

Visual and aural sensations are now routinely transmitted over great distances and increasingly filtered by logical processors in transit. Although television is at present only two-dimensional, various practical methods of creating three-dimensional illusion by stereoscopy (presenting slightly different images to left and right eye) have been developed and tested. Other transmission techniques can create images which are truly three-dimensional, so that the viewer obtains different views by observing from different angles, can move around the image, and so on. Holograms are perhaps the most widely-known method for three-dimensional image presentation. Although full-colour moving holographic images will require data-transmission rates of billions of bits per second, optic-fibre and other communication systems have already been demonstrated in this (gigabit) range. The remaining technical problems of hologram transmission will almost certainly be solved within the decade. Beyond visual and aural sensation, the sensor and effector technology of robotics will also make it increasingly possible to transmit the sense of touch, as well as to recreate actual three-dimensional physical representations of objects sensed at a remote location. Public sensory awareness will continue its historic expansion through broadcast and other public channels. In addition, the increasing density of logic based communication networks via landlines, earth transmitters and satellite links will also greatly expand private dial-up telesensing, for business and personal use.

Interaction

If electronic and logic technology could be expected to go no farther than to give telesensory processes like television and telephony a quantum jump in fidelity, the impact would be substantial. But far beyond the reach of mere high-fidelity sensory transmission, a computer can create the realism of interactive sensation. Although many people may be familiar with interactive computer imagery only through the participatory illusions of electronic games, the importance of interactive sensation is far deeper than its entertainment value. Human-computer interfaces are now being designed and developed to support fundamental new frameworks for interaction between humans and information. Interactive sensory images may be produced in several ways:

- captured under computer-mediated user control (for example, the video images gathered to guide movements of the NASA space shuttle's robot arm);
- managed for display in real time by computer control (interactive videodisc is a contemporary example);
- directly synthesised by the computer from logical models in response to the participant's actions and commands.

From a comfortable and distant location, a telepresent security-guard may 'walk' through a warehouse, a prison or a nuclear reactor, and see exactly those images which correspond to his or her chosen travel and viewpoint. With the computer to act as a realicorder (Deken, 1983), not just the sights and sounds, but the structure of a situation can be captured and stored with high fidelity. When the realicorder's modelled information is 'played back', a viewer/listener/participant will experience the environment in a unique way, depending on his or her own responses to the recreated situation. Buildings, cities and countries may be toured according to each individual's preferred itinerary.

Effects

The final test of presence in any physical situation is the ability to directly change that situation by action. The key to versatile and immediate action-over-distance, tele-operation, is furnished by robotic effectors. These effectors are devices designed to provide physical power in the form of movement, heat, laser energy and so on, controlled not by physical intervention and direction but by their own internal logic and by external command signals.In the near future, undersea miners may work in comfort on dry land, relying on the interactive imagery and other telesensory information transmitted by robots on the ocean floor to locate minerals and materials such as manganese nodules (Ayres and Miller, 1983). The miners' earthbound movements and commands will be transported over great distance and below the ocean to discover, move and retrieve the undersea materials. Modern industrial installations, from sawmills to nuclear power plants, are commonly controlled by electrical devices and electronic switches rather than mechanical links and levers. With logic technology available to render control signals thoroughly transportable and negotiable, the location of the control panel can be independent of the location of the factory. Beyond the best illusions of direct, interactive sensation, logic-driven effectors allow the complete synthesis of artificial presence, translating abstract information into the construction and manipulation of objects in real, physical space (Deken, 1986).

LIBERATED: HUNTERS, ATTENDANTS AND PILGRIMS

Information hunters today, like primitive protein-hunters of the past, are highly constrained to locate and move themselves in search of digestible material. Library, literature, scientific and legal searches are only beginning to be exercises in logical rather than physical agility. This trend will rapidly expand. Another form of information hunting widely practised today is shopping. Despite the element of sport and recreation that some forms of shopping have for many individuals, information hunting by shopping is characteristically time-consuming and inefficient. Another common and economically significant information hunter is the business traveller, who shuttles his or her brain about from one location to another, to make contact with like-minded individuals. Sales presentations, contract negotiations and employee training presently account for billions of miles of travel and untold hours of jet lag, which more effective telepresence will eliminate.

Information attendants today are as abundant and easily recognised as information hunters. Postal-service workers, monitoring and often physically carrying out the movement of information in its historic paper-and-ink format, are perhaps the most obvious. No argument can deny the aesthetic taste of those who savour paper media. The Akkadians were undoubtedly similarly fond of clay tablets. As a practical matter though, all of the features of paper communication—including signatures, security, funds transfer and individualistic graphic style if desired—can be embodied in advanced logical communication systems.

Other occupations which are less obviously those of information attendants will also eventually become far less constrained. Physical goods will hardly disappear, nor will the need to transport them quickly and efficiently vanish. What will liberate truck drivers, airline pilots and others in the tele-operative future is that the responsibility and judgement necessary to manage translocation of valuable goods does not logically require the responsible human director to sit astride the baggage en route. In fact, entirely new modes of shipment in the future, from contained ballistic transport to hypersonic aircraft, will only be feasible if the pilots are physically removed from the cargo and extensively assisted by computer.

As they flock to conventional or traditional fixed locations, many of today's information hunters and attendants behave like classic pilgrims. High real-estate prices have already disturbed many of these pilgrimages, dislocating back-office operations from corporate centres in large cities, and even separating restaurant kitchens by miles from their dining areas. Led by computer software organisations, service and technical operations are gradually being unlinked from the physical necessity of a technician travelling to the site of an ailing machine. Emergency medical technology and computerised medical expert systems demonstrate that the repair technology of physicians may also be transportable and negotiable. The factory at home is more plausible because the superintendent can work there, while the (robotic) labourers and machines toil at a comfortable distance.

Despite free flowing speculation about post-industrial societies, service economies and so on, it is obvious that humans will continue to live in physical space, use physical objects and have physical needs. Logic technology is certainly not magic; no repeal of the law of gravity is in sight. What is in sight though, is a far greater freedom, through powerful telepresence, to negotiate physical circumstances and break historic constraints. Just as humans have not learned to fly by breaking the law of gravity, but by understanding it, we may expect to abolish the tyranny of personal space by using advanced technology to work powerfully within it.

MAKING PERSONAL SPACE A HOME

It would be a fundamental mistake to suppose that the demise of the tyranny of personal space implies an extinction of the role of personal space. In fact, it seems more likely that as unsettling new possibilities for spatial organisation and use emerge, the creation of an immediate, stable and intimate home environment will become even more important.

The practical function of a physical home as a shelter will be enhanced as home devices and architecture increasingly incorporate microprocessor logic. Energy management (including production, in many cases) is a prime target. Robots, although far different in form than the androids of fiction, will take the place of conventional appliances and eliminate much of the tedious overhead of cooking, cleaning, etc., which accrue to home-making.

The home environment created by personal computing will also provide a shelter from the onslaught of raw data generated by an information-leveraged world. In the near future, the personal computer will be not only an access port through which an individual can receive and transmit data, but a powerful filter for screening out noise and interactively interpreting information. Customised information abstracts and summaries will replace the scattershot, superficial coverage of contemporary newspapers and magazines. The realicorder technique of storing information for interactive presentation will replace the rigid linear format of traditional books and open new horizons for personal education, recreation and training at home. The unknown terrain of any new subject area can be mapped like a geographic region, to be navigated by the learner who interactively experiments with it and explores its structure.

The aesthetics of home design and decoration are also likely to grow in importance as personal control of space expands. Shorter working hours, a greater variety and quality of consumer goods, and a vastly expanded potential for customised home-made artefacts and handicrafts will all contribute. The treasured results produced may be home-designed weavings spun at the automated facilities of a community information centre, original and replica furniture pieces, or rare and delicate tropical plants.

An additional, persistent aspect of personal space is the bilaterally-symmetric, inherently obsolescent, physical form called human anatomy. The anatomical facet of personal space management will increasingly evolve toward preventive medicine, health maintenance and physical fitness. Microcomputers, microbots and other artificial-logic devices will contribute to a new age of health maintenance, medical diagnosis and treatment (Deken, 1986). Since microcircuit components can be constructed to function routinely and reliably in extremely small scale or in environments uninhabitable by humans, personal telepresence in the future will extend to far greater physical self-knowledge. Computer 'pills', which traverse the digestive tract and provide a variety of physiological information have already been developed. Such devices are primitive precursors of the medical microbots of the future. Internal telepresence coupled with realicorder modelling techniques for presentation and motivation will make health-building effective and enjoyable. In a compelling realicorder environment designed to recreate ski slope or soccer field, a home exercise machine may show the participant's projected appearance and performance six months or a year in the future, updated automatically as the workout continues.

Telepresence via computer networks will establish intellectual mobility as a routine element of advanced society. It would be unreasonably optimistic now to forecast that the main result of computer networking will be large scale co-operation in business, educational, social and artistic endeavours and an increasingly informed and active public for economic and political processes. It would be defeatist to predict that telepresence technology will merely produce interactive television and a new breed of electronic escapism packaged for easy home-delivery. It is likely that both trends will be identifiable within subcultures of developed countries.

Traditional homes and work spaces now provide most people with a supportive context of neighbours and friends. Rather than removing the social milieu of a physical neighbourhood, the transformation of personal space by telepresence can contribute to the development of multiple, close-knit communities of common interest, sharing and co-operation. Whether the beneficial social impact of information networks will be realised does not depend primarily on technology. Far more important will be government and corporate policies, the influence of organised groups and individual choice exercised within the new architecture of personal space.

REFERENCES

Ayres, R. U. and Miller, S. M. (1983) *Robotics: Applications and Social Implications*, Ballinger Publishing, Cambridge, Mass.

Bolt, R. A. (1984) *The Human Interface: Where People and Computers Meet*, Lifetime Learning Publications, Belmont, California.

Deken, J. G. (1981) *The Electronic Cottage*, William Morrow, New York.

——— (1983) *Computer Images: The State of the Art*, Stewart, Tabori and Chang, New York (also Thames and Hudson, London).

——— (1986) *Silico Sapiens: The Fundamentals and Future of Robots*, Bantam Books, New York.

Papert, S. (1980) *Mindstorms: Children, Computers, and Powerful Ideas*, Basic Books, New York.

Chapter 6

TELECOMMUNICATIONS AND INTERNATIONAL FINANCIAL CENTRES

M. Moss

Communications technologies are often regarded as space-extending phenomena: specifically, they allow individuals and firms to function within a geographically larger set of boundaries (Kellerman, 1984, p. 232). For most observers, the ability to overcome traditional spatial limits implies a weakening of the city (Abler, 1975, p. 133). New technologies, it is argued, allow people to exchange information and ideas without interpersonal contact; thus, the comparative advantage of cities, whose existence has been traditionally based on their role as centres for face-to-face contact, is no longer necessary (Gottmann, 1977). This article argues that advances in communications technologies are not leading to the demise of cities; rather they are strengthening a handful of principal world cities (Hall, 1984, p. 95). Three critical issues are examined to demonstrate the way in which technological change is influencing the pattern of urban development and the emergence of global finance and legal centres.

1. How has the emergence of a global economy led to the creation of major financial centres and what role do these centres play?
2. How will the emerging telecommunications infrastructure influence further patterns of urban development?
3. What types of cities will benefit from telecommunications deregulation and why?

CITIES IN A GLOBAL ECONOMY

Telecommunication systems, by allowing firms to overcome the traditional limits of distance, permit what were once separate economic activities to become highly integrated functions. Multinational firms, for example, although headquartered in one location, produce and sell a diversity of goods and services in numerous countries. More precisely, information and computer systems enable a relatively small number of people to control and co-ordinate production, marketing and financing from geographically-remote points. Because of the widespread decentralisation of manufacturing and assembly operations, there has actually been an increased need for a central headquarters responsible for the policies and financial decision making that allow such dispersion to occur. Noyelle and Stanback (1984) have shown that such corporate headquarters rely extensively on advanced producer-services (such as finance, law, accounting, management consulting and advertising), which are predominantly situated in the central business districts of large cities. The city provides the sophisticated financial and information services that allow a firm to operate globally (Noyelle and Stanback, 1984, p. 67). As a result, cities that specialise in international finance and producer services have witnessed considerable growth in their economic activity (Cohen, 1981).

This emergence of the internationally-oriented financial capital represents a considerable departure from the traditional role of certain large cities as centres for a nation's international trade and commerce. In a report by *The Economist Publications* analysing changes in London's financial markets, the authors highlight London's transformation into a global financial capital:

> The City of London has always been an international financial centre, or at least since the late Middle Ages. However, until recently its cosmopolitanism merely reflected the international scope of British trade and financial interests. British merchant banks . . . traditionally engaged in the finance of British trade, while the "colonial and foreign banks" . . . provided a banking network throughout the Empire and British trading enclaves elsewhere. The international horizons of these institutions were essentially the global horizons of the British political and economic interests which they served. However, the recent globalisation of the City's interests, occurring at a time of economic recession and stagnation in the UK, is a different phenomenon from the internationalisation that occurred in the nineteenth century, when the City financed British trade and economic influence. The central feature of many recent changes has been the development of the City as a centre for an emerging world capital market, as opposed to its more traditional role as a financial base for cosmopolitan interests. (Hewlett and Toporowski, 1985, p. 43).

The rise of the multinational firm has been intimately connected to the globalisation of banking and finance activities. Facilitated by the advent of communications technology, these activities grew for reasons that extend beyond responding to the finance demands created by the rise of the multinational firm. Specifically, the rapid rise and fluctuation of inflation and interest rates, which began in the early 1970s, created conditions of far greater uncertainty and risk. This, in conjunction with an influx of petrodollars, increased nation-state budget deficits, and the lowering of capital barriers between countries, has led to the internationalisation of finance activities and the proliferation of new finance related services and products. Most important, 'Technological change, particularly the advance in computer technology, has altered the environment of financial markets. . . . The increased speed and lower cost of communication have been important to the development and expansion of international markets.' (Germany and Morton, 1985, p. 743).

As a recent report by the Group of Thirty stated: 'Improved telecommunications and the presence of the larger banks in several time zones have created a continuous, round-the-clock market which responds instantaneously to new developments.' (Group of Thirty, 1985, p. 15).This is one of the reasons given for London's continued prominence as an international financial capital: 'London's position in between the US and Far Eastern Time Zones make it a useful centre for arbitrage between financial markets in those zones (chiefly the markets of the USA, Japan and Hong Kong); dealings on all those markets can be orchestrated from London in the course of one deal day.' (Hewlett and Toporowski, 1985, p. 43).In addition, telecommunications systems are now being used to link geographically separate stock and commodity exchanges and leading to considerably longer trading days. For example, the Chicago Mercantile Exchange is linked to a futures exchange in Singapore, the Sydney Stock Exchange has agreed to do joint trading with the New York Commodity Exchange, and the London Stock Exchange and the National Association of Securities Dealers share price information on actively-traded British, American and international stocks (Lohr, 1985a,b).

As Charles Kindleberger has wisely noted, 'The continuous reduction in the costs and difficulties of transport and communication over the last two hundred years has favoured the formation of a single world financial market.' (Kindleberger, 1978, p. 130).The emergence of international finance centres has facilitated the emergence of this global market; however, it has also weakened the role of the small-sized and medium-sized city. The headquarters of the independent firms that once thrived in smaller cities are now subsidiaries of large, multinational companies. As a result, their headquarters have been consolidated within larger financial centres. Thus, communications and information technologies are strengthening a small number of world cities while weakening the traditional autonomy of many smaller cities. More than twenty-five years ago Raymond Vernon predicted that such a process of office consolidation could occur: 'To the extent that the office function grows, therefore, the growth may well occur to a disproportionate extent in the office districts of the larger central cities, at the expense of the regional centres.' (Vernon, 1959).

In the Pacific Rim, Tokyo, Hong Kong, Singapore and, to a lesser extent, Sydney, have emerged as major financial centres. Tokyo's emergence as a particularly powerful financial centre is closely linked to Japan's large capital base and its pre-eminence in the Pacific Rim economy. In 1981, Japan generated 69 per cent of the Pacific Rim's GDP. Japan's economic presence also extends beyond the Pacific Rim. In a recent report by the *American Banker*, Japan was shown to account for 35.5 per cent of the total foreign bank lending to business in the United States, the largest single source of such activity (*American Banker*, 1986).

Tokyo has long been Japan's pre-eminent financial centre. Despite stringent financial regulations that limit its role in offshore activities, Tokyo is the headquarters of eleven of the world's fifty largest banks. Moreover, the Tokyo Stock Exchange is the fastest growing major market in the world (McMurray and Browning, 1986). With the gradual loosening of Japan's finance regulations, and the admission of three major American securities firms as Tokyo Stock Exchange members, Tokyo is becoming an increasingly powerful world financial centre. The emergence of Hong Kong and Singapore as financial centres has been closely linked to their ability to facilitate offshore capital-market activity. Hong Kong has a liberal regulatory environment, and both cities have strong historical ties to international trade and commerce. Most important, an infrastructure exists that adequately facilitates increasing demands for international communication and transport (Kirby, 1983).

In order to determine the location of international financial centres in the Pacific Rim, the ten largest US banks were surveyed. Seven of these banks indicated that they had established a regional headquarters office in the Pacific Rim. Citicorp designated regional headquarters in three cities: Tokyo, Hong Kong and Singapore. Of the remaining six, three banks established regional headquarters in Hong Kong and three in Tokyo. In addition, all banks had their largest operations in Tokyo, Hong Kong and Singapore, while three had equivalently large operations in Sydney. The importance of these financial centres is underscored by a 1983 comparison of world rental levels in which New York led, followed by Tokyo, London, San Francisco, Los Angeles and Hong Kong; Singapore tied with Sydney for seventh place (Rowley, 1984). Langdale (this volume) suggests that a 'loose hierarchy of international cities based on Electronic Information Services' (EISs) can be recognised. He states: 'It is possible to correlate the global reach of a city with its status as an international EIS centre.' He classifies cities into four categories, ranging from major global cities such as New York and London to 'nationally-oriented cities [that] function as international gateway locations for their respective countries'.

As financial centres are drawn to areas with complementary services, the strength of these centres can also be measured by analysing the location of branch offices of leading American law firms. Moreover, the location of these law firms underscores the concept that these areas serve as neutral settings that facilitate face-to-face contact and transactions. While many law firms rely on travel, telephone-based contact and/or correspondent relationships with local counsel, a growing number of firms recognise the need for maintaining a physical presence in the Pacific Rim. Such firms place a premium on being able to readily serve clients without encountering time differences whenever a meeting is necessary. As Table 6.1 shows, Hong Kong houses the largest concentration of American law firms, followed by Singapore and Tokyo. Although one might expect Tokyo to have a larger concentration of law firms due to Japan's economic predominance, Japan's *Practicing Attorneys Act* (1955) severely restricts the presence of foreign lawyers. Thus a regulatory impediment rather than a market impediment is constraining the growth of an international legal community in that country (Weber, 1983).

Table 6.1: Location of 15 Largest US Law Firms in the Pacific Rim

Location	Location of all law firm offices in Pacific Rim	Location of firms with only one office in Pacific Rim
Hong Kong	11	6
Singapore	6	1
Tokyo	5	1
Melbourne	2	1
Shanghai	1	1
Taipei	1	0
Sydney	1	0
Bangkok	1	0
Peking	1	0

Source: Legal Times (1985).

TELECOMMUNICATIONS INFRASTRUCTURE AND CITIES

While the initial development of financial centres has been spurred by cultural, economic and regulatory factors, a new optical-fibre telecommunications infrastructure is being built that will enhance the communications capabilities of national and international financial centres. As stated in *The Economist*,'The world's telecommunications are going on a high-fibre diet. Within two years, optical fibres will carry telephone calls beneath the Atlantic and Pacific.' (*The Economist*, 1986). Although optical-fibre systems have inherent technological advantages over copper wire, satellite and microwave communication systems, the current state of the technology and the economics of fibre favours high-volume, point-to-point communications—from one hub to another hub. As a result, the new optical-fibre systems are initially being built to serve the heavily-used communication routes, typically those linking major cities (Moss, 1986, p. 3). This pattern of development is in sharp contrast with communication satellites, where the economics favour traffic from one point to multiple points or vice versa.

In the United States and other nations, optical fibre systems are being installed along transportation rights-of-way, often following the railroad routes established in the nineteenth century. MCI has built its north-east fibre system along the AMTRAK right-of-way; Cable and Wireless is using the right-of-way of the Missouri-Kansas-Texas Railroad to connect the Texas cities of Austin, San Antonio, Dallas and Fort Worth; and in England, Mercury Communications is building a fibre system on British Rail's right-of-way. It is ironic that the choice of cities to be first served by advanced fibre systems is in part due to the decisions made in the nineteenth century concerning transportation rights-of-way.

The initial comparative advantage fibre optics confer on financial centres is seen by examining New York City and Los Angeles, the United States' pre-eminent east-coast and west-coast financial centres. These cities account for approximately 30 per cent of all overseas telephony emanating from the US. In addition, New York, with 43 per cent, has the largest concentration of foreign bank offices in the US, followed by Los Angeles with 11.8 per cent, and Chicago with 8.4 per cent (*American Banker*, 1986). Further, advanced telecommunications systems have reinforced the comparative advantage of New York and Los Angeles. Within each city, extensive fibre-optic systems are being used to facilitate intra-urban communication flows. New York Telephone has built three fibre-optic networks around Manhattan and an interborough fibre network that links the counties adjacent to Manhattan. In the Los Angeles region, the fibre network built for the 1984 Olympic Games provides an advanced regional telecommunications infrastructure that can support the information-intensive firms in southern California. Clearly, telecommunications has not resulted in the economic decline of the largest central cities in the United States, but is being used to move information in, through and out of such cities with greater speed and efficiency.

Telecommunications and the Pacific Rim

Just as the emergence of fibre-optic networks in the US demonstrates the way in which new technology can strengthen large cities, the fibre and analog networks being built in the Pacific Rim show similar trends. Due to the dramatic increase in trade between Pacific Rim nations and the United States, the current telecommunications infrastructure is being seriously challenged. Because the existing Hawaii-3/Transpac-2 (Haw-3/TPC-2) cable shown in Figure 6.1 is saturated, satellite transmission is being increasingly used to meet communication needs. By the end of 1984, 70 per cent of the transmission in the Pacific Rim was via satellite and 30 per cent via cable. By 1987 it is estimated that 76 per cent of transmission will be via satellite and 24 per cent via cable (Logue, 1986, p. 78).

The Federal Communications Commission considers this dependence on one medium of transmission to present risks and has therefore supported the new Hawaii 4-Transpac-3 (HAW-4/TPC-3) fibre-optic system. By 1991 this system is projected to shift the balance between satellite and

Figure 6.1 North America - Pacific Cables.

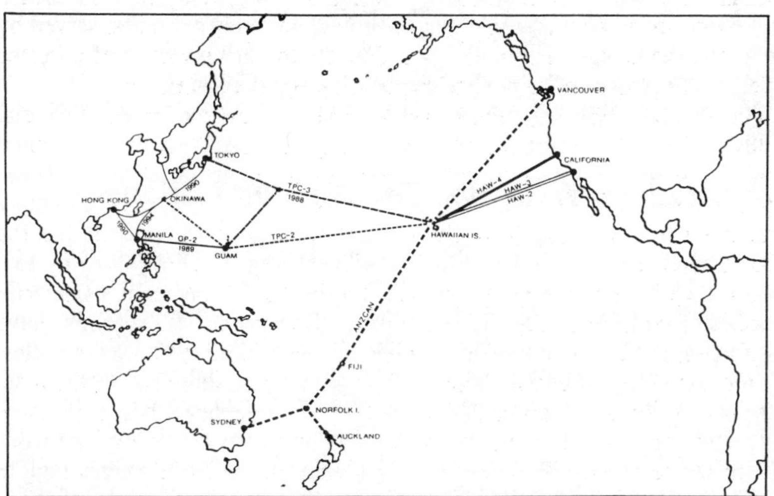

cable traffic to 56 per cent satellite and 44 per cent cable (Federal Communication Commission, 1986). Further, the configuration of this cable provides valuable insights concerning the emerging pattern and location of economic activity. While the initial Hawaii Transpac cables were oriented towards national security interests with direct links from Hawaii to Guam, Figure 6.1 shows how the new HAW 4-TPC 3 cable is designed to accommodate both economic and military linkages. After stopping in Hawaii, the cable extends far into the Pacific and then branches in two directions; the national security link runs to Guam and then on to the Philippines via a branch called GP-2; the economic and trade connection extends directly from Hawaii to Japan. By 1990, other key Pacific Rim centres will be linked via a ring that will connect Guam, the Philippines, Hong Kong, Korea and Japan. By 1994 Taiwan will also be connected.

Although the design of this network appears to favour the northern countries in the Pacific Rim, it is important to note that the ANZCAN cable, shown in Figure 6.1, connects Vancouver, Hawaii, the Fiji Islands, Sydney and New Zealand. With both the Canadian and US cables stopping in Hawaii, this will become an important point for information transfer. Further, it serves to mitigate somewhat the strong comparative advantage that the northern countries would otherwise have received.

As Figure 6.2 shows, there is also an extensive amount of intraregional telecommunications systems being built. The systems shown on the map have been constructed since the early 1980s. As can be seen, these systems link areas that have both cultural and/or economic ties. For example, the Japan-Korean cable shows the important economic relationship between these countries. In the early 1980s, 28 per cent of South Korea's trade was with Japan; in 1980 41 per cent of South Korea's overseas telecommunications messages were with Japan (Kirby, 1983). In addition, the ASEAN cable shows how a new communications infrastructure can further the development of a coalition whose goal is to develop stronger economic and cultural ties.

Urbanisation, Trade and Telecommunications

While the emerging web of cable systems shows that the strength of key centres with an existing comparative economic advantage will be reinforced, Figure 6.3 shows that trade relationships do not completely account for international communication linkages. Although the NICs—South Korea, Taiwan and Hong Kong—show a strong association between trade and communication, Japan's pattern is different. Japan accounts for almost half of all Pacific Rim trade with the United States, yet it has only one-third of the AT&T circuits between the Pacific Rim and the United States. Given the strong cultural ties which also exist between the two countries, one might argue that this is an area where demand for international com-

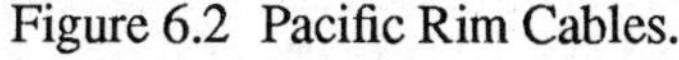

Figure 6.2 Pacific Rim Cables.

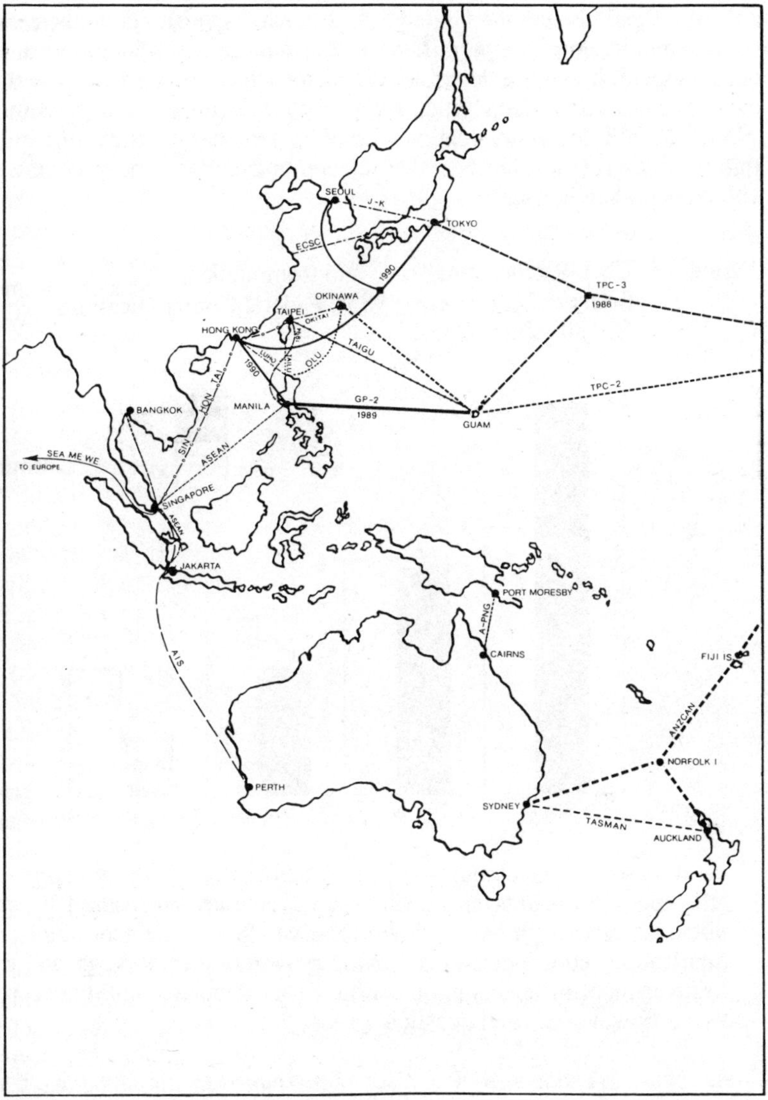

munications should be far higher. Conversely, while ASEAN accounts for 16 per cent of Pacific Rim trade with the US, 22 per cent of AT&T's circuits link ASEAN and the United States. Such disparities demonstrate that telecommunications is a permissive factor that can contribute to locational decisions, but it is not a deterministic factor which, by itself, can generate economic activity (Mandeville, 1983). The Independent Commission for World Wide Telecommunications Development has recently highlighted this complex relationship between telecommunications and economic development when it stated:

Figure 6.3 USA-Pacific Rim: Trade and Communication.
Source: *The Banker* (1985); Federal Communications Commission (1985).

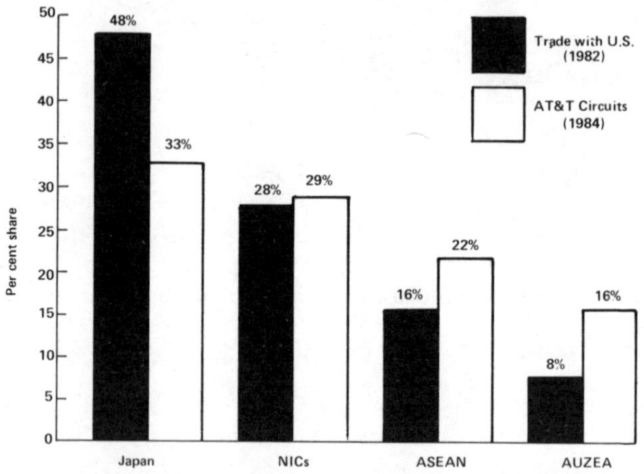

> While a strong correlation has been established between the number of telephones per capita and economic development measured by gross domestic product, it has not been clear whether investment in telecommunications contributes to economic growth or economic growth leads to investment in telecommunications. That there is a link between the two is however beyond question. (1984, p. 9)

As Figure 6.4 shows, GDP statistics for countries within the Pacific Rim show a strong association with number of telephones. As a result, Japan's economic dominance and strong economy translates into a dominance in percentage of telephones as well. For ASEAN, where the greatest variation occurs, this can be largely explained by the lack of urbanisation within member nations. With the exception of Singapore, countries in ASEAN

Figure 6.4 Pacific Rim: Concentration of GDP, Telephones and Population. **Source:** World Economic Indicators (1985); *World Almanac* (1986).

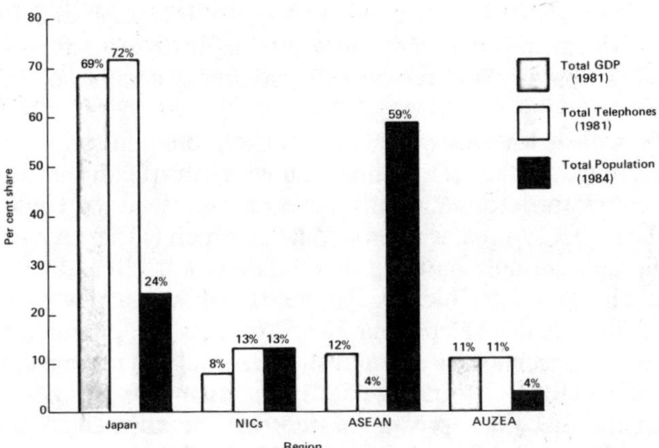

are 25-35 per cent urbanised; this compares to over 80 per cent urbanisation in other Pacific Rim countries. Thus while ASEAN's population is dominant within the Pacific Rim, degree of urbanisation and economic development is a far better indicator of telephone dispersion and demand.

Telecommunications Deregulation and Cities

The design, development and management of the telecommunications infrastructure in all nations (with the exception of the United States) has, until recently, been totally under public control. In the United States the telecommunications infrastructure has been largely built by AT&T, several independent telephone firms and numerous rural telephone companies. Although privately owned all function under close government regulation. In the United States a policy of telecommunications deregulation and the divestiture of AT&T is leading to a profound transformation of the nation's telecommunications infrastructure. Public policy at the Federal level is geared toward letting market forces determine the pattern of telecommunications investment, thus supplementing the policy of universal service, the goal of providing everyone (regardless of location) with low-cost, reliable telephone service. As a result, new telecommunications systems are being built to serve large communications-users, most of whom are located in the nation's largest metropolitan areas (Moss, 1986).

A growing movement to rely on private firms to develop telecommunications systems is occurring in other nations as well. Nowhere, however, is this competitive telecommunications environment as pronounced as in the United States. In Britain, privatisation has occurred with the creation of British Telecom; presently, limited competition in telecommunications infrastructure is taking place between British Telecom and Mercury Communications. In Japan the state-owned Nippon Telegraph and Telephone (NTT) has become a private firm; new private firms, such as the Hughes-Mitsui-CITOH partnership, are now developing systems to compete with NTT.

This movement towards competition in telecommunications is clearly bound to strengthen the telecommunications infrastructure in those cities that are centres for information-intensive industries. In the United States, COMSAT and TRT (a major international common carrier) have formed a partnership and are now building urban gateway satellite stations in San Francisco, Houston and Chicago. Further, they have already established a satellite facility at the Teleport in New York. At the international level, competition and technological innovation are largely responsible for the decision of INTELSAT to deregulate its International Business Services (IBS). For the first time, private carriers will be allowed to build earth stations and lease a portion of INTELSAT's satellite capacity for the purpose of sending data communications. INTELSAT's decision to deregulate a portion of its activity reflects the intense pressures brought by large information users who want to maximise the use of new technologies, such as Ku-band satellite transmission, that INTELSAT has not incorporated into its network. In a survey of four common carriers with plans to offer Pacific Rim IBS in 1986 or 1987, each stated that they were in the process of negotiating or concluding agreements with the PTTs of Japan, Hong Kong, Singapore and Australia. This pattern of development again demonstrates that each of these countries' financial centres is the source of rapid and growing information-related telecommunications demand. Further, the probability that new telecommunications systems will initially serve these areas and reinforce their pre-eminent position is underscored. As a result of these developments, planners and policy-makers concerned with regional development will need to give heightened attention to the availability of telecommunications systems in cities.

CONCLUSION

This paper has examined the relationship between advanced communication and information technologies and the future pattern of urban development. It has sought to demonstrate that:

1. telecommunications technologies are facilitating the globalisation of world financial markets and increasing linkages among principal world cities;

2. the emergence of fibre-optic systems for international and inter-urban communications is strengthening the telecommunications capacities of large metropolitan centres; and
3. increased competition in telecommunications will require policy-makers to give greater attention to the private sector's role in providing advanced telecommunications services in cities.

ACKNOWLEDGEMENT

The author would like to acknowledge the valuable research assistance of Andrew Dunau in the preparation of this chapter.

REFERENCES

Abler, R. (1975) 'Effects of Space-Adjusting Technologies on the Human Geography of the Future', in R. Abler, D. Janelle and A. Philbrick (eds) *Human Geography in a Shrinking World*, Duxbury Press, Scituate, Mass.

American Banker (1986) 'Foreign Banking in the US', 14 February, 1A.

Cohen, R. B. (1981) 'The New International Division of Labor, Multinational Corporations and Urban Hierarchy', in M. Dear and A. J. Scott (eds) *Urbanization and Urban Planning in Capitalist Society*, Methuen, London.

Federal Communications Commission (1985) *Memorandum Opinion, Order, and Automation*, File No. I-T-C-85-219, 7 June, Washington, DC.

Germany, J. D. and Morton, J. E. (1985) 'Financial Innovation and Deregulation in Foreign Industrial Countries', *Federal Reserve Bulletin*, October, 743-53.

Gottmann, J. (1977) 'Megalopolis and Antipolis: The Telephone and the Structure of the City', in I. de Sola (ed.) *The Social Impact of the Telephone*, MIT Press, Cambridge, Mass.

Group of Thirty (1985) *The Foreign Exchange Market in the 1980s*, New York.

Hall, P. (1984) *The World Cities*, Weidenfeld and Nicolson, London.

Hewlett, N. and Toporowski, J. (1985) *All Change in the City*, Special Report No. 222, Economist Publications, London.

Independent Commission for World Wide Telecommunications Development (1984) *The Missing Link*, Geneva, Switzerland.

Kellerman. A. (1984) 'Telecommunications and the Geography of Metropolitan Areas', *Progress in Human Geography*, 8(2), 222-46.

Kindleberger, C. P. (1978) *Economic Response: Comparative Studies in Trade, Finance, and Growth*, Harvard University Press, Cambridge, Mass.

Kirby, S. (1983) *Towards the Pacific Century: Economic Development in the Pacific Rim*, Special Report No. 137, The Economist Intelligence Unit, London.

Legal Times (1985) '500 Largest Law Firms', 8 (15 September), 15-7.

Logue, T. (1986) 'Recent Major US Facilities-Related Policy Decisions: Letting a Million Circuits Bloom', in D. Wedemeyer and A. Pennings (eds) *Pacific Telecommunications Council '86 Proceedings*, University of Hawaii Press, Hawaii.

Lohr, S. (1985a) 'Global Stock Trading Near', *The New York Times*, 27 November, D 1.

——— (1985b) 'The Global Exchange', *Fortune*, 4 February, 9.

Mandeville, T. (1983) 'The Spatial Effects of Information Technology', *Futures*, 15(1), 65-72.

McMurray, S. and Browning, E. (1986) 'Merrill Lynch, Other US Firms, Expect Big Things from Tokyo Exchange Seats', *Wall Street Journal*, 31 January, 24.

Moss, M. L. (1986) 'Telecommunications and the Future of Cities', *Land Development Studies*, 3(1), 1-12.

Noyelle, T. J. and Stanback, T. M. Jr. (1984) *The Economic Transformation of American Cities*, Rowman and Allenheld, Totowa, New Jersey.

Rowley, A. (1984) 'Where Does the Market Go From Here?', *Far East Asian Review*, 8 March, 45.

The Banker (1985) 'The Pacific Arrives', 135(713), 16-23.

The Economist (1986) 'Optical Fibres Straddle the Globe', 289(7438), 84.

Vernon, D. (1959) *The Changing Economic Function of the Central City*, Supplementary Paper No. 1, Area Development Committee of CED, New York.

Weber, D. (1983) 'The Asian Connection: Lawyers in the Financial Centres of the Far East', *California Lawyer*, 3(11), 28-9, 52-3.

Chapter 7

TELECOMMUNICATIONS AND ELECTRONIC INFORMATION SERVICES IN AUSTRALIA

J. Langdale

The emergence of the information economy in industrialised economies has created considerable public awareness. While most attention has been directed towards examining the growth of information-equipment industries and the underlying technological changes in electronics and computers, there is a growing level of interest in those service industries whose development is reliant on the adoption of electronic information systems.

Information services are provided by firms involved in the collection, processing and/or transmission of information (Langdale, 1985b). Electronic information services (EISs) represent a subset of this group in that the collection, processing and/or transmission is primarily in an electronic form. It is difficult to classify information services into one category or another since virtually all firms handle information in both its electronic and non-electronic forms. The distinction relies on the degree of importance of electronic information systems for service firms. Traditionally, non-EISs include those firms that transfer information in the form of personal skills (advertising, legal services and accounting); on the other hand, EISs tend to be found in the banking, finance and computer services areas. However, the distinction is becoming increasingly blurred as a number of non-EIS firms are expanding their usage of electronic information systems.

This chapter focuses on services that are heavy users of electronic information systems, such as banking, finance and business, and financial information services. However, it also includes some consideration of other services (accounting, advertising and legal services) which are less heavy users of electronic information systems; these services are also referred to under the general heading of EISs.

To further complicate matters, EISs may be provided by specialist EIS firms or they may be provided by other firms such as those in manufacturing or mining. Firms in the former category are classified as primary EISs and include banking and finance, advertising, computer services and media firms. Firms in the secondary EIS category are those whose main function is in such industries as manufacturing or other services but which provide EISs on an intra-company basis.

INTERNATIONALISATION OF EISs

The rapid growth of EISs internationally and within Australia needs to be seen in the context of internationalisation of production in some manufacturing and service industries. A number of transnational corporations (TNCs) have integrated their production on a worldwide basis and have upgraded their international information systems, which are necessary to link the various parts of a global organisation. These developments have led to TNCs devoting more of their resources to secondary EISs in terms of intracorporate financial, accounting and computer systems operations. Intracorporate telecommunications linkages on a world-wide basis are important to the extent that they allow TNCs to co-ordinate their international operations more effectively. The internationalisation of production has also stimulated growth of primary EISs. Many TNCs in manufacturing and other service industries require better services from primary EIS firms. Thus banks, computer services and accounting firms have had to improve their own international information systems in order to adequately service their clients' requirements.

National Perspectives

The shift to an international information economy is increasing the level of interdependence of economies at different geographical scales. The location and functions of Australian EISs and the geographical patterns of business telecommunications-flows within Australia are closely interrelated with international factors.

A number of the processes discussed previously at the international level can also be identified at the national level. Large organisations (both public and private) are increasingly utilising secondary EISs in their activities; intra-corporate telecommunications are integrating their operations on a nationwide basis. Similarly, many primary EIS firms are geographically expanding their operations to achieve a national coverage and diversifying their range of services so as to meet the demands of client firms.

EISs AND THE INTERNATIONAL URBAN HIERARCHY

A loose hierarchy of international cities based on EISs can be recognised. It is possible to correlate the 'global reach' of a city with its status as an international EIS centre. Major global cities such as New York and London are key decision-making centres for international capital and have concentrations of specialist international EIS firms. Lower-ranking cities (Tokyo, Chicago, Paris, etc.) also have these functions, but do not have the diversity and specialisation of functions present in global cities.

Further down the urban hierarchy regionally-oriented cities such as Singapore, Hong Kong and Bahrain are emerging as EIS centres for their surrounding regions, especially in the banking and finance area (Langdale, 1984a,b). These cities perform an important focal role: they link particular regions to major international cities. Singapore performs such a function for the South-East Asian Region. It has attracted numerous offshore banking and financial institutions and at present is attempting to attract other information services (legal, computer and accounting services) to complement its banking and finance role. In addition, Singapore functions as an important regional head office for TNCs in a variety of industries; many of these firms generate heavy telecommunications linkages with their subsidiaries throughout the South-East Asian region as well as with their head offices, usually located in major international cities.

Finally, at the lowest level in this international hierarchy, nationally-oriented cities function as international gateway locations for their respective countries. EISs located in these cities are linked to other cities within the country and to regional, international and global cities. These cities are frequently the primate and/or capital cities for their respective countries and have national concentrations of corporate head offices and specialist EISs.

The key role of international cities in EISs stems from their function as decision-making centres for large TNCs, combined with their concentration of specialist EIS firms. There is a strong complementarity between these two functions (Cohen, 1981; Sassen Koob, 1984). Head offices of TNCs require specialist taxation, financial and legal advice from EIS firms; frequently there is a need for close communications between these firms. There has been some research examining the nature of these linkages in the office location literature (Goddard, 1973; Daniels, 1982). The linkages are not simply between EIS firms and their clients, but also between the service firms themselves.

Unfortunately, few studies have examined the role of EISs in international cities in the context of international, national and intra-urban telecommunications linkages. Moss (1984) has considered aspects of these linkages for New York City, however, there has been little attention given to the location of EIS firms in the context of their overall information collection, processing and transmission functions. This study provides an exploratory study of Australian EISs in terms of their international and domestic telecommunications flows. However, because of data limitations, the study's conclusions must be regarded as tentative.

AUSTRALIAN CITIES AND INTERNATIONAL EISs

Sydney and Melbourne as International Telecommunications Gateways

Sydney, and to a lesser extent Melbourne, are key locations for Australia's international EISs; they act as international-telecommunications gateways between Australia and the rest of the world. Sydney is particularly dominant for business-oriented services. For outgoing international telex traffic in March 1981 over 50 per cent of traffic originated in the Central Coastal Region of NSW (which includes Sydney, Newcastle and Wollongong), compared with 30 per cent for Melbourne and 14 per cent for all other capital cities (Langdale, 1982, p. 79).

Several factors account for Sydney and Melbourne's dominance. They are the location of most of the head offices of large firms in Australia (Taylor and Thrift, 1980). These offices, especially those of TNCs, generate a large volume of international telecommunications traffic. The Sydney-Melbourne head office functions as an international information gateway: information from the firm's Australian operations is channelled through the head office; it is sifted and repackaged and then sent to the overseas head office. Conversely, the Sydney-Melbourne office functions to filter information coming into Australia from overseas sources (Langdale, 1985b).

Sydney and Melbourne have a large number of specialist EIS firms. The banking and finance industry in particular generates large international telecommunications-flows and is concentrated in these cities. Similarly, other EISs, such as business and financial information providers, media firms and computer service bureaux are also concentrated in Sydney and Melbourne. Other state capital cities have much smaller international telecommunications flows and less specialised EISs, reflecting their primary orientation towards their state economies. Canberra is an exception with relatively strong international linkages: this results from the concentration of Australian Government departments, research institutes (e.g. CSIRO), the National Library and the Australian National University.

Discussion of Australia's international telecommunications linkages is initially focused on secondary EISs, those intra-organisational EISs associated particularly with large public and private organisations in manufacturing and service industries. Attention is then directed towards primary EISs, particularly in the banking and finance area.

Secondary EISs: The Role of TNCs

It is likely that a substantial component of Australia's international business telecommunications is generated and/or received by the Sydney-Melbourne head offices of Australian subsidiaries of foreign-based and Australian-owned TNCs. Unfortunately, there are few data to support this assertion. However, studies in other countries of international telecommunications traffic suggest that a small number of TNCs generate a large percentage of business-oriented traffic (Gassmann, 1985; United Nations,1982). It is likely that much of this intracorporate traffic is between Sydney and Melbourne and the TNCs' head offices located chiefly in such cities as London, New York or Tokyo. However, it would appear that more complex patterns of linkages are found in some large TNCs which have more fully internationalised their operations: a number of these firms have established regional offices (Singapore or Hong Kong for the Asia-Pacific Region) and have integrated production plants and offices throughout the world into global networks.

There is little evidence as to the number and geographical extent of these networks. A partial indicator is provided by the number of international leased circuits between Australia and other countries (Table 7.1). Many large organisations operate domestic and international leased networks for their internal traffic requirements; these networks are rented from telecommunications carriers. While large organisations are not the only users of these leased circuits, they dominate this type of service; the larger the organisation, the greater the possibility of fully utilising the capacity of the leased circuits. Numerous TNCs are rapidly expanding the geographical extent and sophistication of their international leased-circuit networks (Langdale, 1985a).

Hong Kong has the largest number of international leased circuits with Australia (Table 7.1). While Hong Kong is of increasing importance to Australia as an international banking and finance centre, the concentration of leased circuits between the two countries reflects Hong Kong's favourable international telecommunications rates. Many TNCs use a Sydney-Melbourne to Hong Kong leased circuit to connect with their global telecommunications networks.

The US has the largest number of high-speed voice and voice-data links (Table7.1).This reflects the greater technical sophistication of international information systems of US-based TNCs, as well as the fact that many are concentrated in high-technology industries (automobiles, aerospace and computer equipment). This is supported by a study of the diffusion of international data communications technology in US and European TNCs

Table 7.1: Number of Australian International Leased Line Circuits, 1983

	Telegraph	Voice/ Voice, Data	TOTAL
Hong Kong	88	13	101
New Zealand	59	18	77
US	38	40	78
UK	30	4	34
Japan	32	0	32
Singapore	17	2	19
Papua New Guinea	18	2	20
Philippines	14	1	15
Fiji	5	1	6
Others	60	8	68
TOTAL	361	89	450

Source: Overseas Telecommunications Commission (1983, p. 50).

(Antonelli, 1985). US-based TNCs with in-house telecommunications skills were the fastest adopters. In contrast, European TNCs were slower in adopting the new technology.

Primary EISs

The rapid internationalisation of primary EISs is having a major impact on the growth, industry structure and international linkages of these services in Australia. As was the case with secondary EISs, the impact has been concentrated in Sydney and Melbourne, although a number of TNCs have extended their branch networks to other capital cities and, in some cases, to country towns. Such an expansion strategy has reinforced the role of Sydney and Melbourne as international and domestic foci in the EIS area.

Banking and Finance

The banking and finance industry illustrates a number of these points. The round of mergers amongst Australian banks in the early 1980s reflected the need for Australian banks to be of sufficient size to compete with foreign banks. The large Australian banks have rapidly internationalised their operations, partly in response to the international expansion of their client firms, but also to meet the competition from foreign banks. It was also necessary for the Australian banks to provide electronic funds transfer (EFT) services for their corporate clients in Australia and in as wide a range of countries (especially in the Asia-Pacific region) as possible.

The entry of foreign banks and financial institutions has reinforced Sydney's—and to a lesser extent Melbourne's—international financial status. Sydney dominates the head office locations of merchant banks in Australia with 73 per cent of the total; Melbourne has 25 per cent and other states 2 per cent (Peat Marwick Mitchell, 1983, 64). However, the fourteen largest merchant banks, which account for approximately 65 per cent of the market, have their head offices evenly divided between the two cities and have offices in both (Henderson, 1983, p. 46). All except two of the seventeen foreign banks recently given permission to enter Australia are located in Sydney and Melbourne. Sydney has eight of the head offices, Melbourne has five (two banks have a joint Sydney-Melbourne head office) and one each is located in Perth and Adelaide.

The international telecommunications linkages of domestic and foreign banks and financial institutions are quite extensive, particularly those of large US financial institutions which operate extensive global information networks (Citicorp, Bank of America and Bankers Trust). For example, Citicorp has a medium-speed leased circuit (9.6K bits/sec.) between Singapore and Sydney, which in turn is linked to other leased circuits to New Zealand and Australian capital cities. From Singapore messages are routed to Citicorp's global network. The move to full banking status for a number of foreign banks is likely to expand the volume of international traffic, both on public switched services (telephone and telex) and on each bank's leased network.

AUSTRALIAN CITIES AND DOMESTIC EISs

Sydney and Melbourne as Domestic Telecommunications Nodes

In addition to their international gateway role, Sydney and Melbourne are also national telecommunications centres with extensive linkages throughout the country. The other state capital cities also have a significant role in the national economic system, although their linkages are primarily with their respective states: Canberra, as the national capital, has significant linkages throughout the country; Sydney and Melbourne have large reciprocal flows; Sydney is strongly linked with Canberra, Brisbane and Perth; while Melbourne is closely connected with Tasmania. Adelaide's traffic tends to be fairly evenly split between Sydney and Melbourne (Langdale, 1979).

The concentration of business subscribers in the capital cities accounts for Telecom's introduction of new business-oriented services in these cities; at a later date, the services will be introduced in smaller metropolitan centres and country towns. Expansion of new, business-oriented networks (packet-switched and digital data) have followed this pattern. In addition, expansion of telecommunications flows between capital cities has led to the laying of a high-capacity fibre-optic cable between Sydney, Canberra and Melbourne to be completed by 1988; other state capital cities and smaller centres will be progressively connected to this network in the 1990s.

Secondary EISs

It has already been argued that major public and private organisations generate and/or receive a large share of business traffic. While detailed statistics are not available in Australia, in the UK 4.6 per cent of business customers generate 50.7 per cent of revenue for long-distance business traffic (Post Office Engineering Union, 1981), and in the US 4 per cent of business customers generate 62 per cent of revenue for long-distance business traffic (Telecommunications Industry Task Force, 1977).

There are few Australian statistics in this area. Information on the industry composition and revenue for Telecom's forty largest customers is available for 1981 (Table 7.2). Somewhat surprisingly, Federal Government departments (37.0 per cent) and authorities (13.2 per cent) dominated revenues, with State Government departments accounting for a further 11.2 per cent. Of the remaining users, banks (13.8) and airlines (7.6) were heavy users with several large manufacturers (BHP, ACI, CSR and ICI) quite prominent.

Information on the geographical distribution and purpose of the telecommunications flows generated by large organisations is not publicly available. It would be expected that much of this traffic is for intra-organisational purposes. We have already seen in the international section that corporate head offices are telecommunications intensive. Such a situation is repeated at a domestic level for both large Australian organisations and TNCs. Head offices of organisations located in Sydney and Melbourne have strong linkages to branch plants and offices in other locations. While these flows are particularly strong within the organisation, they may be quite substantial to closely-related industries (major suppliers and clients).

The geographical pattern of control in the corporate sector is concentrated in Sydney and Melbourne (Taylor and Thrift, 1980). While the evidence is not conclusive, it would appear that Sydney has increased its share of head offices of large firms because of its stronger international

Table 7.2: Telecom's Revenue from its Forty Largest Customers, 1981

	Number of Customers	Revenue ($'000)	% of Revenue from the 40 Largest Customers
Federal Government Departments	10	65,715	37.0
Federal Government Authorities *	4	23,488	13.2
State Government Departments and Instrumentalities	6	19,964	11.2
Banks	6	24,421	13.8
Manufacturers	4	10,851	6.1
Airlines	3	13,545	7.6
Media	2	4,867	2.7
Others	5	14,764	8.3
TOTAL	40	177,615	99.9

Source: Australian Financial Review, 18 January 1982, pp. 1, 16, 18.

* The Overseas Telecommunications Commission (OTC) accounted for $7,519,000 revenue (or 4.2 per cent of total revenue of the 40 largest customers). Much of this would be the domestic component of international telecommunications traffic generated by business and private subscribers.

connections, and because it has attracted firms in more dynamic industries. However, Melbourne retains the headquarters of thirteen of the twenty largest companies; they are concentrated in mining and manufacturing and include some of the largest foreign-owned firms (Macquarie Atlas, 1984, p. 192). This pattern of corporate control suggests that the major telecommunications flows associated with these firms are focused on Sydney and Melbourne; it is also likely that the most sophisticated business telecommunications services would be first adopted by these head offices.

Primary EISs

There is a strong concentration of primary EISs in Sydney and Melbourne with smaller concentrations in other capital cities. Unfortunately, there is little published information which provides an overview of the growth and geographical location of these services. The banking and finance industry is of considerable interest because of its size, central importance in the economy and rapid development of EFT services. We have already seen that the industry is heavily concentrated in Sydney and Melbourne. Its growth is reviewed later in this section. Other primary EISs are also heavily concentrated in Sydney and Melbourne; the legal-services area illustrates this dominance. Amongst the ten largest legal firms seven had their head offices in Sydney and three were Melbourne based (Walter, 1985).

Banking

The banking industry has been a heavy user of domestic telecommunications for some time. An early stage of the banks' computerisation program was the introduction of centralised computer-communications networks linking their branches throughout Australia to their respective state regional offices, which in turn were linked to the national head offices in Sydney and Melbourne. This early computerisation was primarily aimed at internal administrative functions. EFT developments from the early 1980s have been oriented towards retail (individuals and small business) and wholesale (government and corporate banking) areas.

Retail-based EFT developments have allowed customers to operate their accounts from any branch in Australia. Automatic teller machines (ATMs) have been widely accepted by customers and point of sale (POS) terminal networks are being extended into retail outlets. In addition, home-banking services which utilise Telecom's videotex service, VIATEL, are being introduced.

It is not possible to fully review the range of wholesale EFT services oriented towards the corporate and government sectors. In common with the retail area, they are providing customers with greater diversity of electronic-based services and are allowing firms to manage their financial transactions more efficiently. The urban and regional implications of these changes are quite significant. EFT services have facilitated operation at the level of the national urban system. At a national (and international) level, corporate treasurers in the Sydney and Melbourne head offices are able to use corporate cash-management services to transfer funds from one subsidiary's account to another. Head office's financial control is thus enhanced by these technological developments.

Retail EFT developments are also likely to have significant impacts at an intra-urban level. The geographical diffusion of ATMs and POS terminals in suburban retailing outlets combined with increased acceptance by customers will have a substantial impact on travel behaviour and on the suburban retail property market. Banks may close some small branches with the possibility of opening unattended ATMs in shopping centres; the greater utilisation of POS terminals may reduce the demand for some banking services at branch locations. Some commentators have speculated that home-banking services provided by videotex will have an important impact on travel behaviour. However, given the low penetration of the videotex service, it is unlikely that home banking will have a significant impact in the immediate future.

Other EISs

Other EISs also generate heavy telecommunications flows, although their impacts are less significant compared with the banking industry. Virtually all the major media firms (newspapers, radio and television broadcasting) are rapidly expanding their investment in electronic information systems. Much of their expenditure in the 1970s was concerned with computerising their newspaper operations. More recently they have expanded their use of domestic and international telecommunications so as to receive and disseminate information more quickly and cheaply. The Fairfax organisation was ranked thirty-sixth in Telecom's major customers' list in 1981; News Limited was close behind at thirty-eighth (Table 7.2).

Communication satellites are becoming more important for media firms' operations, although their use of the AUSSAT domestic satellite has been plagued with controversy and has not as yet been resolved. On an international basis the Seven and Nine networks have earth stations in Sydney which receive programs beamed up to the Pacific Ocean INTELSAT satellite from Los Angeles. These programs are then distributed using Telecom's terrestrial network and AUSSAT to television stations in other locations.

Australian Associated Press (AAP) plays an important role in supplying media firms with information. AAP is headquartered in Sydney and employs about 140 journalists, has bureaux in every Australian capital city as well as overseas correspondents. The Sydney data-processing centre, to which all reporters file, provides world and Australian news to newspapers and commercial radio and television stations throughout the country (Woolford, 1985). In common with overseas firms such as Reuters, AAP has diversified away from supplying information to media firms. It provides a wide variety of financial and business information to corporations and government departments; about half of its income comes from such services. Most of its subscribers to these business-oriented services are located in the major decision-making centres of Sydney and Melbourne, although Canberra and to a lesser extent other capital cities are significant markets. Access to domestic and international business and financial information is becoming increasingly important for decision makers. Consequently, the web of telecommunications-flows linking information providers with corporate and government head offices is likely to intensify in the future.

Some primary information services (legal, accounting and consulting firms) are characterised by relatively small firms. While they may have heavy face-to-face communications and/or telecommunications for their size, in aggregate they are not major generators of telecommunications traffic. However, a number of the larger legal and accounting firms are rapidly adopting electronic information systems which link together their Australian and international operations, although the bulk of information transferred is likely to be on a local basis. The large accounting (particularly the Big Eight) firms are expanding their work on computer systems, rather than confining themselves to auditing. It is likely that there will be substantially increased telecommunications flows between accounting firms and their clients as these computer services are provided on an on-line basis.

Other information services are characterised by larger firms (insurance and advertising firms), but do not appear, on the limited evidence available, to be heavy users of domestic or international telecommunications. However, this situation appears to be changing as these firms computerise their operations and diversify in terms of the range of services provided to their customers.

Industry Structure of Primary EISs: Impact of Mergers and Takeovers

There are a number of changes taking place in the industry structure of primary EISs which are having a significant impact on the nature of domestic (and international) telecommunications linkages for major metropolitan areas. One of the most significant of these is the increased level of merger and take-over activity involving Australian EIS firms. In part this is related to the internationalisation of EISs: foreign-owned EIS TNCs are expanding into the Australian market by establishing new branch offices and by taking over or merging with Australian firms. In addition, Australian EIS firms are engaged in merger and take-over activity, partly because of the increasing level of international competition, but also because of the need to achieve economies of scale. The computerisation of EISs has enhanced this trend since it allows larger firms to spread fixed overheads.

Regulatory changes have also been a significant influence on the level of mergers and take-overs in a number of EISs. For example, the Ministerial Council for Companies and Securities has recently approved an increase in the maximum size of law partnerships from 100 to 200 partners (Davies, 1985). This move, in conjunction with other trends, such as the internationalisation of legal services and the computerisation of law firms, is likely to accelerate the number of mergers and rationalisations which have taken place in recent years. Large legal firms based in Sydney and Melbourne are likely to expand the number of branch offices in other states and overseas as a result of these changes.

EIS firms are under pressure from clients to extend the geographical coverage of their operations so as to adequately service their client firms' needs in various parts of Australia (and overseas). Increasingly, it is necessary for the EIS firm to support such domestic and international operations with improved information systems. Such moves are enhancing the degree of integration of the national and international urban systems.

CONCLUSIONS

A key theme in this chapter is the importance of the interrelationships between various geographical scales of analysis when considering telecommunications and EISs. While it is true that the volume of domestic telecommunications is far larger than that at the international level, it is important to consider the content of the information flows. This is especially significant given the high degree of foreign control in many Australian industries. International information flows in TNCs need to be considered in the context of the management structure of these firms. How much autonomy does the Australian subsidiary of a foreign-owned TNC have? What types of information are transmitted in TNCs as a result of different administrative structures? Similar types of questions need to be considered at a domestic level. What is the nature of the control exerted by the head office of large firms based in Sydney and Melbourne over the rest of the economy? How is this control reflected in the nature of domestic information flows?

There is a need for more data both on the growth and the geographical distribution of EISs in Australia. More information is also needed on the nature of telecommunications linkages generated by primary and secondary EISs, both at an international and at a domestic level. Such information is of particular significance given the growing importance of these services to Australia's domestic economy and its international trading relationships.

REFERENCES

Antonelli, C. (1985) 'The Diffusion of an Organisational Innovation: International Data Telecommunications and Multinational Industrial Firms', *International Journal of Industrial Organization*, 3, 109-18.

Cohen, R. B. (1981) 'The New International Division of Labor, Multinational Corporations and Urban Hierarchy', in M. Dear and A. J. Scott (eds) *Urbanisation and Urban Planning in Capitalist Society*, Methuen, London.

Daniels, P. (1982) *Service Industries: Growth and Location*, Cambridge University Press, Cambridge, England.

Davies, A. (1985) 'Law Firms With 200 Partners Approved', *Australian Financial Review*, 14 May, 1.

Gassmann, H.-P. (1985) *Transborder Data Flows*, Proceedings of an OECD Conference held in December 1983, North-Holland, Amsterdam.

Goddard, J. (1973) *Office Linkages and Location: A Study of Communication and Spatial Patterns in Central London*, Pergamon Press, Oxford.

Henderson, C. G. (1983) *Foreign Banking in Australia*, Unpublished-Honours Thesis, Macquarie University, Sydney.

Langdale, J. (1979) 'Telex and Data Transmission and the Australian Economic System', *Proceedings of the Tenth New Zealand Geography Conference*, Auckland.

——— (1982) 'Telecommunications in Sydney: Towards an Information Economy', in R. Cardew, J.Langdale, and D. Rich (eds) *Why Cities Change: Urban Development and Economic Change in Sydney*, Allen and Unwin, Sydney.

——— (1984a) 'Computerization in Singapore and Australia', *The Information Society*, 3(2), 131-53.

——— (1984b) *Information Services in Singapore and Australia*, ASEAN—Australia Economic Papers No. 16, Australian National University, Canberra.

——— (1985a) 'Electronic Funds Transfer and the Internationalisation of the Banking and Finance Industry', *Geoforum*, 16(1), 1-13.

——— (1985b) *Transborder Data Flow and International Trade in Electronic Information Services: An Australian Perspective*, Report to the Department of Communications, AGPS, Canberra.

Macquarie Atlas (1984) *The Macquarie Illustrated World Atlas*, Macquarie Library, Sydney.

Moss, M. (1984) 'New York Isn't New York Any More', *Intermedia*, 12(4/5), 10-14.

Overseas Telecommunications Commission (1983) *Annual Report*, Sydney.

Peat Marwick Mitchell (1983) *Sydney: Financial Growth Centre for the Pacific*, A Study on Behalf of the NSW Department of Industrial Development and Decentralisation, Sydney.

Post Office Engineering Union (1981) *The POEU Submission on the Beesley Report*, London.

Sassen-Koob, S. (1984) 'The New Labor Demand in Global Cities', in M. P. Smith (ed.) *Cities in Transformation: Class, Capital and the State*, Urban Affairs Annual Reviews, Volume 26, Sage, Beverly Hills.

Taylor, M. and Thrift, N. (1980) 'Large Corporations and Concentrations of Capital in Australia: A Geographical Analysis', *Economic Geography*, 56, 261-80.

Telecommunications Industry Task Force (1977) *The Dilemmas of Telecommunications Policy*, New York, (*ad hoc* committee).

United Nations (1982) *Transnational Corporations and Transborder Data Flows: A Technical Paper*, Centre on Transnational Corporations, New York.

Walter, B. (1985) 'Why Legal Firms Are Tying the Knot', *Australian Business*, 20 February, 38-41.

Woolford, D. (1985) 'The Nation's News Agency Celebrates its 50 Years', *Canberra Times*, 29 May, 27.

Chapter 8

TRADED SERVICES, REGULATED MARKETS AND TECHNOLOGICAL CHANGE

K. Tucker

In a post-industrial society, the spatial impact of innovation and technological change will depend to some extent on the degree to which services are traded across regional or international borders and are supplied from or into regulated markets. To begin with, some evaluation of the significance of traded services is required. Then it is necessary to consider whether the services traded originate in or are destined for markets which are regulated by tariff or non-tariff barriers to trade or are monopolised by key suppliers or users. Finally, the role of innovation or technological change may be to overcome the impediments presented by such regulated markets. By contrast, a traditional view is that the role is to reduce the resource inputs required in a production or distribution process or augment the benefits to users by means of additional characteristics supplied with the products or services.

TRADED SERVICES: THEIR SIGNIFICANCE

Daniel Bell (1974) commented as follows:

> A post-industrial society is based on services. Hence it is a game between persons. What counts is not raw muscle power, or energy, but information. The central person is the professional, for he is equipped by his education and training, to provide the kinds of skill which are increasingly demanded in the post-industrial society.

Bell's view of the world suggests that spatial impacts depend very much on the information content of services and the nature of skills required by professionals associated with the supply of these services. Therefore, the supply of suitably-trained and skilled-labour inputs is a critical element. Where these educational and know-how providing institutions are located and how they satisfy their client-group needs is not the key focus of this chapter (for discussion of this topic see Tucker and Chataway, 1983). Rather, the emphasis is on those activities with a high degree of 'service intensity' in the economy and the ways in which these are sold to inter-regional or international customers.

It is now more commonly accepted that services in any economy are neither parasitic nor luxuries but are essential ingredients in the functioning of both developing and advanced economies (Shelp, 1981). Further, it is evident that the output of service industries is widely traded, especially across national boundaries. Estimates of directly traded services suggest they are about 25 per cent of total international trade. Indirectly traded services, that is, intermediate or embodied services, may also be a significant addition for some countries to this international trade in services (estimates are provided in Tucker *et al.*, 1983, and Bureau of Industry Economics, 1980). Figure 8.1 shows how some of these links operate through the balance of payments.

Figure 8.1 Traded Services: Sectoral and International Relationships. **Source:** Tucker *et al.* (1983).

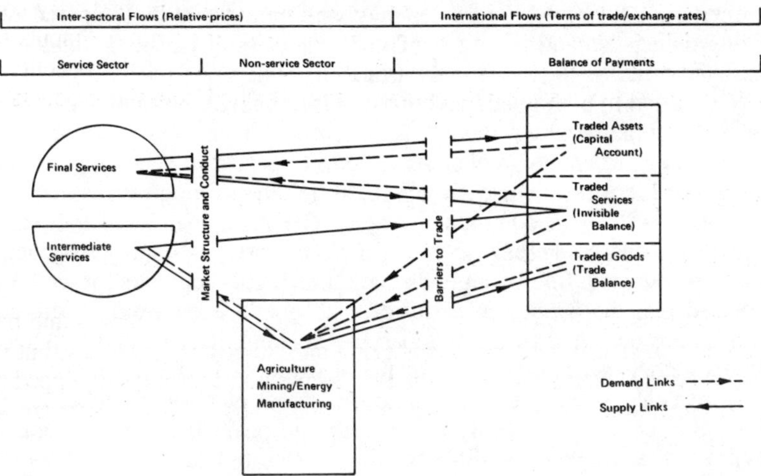

A recent examination of the service intensity of traded goods for three countries—Australia, Singapore and Thailand—revealed that the ratio of indirectly traded services to directly traded services was as high as 50 per cent for Australia with the services content of manufactured exports being 33 per cent. By contrast, the figures for Singapore were 18.5 per cent and 7.5 per cent respectively (Tucker and Sundberg, 1985).

REGULATED MARKETS: THEIR IMPACTS?

The question arises, therefore, as to whether the share of output of the service sector that is traded and the service intensity of goods trade is in any way related to the degree to which markets are regulated by government intervention (for example, tariffs or other barriers) or by the concentration of buyers or sellers (that is, monopoly or restrictive practices).

It may be argued, of course, that the spatial distribution of customers and technological change associated with the design, production and distribution of goods and services are the determinants of these key ratios. However, it also could be asserted that the very existence of regulated markets, besides the tyranny of distance or the relative scarcity of factors of production (labour, energy, capital, etc.) is a major catalyst for innovation and technological change. In other words, rather than technology augmenting factor availability or equalising the access of customers to products or services in any one location (that is, removing transport costs), the unbundling or splintering of services from goods and the inter-regional or international supply of services may well be induced by the perception of unsatisfied demand in any regulated market (that is, the equivalent of restricted or unsatisfactory supply arrangements).

When trade in both goods and services is to heavily-protected markets it will be limited, thereby providing an incentive for innovation and technological change to create an alternative form of access to such markets. The concept of footloose industries supported by direct foreign investment behind tariff walls is a common example. The phenomenon of tourists as itinerant consumers getting duty free and other price or quality advantages is well recognised. Should a goods market be highly regulated, it would be expected that the service intensity of such goods trade would decline as services are detached or made separable in supply and demand, to avoid the regulatory regime. One international example is the development of offshore banking services, supported by global electronic networks (see Moss, this volume). Not only is it possible to provide by these means an integration of markets in different time zones but taxation and foreign ownership regulations are also overcome. At a more local level, the incentives to innovate in financial markets by providing home-banking services recognises the restrictions on trading hours facing customers. Similarly, where advisory services (such as an R & D division) provided from within a company are offered at prices above market rates, the likelihood of contracting out such services increases.

TECHNOLOGICAL CHANGE: WHAT IS ITS ROLE?

The role of innovation and technological change may, in the context of the above discussion, be better viewed as a device for overcoming existing regulated market structures. In other words, the 'natural' imperative of economising on resources and reducing transport or consumption costs may have been supplanted by the 'synthetic' objective of counteracting the barriers to trade erected by suppliers, or consumers having excessive market power, or governments whose intervention was designed to protect specific interest groups. If this view is accepted, then the role of technological change and innovation can be seen as affecting more crucially consumption location than production location. In other words, technology essentially breaks the influence of regulated markets and makes consumers of products or services less dependent on local production capacity in formal market arrangements.

Taking this argument even further, perhaps our concern about the long-term adverse effects of mechanisms to protect existing market suppliers or preferred distribution channels (for example, tariffs, quotas, regulated entry) may be unwarranted if innovation and technological change can sooner (preferably) or later neutralise these barriers or restrictions. Furthermore, the immediate concerns we may have for measuring inequalities or differential impacts in a spatial context may only be of a temporary nature. These are transitory outcomes of a causal process something like that described above. They are not discrepancies capable of welfare solutions or further intervention. If that occurs, the incentive for innovation to remove the effects of such will disappear.

REFERENCES

Bell, D. (1974) *The Coming of the Post-Industrial Society—A Venture in Social Forecasting*, Heinemann, London.

Bureau of Industry Economics (1980) *Features of the Australian Service Sector*, Research Report 5, AGPS, Canberra.

Shelp, R. K. (1981) *Beyond Industrialisation: Ascendancy of the Global Service Economy*, Praeger, New York.

Tucker, K. A. and Chataway, G. (1983) *Training Needs in the Application of Micro-Electronics: The Service Sector*, Workshop on Micro-electronic Training Needs, National Training Council, September.

———, Seow, G. and Sundberg, M. (1983) *Services in ASEAN-Australian Trade*, Economic Paper 2, ASEAN-Australian Joint Research Project, Kuala Lumpur and Canberra.

——— and Sundberg, M. (1985) *Comparative Advantage and Service Intensity in Traded Goods*, Economic Paper, ASEAN-Australian Joint Research Project, Kuala Lumpur and Canberra.

Chapter 9

INSTITUTIONAL ECONOMICS AND TECHNOLOGICAL CHANGE IN THE AUSTRALIAN RETAIL INDUSTRY: THE SPATIAL RAMIFICATIONS

P. Fisher

Wholesale and retail trade is Australia's largest service employer and small business sector and a major source of employment. Along with community services, it is the fastest-growing avenue of part-time casual labour and one of the major employers of females. Australia's labour market is, however, tightly regulated. Wages and conditions of employment are generally set by Awards negotiated between unions and employer organisations under Federal and State arbitration systems. Apart from certain exempt categories, the retail trade also incurs stringent restrictions on its trading hours. Most States limit evening trade to one designated day per week and, with the exception of declared resort areas and certain categories of business in New South Wales, all currently prohibit trade on weekends after 1 p.m. Saturdays.

Labour costs, particularly those of an indirect nature, are high relative to, say, the United States which has a similar pattern of retailing. In that country, fringe benefits assumed as-of-right in Australia are scarcely contemplated. Awards further restrict how labour can be used. There are, too, indications that additional restrictions may be placed on conditions of employment. (A recent agreement negotiated between large retailers and the union in one State effectively precludes the engagement of juniors under 16 years of age.) In essence, the industry has experienced a blow-out in costs, both labour and general overheads, and this is reflected in a significant decline in profitability (e.g. gross operating surplus of GDP was 38 per cent in 1971 and 33 per cent in 1980). The response by major elements of the industry has been to increasingly hire junior casual labour and convert to self-service format; packaging and labelling has assisted this transition.

Since that time new technology has been developed in the form of electronic-funds transfer and point-of-sale (EFT-POS) systems which are now being introduced into major outlets. This technology will inevitably lead to further adjustments in the labour-capital mix within both stores and warehouses. Such developments will be locally accentuated by recent mergers between Australia's major retailers, Myer and Coles, and Woolworths and Safeway. The new corporate strategies of these large organisations and their concomitant push into new technology, carry enormous implications for customers, unions, landlords, employees, suppliers and the service organisations that deal with them. This chapter accordingly explores technological change in the tightly regulated environment of the Australian retail industry, having particular regard to the spatial ramifications of this change.

MAJOR FACTORS AND TRENDS TO THE PRESENT

In the last decade or so, major changes have occurred in employment and trading hours within the Australian retail industry. As elsewhere, other changes have taken place in packaging and marketing, including the adoption of credit-card facilities and utilisation of electronic cash registers. Over the course of the decade, employment in wholesale and retail trade remained a steady 20 per cent of total employment and has now displaced manufacturing as the major source of employment.

Labour Intensity

In Table 9.1, ten major sectors of the Australian economy are ranked according to their labour intensity as measured by employment per $1 million of output. This shows that wholesale and retail trade was the third most labour-intensive industry in 1976-77. Whereas it reduced its labour intensity during the period 1966-67 to 1976-77 by 20 per cent, it merely kept pace with the other sectors as there was an overall decline of 21 per cent. There can be little doubt that such reductions represent a response to the disproportionate increase in labour costs characteristic of that period (Golledge, 1982).

Retail census data suggest that, in respect of employment numbers, no significant change occurred in the mix of retail types between 1968-69 and 1979-80. However, the average employment per establishment rose by almost one-quarter from 5.4 employees to 6.7 employees. This is largely explained by the substantial growth in employment per establishment of de-

Table 9.1: Industry Ranking by Labour-Intensity

Industry	1976-77 Empl. per $1M Output	Change in Labour Intensity from 1966-67 to 1976-77
Community services	113.1	-9
Recreation and personal services	112.4	-3
Wholesale and retail trade	105.1	-20
Construction	95.5	-15
Manufacturing	81.4	-26
Agriculture, etc.	80.9	-28
Finance and property services	68.1	+1
Transport and services	65.1	-34
Electricity, water and gas	48.6	-42
Mining	33.7	-41
All Industries	86.0	-21

Source: Ministry for Economic Development (1982)

partment, variety and general stores and food stores, which together accounted for almost half of the total employment in the retail trade. In particular, the average employment per establishment in department, variety and general stores increased from 41.2 to 116.2 employees over the same period while for food stores, it increased from 4.1 to 6.6 employees.

Two developments have contributed to this growth. Firstly, large retail enterprises, which are prominent in these two categories have, by exercising scale economies, generated an environment of intense price competition. In particular, self-service operations, a measure of automation, and changes in packaging and labelling, were tailored to meet the demands of the larger enterprises. Consequently closures have outstripped openings of small, independent, owner-managed shops. Secondly, large retail enterprises, in the course of expanding their operations, have amalgamated their own and take-over-acquired units into larger entities (e.g. closure of street-side variety stores in favour of discount department stores in free-standing situations) and have been constructing much larger stores after the fashion of British and South African type hypermarkets. The decline in labour intensity in the wholesale and retail trade can therefore be attributed to the consolidation into spatially larger units.

Employment Characteristics

The retail industry closely follows behind entertainment and recreational services and community services as a major provider of jobs for women. Australian Bureau of Statistics (ABS) figures show that since 1966, the female:male employment ratio has slowly risen from 38:62 to the present-day figure of 43:57. This increase is a direct result of growing importance of part-time and casual employment in respect of which women have long been dominant by a factor of 4 :1.

Growth in Costs

Given the significant level of female employment, the phasing in of equal pay in the early 1970s, whereby female award rates doubled in three years, resulted in a major shock to the cost structure of the industry. ABS figures on the retail labour content of sales on a mark-up basis show that when the equal-pay provision was phased in, the decline in the retail labour content of mark-up on sales (viz. 6.4 per cent in 1963 to 5.1 per cent in 1969) which originated from the steady introduction of self-service operations during the sixties, was reversed (viz. 5.8 per cent in 1975). But by the late seventies, with the consolidation of many stores into larger, self-service operations, and the adoption of some automation, the decline was again resumed (viz. 5.4 per cent in 1978). Later in the decade, trading hours were extended beyond 9.00-5.00 Monday-Friday and 9.00-12 noon Saturday, by the introduction of late Thursday/Friday night shopping, accompanied by new industrial awards which incorporated a penalty loading of 25 per cent for the Saturday morning or weeknight after 6.00 p.m.

In addition to the increase in direct labour costs as a result of the equal pay decision and the extension of trading hours, the retail industry has, in common with other Australian industry sectors, incurred significant increases in indirect costs over recent years. Employers' liability insurance premiums, for instance, after a period of intense competition between insurance companies during the late 1970s, have now increased twofold to threefold. Enterprises with more than about ten employees further incur a State-levied payroll tax, which has risen from 2.5 per cent of wages to 6 per cent of wages currently. Finally, holiday loadings introduced in 1974 add 17.5 per cent to the basic four weeks' recreational-leave pay entitlement.

Employment Outcomes

The response by the industry—notably major supermarket, department and chain-store enterprises—to these significant increases in labour costs has been dramatic. As foreshadowed above, it has passed over full-time employees in new hirings in favour of part-time and casual employees most of whom are junior and female. According to ABS figures, the respective proportions of full-time to part-time casual employment have changed from 85-15 per cent, ten years ago to 75-25 per cent currently. Along with community services, the wholesale and retail trade is the fastest-growing source of part-time and casual employment in the Australian economy. This is reflected in the reduction in average hours worked per employee in the industry over recent years.

Youth Wages Impact

Since 1974 full-time juniors have been renumerated at rates which range from half the adult wage at 16 years of age to full adult wage at 21 (viz. 15-16 years, 50.0 per cent of adult wage; 17 years, 55.0 per cent of adult wage; 18 years, 67.5 per cent of adult wage; 19 years, 80.0 per cent of adult wage; 20 years, 90.0 per cent of adult wage; and 21+ years, 100.0 per cent of adult wage). Part-time and casual rates are pro rata, but casual employees have no entitlement to fringe benefits. However, for the first twenty hours, they must be paid a 33.3 per cent loading reflecting the entitlements of permanent employees. In certain sections of the industry this loading is being progressively supplemented. Obligations to provide substantial fringe benefits for the permanent employee and the need to meet unforeseen shifts in demand, clearly constitute a strong incentive to hire casual junior labour, particularly in the 15-16 year age group. Further, they provide cause to shed the least skilled of this labour at 17 years of age when a wage rise becomes obligatory. In fact, an analysis of the part-time and casual employment experience of youth born in the five years 1963-67 respectively, over their fifteenth, sixteenth and seventeenth years reveals that for each cohort, part-time/casual employment increases in the sixteenth year and falls away in the seventeenth year by as much as 20-30 per cent.

FUTURE CHANGES AND SCENARIOS

Institutional Economic Factors

There is an expectation that both direct and indirect labour costs will continue to increase in real terms. In respect of direct labour costs, a revision of the awards can be foreseen to allow for a shorter working week (e.g. 35 hours) and to accommodate an extension of trading hours during the weekends. In the first case, the hourly wage rate will increase and unless remedial action is taken, that is, varying the capital-labour mix, it will be necessary to hire additional employees to perform the same amount of work, therefore increasing the total wages bill. In the second case, additional employees would normally need to be hired to perform the extra amount of work, again increasing the total wages bill. And, if a recent agreement concluded between five major retailers and the Shop Distributive and Allied Employees' Association in one State is any guide, penalty rates are likely to be further increased, and the under-17-years wage rate eliminated, thus sealing off an avenue for a 10 per cent reduction in labour costs.

In respect of indirect labour costs, any increase in the total wages bill will increase the amounts paid in employers' liability insurance since these are determined as a proportion of payroll. Although premiums can be expected to rise over the short-run, the introduction of legislative occupational health and safety provisions will eventually reduce them by lowering the incidence and seriousness of accidents, leading to fewer and lesser claims. On the other hand, these provisions require some small retail enterprises to more adequately cover themselves against employer's liability claims, significantly lifting their overheads. Some may also have to provide and pay for sick room facilities. However, both of these imposts will be one-off.

Finally, unions can be expected to press claims for increased benefits such as technology-inspired job redundancy, retirement, pension and long-service leave provisions in their awards. If granted, these benefits will significantly add to labour costs. The degree to which increased turnover from extended trading hours will offset the ballooning in labour costs for which it would be partly responsible, remains unresolved. The populace now have a far wider range of choices as to how to occupy their leisure time—with VCRs, indoor sports centres, aerobics, etc.—a feature of the 1980s' lifestyle. These activities are emerging as alternative recreational pursuits to shopping.

Technology

Recent years have seen substantial advances in computer-related (information handling) technologies within the commercial sector. The EFT system now being locally implemented aims for a substitution of ordinary paper-transmitted orders by machine-processable electronic impulses. Many large Australian stores already deploy POS equipment in association with their own computers to monitor stock level and cash in hand. To provide an EFT-POS facility to the customer, therefore, this equipment need only be linked in with the bank or financial institution's computer.

Internally, POS offers efficiencies at the cashpoint, at the store, and at the enterprise or group level for chains. At the cashpoint, the main benefits accrue at the store level with a reduction in manual stock-control, cash audit and sales returns. Security is also markedly improved with the cost of transferring cash from stores to banks eliminated. At the enterprise or group level, the system, in providing detailed transaction information, permits lower stock level, better buying, promotion planning and space allocation.

Automatic stock-monitoring and price assignment are facilitated by product-marking systems. Under these systems, information about the product (colour, size, supplier, etc.) is read by machine and transmitted by electronic impulse to computer for processing. At this stage the price can be identified by reference to a price-by-product memory file. The bar-coding system, which is especially designed for retailing, deploys laser scanning equipment. Laser scanning enables daily adjustments to the price of items, a boon in the supermarket trade where selective discounting is commonplace. It specifically offers a faster and more accurate check-out operation and the elimination of price marking.

POS and laser equipment obviously form important complements and, as such, are being increasingly integrated. Together they offer an extremely powerful means of rationalising the buying and selling of stock as well as reducing the impost of hiring labour and overheads in general (Musaniff, 1984). They can also facilitate in-store video-disk selling and construct detailed social and economic profiles of the store's clientele. This integration is about to be taken further: an information management systems approach is being developed whereby the computer system not only facilitates access to data, but also recommends action such as re-ordering stock. By linking back to the enterprise's main data-processing facility it will then be possible to control goods from receipt at a central warehouse through store stockroom to checkout. Computer controlled conveyor (stacking) retrieving systems in the central warehouse are integral to this advance. Large supermarket, department and chain-store enterprises are likely to be first to deploy these developments just as they have been quick to adopt the bank-promoted EFT-POS system. However, regardless of whether they have a similar understanding of what this technology can do for them or not, the cost factor will deny small shopkeepers technology-derived efficiencies over the short run. But equipment manufacturers will have to capture the small-business market in order to fully recoup their R & D or product-development outlay. This requirement is becoming more critical because of the shortening economic life of electronic products.

One can confidently predict that the unit costs of such new technology, as opposed to training and implementation costs, will decrease over time aiding its widespread adoption. This will, in turn, be pushed along by the rising usage of the debit card and the consumer's expectation of being able to employ it at virtually any retail outlet. Looking further into the future, teleshopping is destined to become more common for grocery items, but personal shopping seems certain to remain the preference for apparel. Meanwhile, developments in hardware and software will continue to enhance the attractiveness of the POS terminal which can be expected to become cheaper, have increased flexibility and greater memory.

By the 1990s, the new micro-mini computers, word processors and view-data systems can be expected to play a major role in the retail enterprise, undertaking functions such as profitable space allocation, merchandise allocation, automatic stock-control and merchandise re-ordering, electronic purchase-order management, EFT, staff scheduling, market research, energy management, and store design and layout. In short, a fully-fledged automation covering sensation, logic and action phases will exist. A similar rationalisation is already underway in the manufacturing sector where design, manufacture, marketing and sales functions are being merged.

The Meshing of Technological and Institutional Economic Factors

Although new technology is commonly viewed as the agent of change in the retail industry the preceding discussions suggest that in regulated labour markets typified by Australia institutional economic factors, especially revisions to award provisions and consequent increases in labour overheads, are equally significant. Most importantly, these two factors bear an interactive relationship to one another.

In the period over which the cited developments in new technology will be released onto the market, real direct-labour costs can be expected to further increase through revision of awards to allow for a shorter working week (e.g. 35 hours), job redundancy provisions, and to accommodate an extension of trading hours during weekends—lent impetus in turn by EFT-POS systems which provide for unlimited discretionary expenditure outside banking hours. A continued rise in real labour overheads also seems likely with unions pressing claims for retirement pensions and long service leave provisions, while over the short run, employers' liability insurance premiums can be expected to increase too. In the United States, Roessner (this volume) envisages these sort of demands, including the formation of white-collar unions, could only occur under very rapid rates of market penetration of office technology.

In order to contain their labour costs therefore, Australian retailers have, in the context of re-examining their organisational practices, two immediate options: they can increase the proportion of casual to permanent labour (Macken, 1983), and/or increasingly deploy new technology. If specific agreement between unions and employers precludes the former course of action, staffing levels will be eroded by an accelerated recourse to new technology and in the process, avoidance of the onus attaching to employing persons.

The POS/laser-scanning equipment and advanced software and

hardware suites, along with the revision of organisational practices which are invariably the context for their introduction, have the capacity to substantially check labour costs and operating overheads, leading to considerable price discounting.

SPATIAL RAMIFICATIONS

The extent to which these organisational and technological developments will affect competitiveness within the same line of business, and therefore the geography of retailing, can be evaluated at both the enterprise or group level and at the establishment or store level. Over the short run, it will be the small enterprises which will tend to be the most affected because of the slower take-up of the new technology and limited scope for revision of organisational practices on account of inflexibility of divesting their capital stock.

The majority of small enterprises consist of a single establishment or store and therefore have no opportunity to dissipate the effects of intense price competition at one location over a store-wide operation. In short, single-store enterprises are likely to prove even more vulnerable to competition from multiple store enterprises than is currently the case. However, the severity of impact will depend upon how rapidly the Australian public accepts the EFT-POS system which in turn will have some bearing on the rate of installation of these facilities. Australians have hitherto taken to new consumer-oriented technologies, such as the VCR, with relish, market saturation occurring very quickly.

However, to be truly successful the EFT-POS system will require automatic crediting of employee bank accounts by their employers. For Australian low income and welfare groups there continues a practice of being paid wages in cash or by cheque. Consequently, the rate of POS transactions—or even installation of POS facilities—may be significantly less in low-income neighbourhoods because of a slower take-up and use of the debit card. As a result, small shopkeepers in these places could be shielded from any adverse effects for some time.

The age element may also be important. EFT systems have yet to gain wide acceptance in the United States among the wealth-holding, older age groups who maintain a suspicion about non-traditional approaches to financial transactions. The young, by contrast, do not, generally share their reservations—although a high-tech backlash against machine depersonalisation does extend into this group. In respect of banking, customers would seem to feel more comfortable writing cheques and where possible even going to the same teller each visit to the bank. Nonetheless, one has the distinct impression that the young are attracted to automated-teller machines and use them even when the banks are open: perhaps these devices are seen as high-tech chic? The young constitute an obvious target for large retail enterprises adopting POS and new-tech facilities in general. It is also destined to become the dominant group as successive generations, weened on PCs and taking computer-related technology in their stride, attain adulthood, although Roessner (*op. cit.*) anticipates some literacy problems. Small retail enterprises to match the variety and quality of services, will need to follow suit in line with the age and income characteristics of their clientele or concentrate even harder on the personalised service aspect. Roessner (*op. cit.*) has further observed that the adoption of new information-handling technology by a firm may force rapid adoption by its competitors even if this generates net losses. But many small retail enterprises will need to resort to some automation in any case, to check the growth in their own labour costs and overheads. Fully fledged automation would, however, appear inappropriate to small, isolated situations (Cohen, 1985).

It would appear that the ownership of the Australian EFT-POS system will be vested in the banks and other financial institutions rather than the retailers. It remains to be seen, however, whether there will ultimately be a single system: at present the system is fragmented, with each bank cultivating a different set of retail outlets. There is, too, the prospect of large retailers entering joint ventures with banks; for example, Coles-Myer and BA-Australia have announced plans for retail banking from K-Mart stores, further blurring the distinction between banking and retailing functions. These joint institutional-technological developments appear certain to accelerate the demise of the traditional shopping strip. Larger stores and chains have already relocated in large numbers from streetsides to free-standing shopping centres. It will be these stores which first adopt the new technology, further enhancing their cost edge over the small enterprise stores which predominate along the strips. In at least one capital city, the demise of the strip is being hastened by local street closures which have resulted in increased arterial traffic. Fixed-rail transit separation may also become operative in this general decline.

Unfortunately, part-time or casual work is assuming a role as the only source of income for significant numbers of people in the Australian economy. Because of the nature of this work and other commitments of those undertaking it, most part-time labour market participants find it necessary that such employment be close to home. ABS data reveal that in relation to women who are not in the labourforce who want a job but are discouraged from looking, 37 per cent report the reason as there being no jobs in the locality or line of work. Even among males, 27 per cent report this as their reason for being discouraged. These statistics suggest that a consolidation of commercial-retail facilities into selected centres is out of kilter with developments in the labour market and likely to result in inequities.

CONCLUSION

This chapter has sought to demonstrate that, in the highly regulated environment of the Australian retail industry, institutional economic changes and the adoption of new technology are closely interconnected and will critically affect the profits and/or viability of retail enterprises and thus the scale and location of stores.

Rises in direct and indirect labour costs as a result of the possible introduction of a 35-hour week, long-service leave, pension benefits, redundancy provisions, increases to employers' liability insurance premiums, and the extension of trading hours through Saturday, etc., will fuel automation and/or the further employment of casuals over permanent staff, at least in those sections of the industry which offer no substantial fringe benefits above the casual wage loading of 33.3 per cent. New technology, in the form of EFT, where the debit card can provide for spending on a 24-hour day, seven-day-a-week basis, will be viewed by large retailers as an opportunity to capture an even larger proportion of the consumer's discretionary expenditure. The drive for total deregulation of trading hours will thus be given further impetus.

Such developments will engender a major restructuring of the retail industry. As transfer of this new technology is well underway in large retail enterprises but scarcely contemplated in their smaller counterparts, an accelerated consolidation of shopping facilities into larger and fewer outlets operating from regional complexes, would appear inevitable, especially if the hypermarket concept proves successful locally. A similar restructuring is underway in Britain (Taylor, 1984). In view of the increasing importance of part-time work in such establishments, this consolidation will further compound community employment problems. As matters stand, by the time the present data-capture technology has diffused through to small retailers, large retailers will be in the process of adopting the next generation of technology—a technology capable of optimising many facets of their operations. In contrast to the earlier adoption of the data-capture technology, this will be accompanied by a shift of emphasis toward considerations more directly related to boosting productivity than checking labour costs.

REFERENCES

Cohen, S. (1985) *Telecom and Service Industries: the Future for Employment*, The Berkeley Round Table on International Economics, University of California, Berkeley.

Golledge, R. G. (1982) *Small and Medium Business Districts and Owner-Managed Enterprises*, Report to Department of Planning, Melbourne.

Macken, J. (1983) Report to Minister for Industrial Relations, Industrial Commission of New South Wales, Reference No. 1074/1982, October.

Ministry for Economic Development (1982) *Economic Development Strategy for Victoria: Paper 3 Evaluation of Strategy Options and Recommendations of Strategy*, Ministry for Economic Development, Melbourne.

Musaniff, N. Y. (1984) *Chips in Retailing*, Special Report No. 138, Economic Research Unit, London.

Taylor, A. (1984) 'The Planning Implications of New Technology in Retailing and Distribution', *Town Planning Review*, 55(2).

Chapter 10

MARKET PENETRATION OF OFFICE AUTOMATION EQUIPMENT IN THE UNITED STATES

J.D. Roessner

Eventually, the revolution in office technology based on the microprocessor will produce fundamental changes in what white-collar workers do, how they should be managed, and how the activities of their employers are organised. If office-automation (OA) technology spreads rapidly, the revolution in technology could result in major social disruptions such as the dislocation of large numbers of clerical workers or new demands on educational institutions that cannot be met. There is some evidence, however, that in the US acceptance of new office automation technology is slower than would be expected, given its cost and performance features (Uttal, 1982). The rate at which office automation technology penetrates the market will, it appears, be largely the result of non-technological forces. This chapter is based on the assumption that constraints on the rate at which office automation technology spreads will arise not from the need for technical breakthroughs, but from a host of economic, organisational and behavioural factors that interact in complex ways. In the chapter I identify some of these non-technological factors and examine how they are likely to shape the rate and pattern of market penetration of OA technology in the United States.

FACTORS LIKELY TO INFLUENCE THE MARKET PENETRATION OF NEW OFFICE TECHNOLOGY

Most empirical research on the spread of new technologies in industry has focused on industrial process innovations in competitive industries. Technology diffusion and market penetration researchers have devoted very little attention to the service industries or to technologies such as computers that constitute the production equipment of service industries. (One exception is Malecki, 1977.) The available literature is useful for identifying factors that influence the pattern and rate of spread of new technology, but it lacks verified models that would enable one to predict the relative importance of these factors and the penetration rate for specific technologies and industries (Roessner, 1980). The following list of factors likely to influence the spread of OA technology has been generated from two sources: the empirically-based, scholarly literature on the diffusion of innovations in industry, and the more recent, anecdotal material on office automation. Key references of the former type include Nelson, Peck and Kalachek (1967); Mansfield (1968); Mansfield *et al.* (1977); Nabseth and Ray (1974); Utterback (1974); and Hurter and Rubenstein (1978). The latter include articles in the trade press and in business magazines such as *Business Week* and *Fortune*. The factors are:

1. cost of the new technology, including initial capital cost and operating and maintenance costs;
2. availability and cost of labour, both as a possible substitute for the new technology and as users of it;
3. industry structure and competitive environment;
4. organisational factors such as corporate and managerial strategies;
5. behavioural factors such as fear and inertia; and
6. macro-economic conditions.

In the following section I detail the form that each of these factors takes in the case of OA technology and present evidence, where available, of the relative significance each factor is likely to have on the spread of new office technology.

Analysis of the Significance of These Factors for the Spread of Office Automation Technology Cost

Firms will expend capital to automate their office functions in two major categories: equipment (including software) and implementation. Implementation costs include the costs of training employees to use the new equipment, the costs of re-organising work to realise the full benefits of OA beyond its impact on clerical functions, and some of the costs of retraining workers displaced by the new technology. A firm's decision to automate its offices does not, of course, lead to a single, lump equipment purchase. Capital expenditures, maintenance costs and support costs (e.g. training, software updating) will be ongoing. From a budgetary perspective, office automation involves deciding what percentage of the budget will be devoted to OA over an extended period. Implementation costs may equal or exceed the costs of hardware, and continuing support costs may be equally large. Industry's past rates of investment in new equipment offer another possible indicator of future constraints on the market penetration of OA equipment. Figure 10.1 shows trends in US annual investment in new equipment in manufacturing, wholesale and retail trade, and finance, as a proportion of each industry's GDP (in current dollars). For comparison, data on total investment in producers' durable equipment as a proportion of GNP are presented as well. During the decade 1970-1980, this ratio increased from 6.5 per cent to 9.7 per cent in manufacturing, an

Figure 10.1 Investment in Equipment as a Proportion of Industry Product (current dollars).

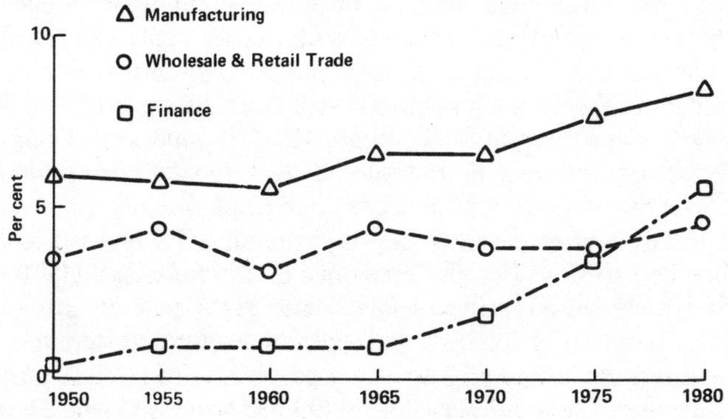

Figure 10.2 Investment in Equipment as a Proportion of Industry Product (billions of constant 1972 dollars).

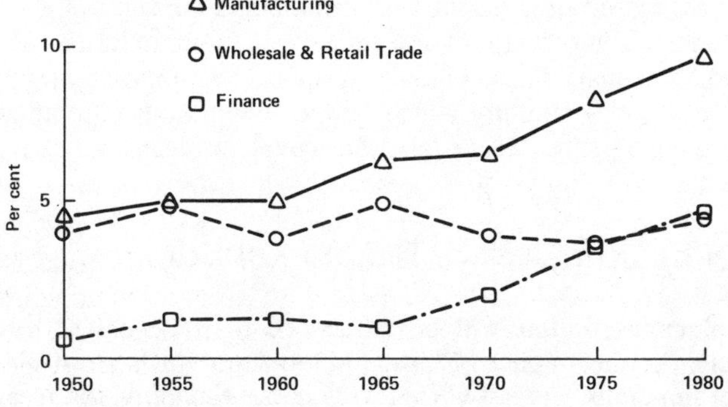

average annual rate of 4.1 per cent; in wholesale and retail trade it rose from 3.9 per cent to 4.5 per cent (1.4 per cent annually); in finance the ratio more than doubled, increasing from 2.0 per cent in 1970 to 4.7 per cent in 1980 (8.9 per cent annually). There is no particular reason to expect that the ratio of annual investment in new equipment to GDP should be similar across industries, although it is entirely possible that office-dominated industries such as finance will approach the level of investment exhibited by manufacturing. Projections of investment in new equipment by industry and for the nation, coupled with estimates of the proportion of that investment spent on OA equipment, offer one basis for setting the bounds within which market penetration forecasts for OA should fall. Figure 10.2 illustrates the same trends as Figure 10.1, expressed in constant 1972 dollars.

Although the ratio of US investment in new equipment to GNP has remained between roughly 6.5 per cent and 7.5 per cent since 1970, and is projected at between 7.5 per cent and 9 per cent, the proportion of that investment devoted to office automation has doubled since 1970 from 6 per cent to nearly 15 per cent in 1983 (Figure 10.3). It appears that this steeply-rising investment curve will continue, at least for the next several years. The penetration of OA technology will depend directly upon decisions about the relative proportion of total investment in all types of equipment to be allocated to OA. For the economy as a whole, the US Bureau of Labor Statistics (BLS) projects GNP and gross private investment in equipment (Table 10.1) for 1990 and 1995 under three different economic scenarios. Investment in equipment as a proportion of GNP is projected to grow to between 7.5 and 9 per cent by 1995 (current dollars). These ratios can be used to project expenditures for office automation equipment under different assumptions about changes in the ratio of investment in OA to total investment in equipment. As we saw, the ratio is now about 15 per cent and rising steeply; it could easily attain 20 per cent or even 25 per cent in the next decade. The results of such calculations are shown in Table 10.2 expressed in constant 1972 dollars and current dollars, assuming an average of 6 per cent inflation since 1980. Other inflation rates could be assumed, of course. These kinds of estimates can be used to generate a series of 'envelopes' within which office automation equipment sales forecasts might be expected to fall. The envelope using the estimates of Table 10.2 appear in Figure 10.4.

Availability and Cost of Skilled Labour

The rate of capitalisation will be influenced to some extent by overall wage levels. As the impact of office automation shifts from clerical to managerial functions, overall wage levels in the economy, not just clerical wages, will be relevant. However, a decision to automate an office func-

Figure 10.3 Ratio of Investment in Equipment and OA Equipment to Various Bases (current dollars).

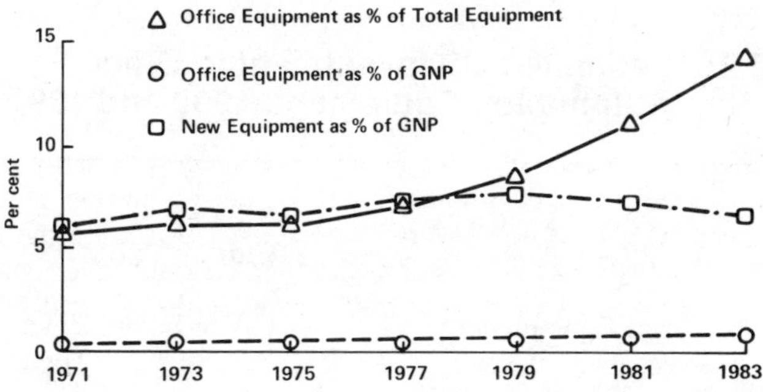

Table 10.1: Estimated GNP and Gross Private Investment in Equipment, 1990 and 1995

Year	GNP (constant 1972 dollars)	Gross Private Investment in Equipment (billions of constant)	Investment as % of GNP (1972 dollars)	GNP (billions of current dollars)	Gross Private Investment in Equipment * (billions of current dollars)
1982	1485.4	112.7	7.6	3069.3	–
1990 (LO)	1857.9	132.4	7.1	–	347.3
1990 (MOD)	1915.5	149.1	7.8	–	381.6
1990 (HI)	2004.2	166.2	8.3	–	406.0
1995 (LO)	2126.7	159.6	7.5	–	491.0
1995 (MOD)	2166.9	177.2	8.2	–	536.8
1995 (HI)	2264.6	202.8	8.95	–	585.9

Source: US Department of Labor (1984, p. 14)

* Assumes 6 per cent average annual inflation rate after 1982.

tion is not a simple trade-off between the cost of the equipment (and its implementation) and the wages paid to workers who perform that function. Rather, it is increasingly clear that firms do not invest in office automation simply to reduce labour costs or increase efficiency (Curley and Pyburn, 1982; Kettinger, 1983). As the International Data Corporation put it: 'Justification for office systems can come only in part from direct labour savings. The rest must come from better turnaround times, increased responsiveness to customers, and smarter decisions.' (IDC, 1983a,

Table 10.2: Estimated US Investment in Office Automation Equipment, 1990 and 1995

	Billions of Constant 1972 Dollars Office Equipment as Percentage of Total Investment in Equipment:			Billions of Current Dollars* Office Equipment as Percentage of Total Investment in Equipment:		
Year	15%	20%	25%	15%	20%	25%
1990 (LO)	19.86	26.48	33.1	52.1	69.5	86.8
1990 (HI)	24.93	33.24	41.55	60.9	81.2	101.5
1995 (LO)	23.94	31.92	39.9	73.65	98.2	122.75
1995 (HI)	30.42	40.56	50.7	87.9	117.2	146.5

Source: Calculated from data in Table 10.1

* Assumes average 6 per cent inflation rate after 1982.

p. 102). They punctuate this comment by estimating that an expected level of investment in OA of $195 billion over the period 1983-1987 will yield labour savings of only $110 billion. Expenditures for OA must include the wages of skilled operators and trainers. Various sources seem to agree that implementing OA equipment costs at least as much as the equipment itself. Past trends in industry expenditures for training offer some guide as to what levels might be expected in the future. The availability of workers possessing the requisite fundamental skills such as analytical sophistication, problem solving, ability to synthesise disparate materials, and communicate effectively could place serious constraints on the pace of market penetration. Industry representatives agree that all office workers, managers and clerks alike, will need to acquire numerous new skills that are not currently needed to perform their duties (Roessner *et al.*, 1985).

Many employers are relying upon vendor-provided training during this early period of penetration of OA equipment. Vendors provide training for

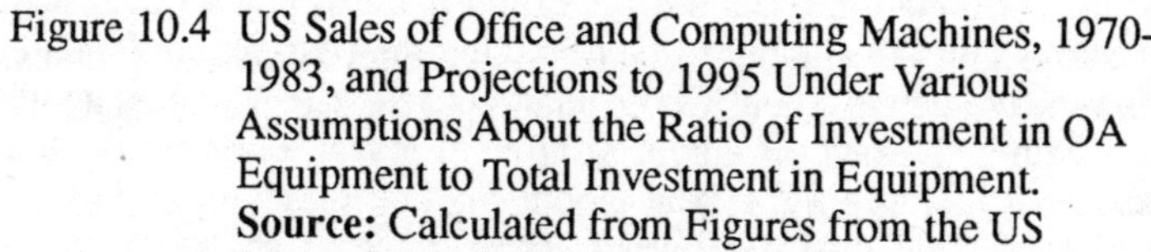
Figure 10.4 US Sales of Office and Computing Machines, 1970-1983, and Projections to 1995 Under Various Assumptions About the Ratio of Investment in OA Equipment to Total Investment in Equipment.
Source: Calculated from Figures from the US Department of Labor (1984, p.14).

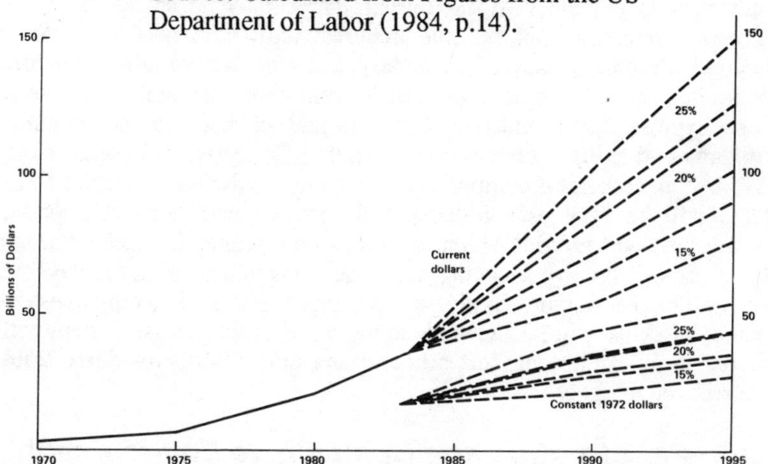

their dealers, and some offer seminars and classes at their local offices or at dealer offices. As vendors seek to shift training costs to purchasers of their equipment, self-paced training packages (cassettes, manuals, exercise workbooks, etc.) are becoming more popular (Friedman, 1982). But evidence is growing that this approach will not suffice, particularly as the focus of training programs moves from clerical workers to managers and other professionals (Bikson and Gutek, 1983, 1984).

Casual estimates of the cost of implementing OA tend to approximate the cost of OA hardware. This means that US employers may have spent as much as $20 billion in 1982 to train employees to use OA equipment and to implement OA systems. If this pattern continues, and vendors successfully transfer the bulk of training costs to end-users (as seems likely), then training and implementation costs could have a significant retarding effect on the rate of penetration of OA equipment. But whether this actually occurs will depend on many factors, including whether training costs are included in a firm's initial decision to purchase OA equipment, the degree of sophistication of the software (i.e. its friendliness), and whether cost savings rather than competitive pressures based on quality of service are the driving force behind the decision. The availability of persons in the labourforce with requisite skills may, in some regional labour markets, prove significant for the spread of office automation. On the one hand, demographics alone may stimulate some companies to automate. In the US, the combination of labour migration patterns and a declining rate of new entrants into the labour force following the post-World War II baby boom may leave certain local labour markets (e.g. the Midwest and New England) with a shortage of labour at all skill levels. Large firms, unable to relocate but facing a tight local labour market, may choose OA as the only way to expand business output. There may be an opposing force, however, that may not become manifest until well into the 1990s and beyond: the inadequacy of secondary and high school education for the computer age. The issue goes well beyond computer literacy to the concern, expressed in recent reports on the state of American education, that fundamental skills such as communication, analysis, and common sense will become crucial if computers are to be used effectively. Future OA systems will be very user friendly, will possess enormous computational power, and will provide volumes of data on command. The central issues for users will involve knowing what questions to ask, what the data mean, and when the output simply doesn't make sense. Teaching these fundamental skills is not easy, yet locating workers who possess them will be increasingly important for office managers. Short, remedial training courses are unlikely to suffice.

Industry Structure and Competitive Environment

The rate and pattern of market penetration of OA equipment will vary widely across and within industries. These variations will be a function of the marketing strategies of vendors and a host of factors that influence the spread of innovations. Here I discuss three factors related to industry structure and interfirm competition.

1. *The contribution office work makes to the industry's production or service delivery function.* Indicators of the significance of office work include the proportion of the industry's employees who are white-collar workers, the ratio of total white-collar wages to blue-collar wages, or the ratio of office costs to total costs of doing business. Office work, and workers, are concentrated in the US in the finance, services and trade industries and government. On this basis, OA would appear initially in the wholesale trade and finance industries. Within white-collar industries, those with a relatively large proportion of clerical workers have been the first to automate because of the initial emphasis on automating clerical functions. The popular and trade literature indicate that banking and insurance have been the bellwether industries for OA.
2. *Whether firms in the industry compete primarily on the basis of the price of the product or service delivered, or on the basis of the type, variety and quality of product or service.* In a typical manu-

facturing industry, new production technology displaces human labour in order to reduce the costs of production. Here, automation is justified on efficiency grounds. The cost of production declines as the amount of capital per worker increases, and firms compete on the basis of product price. In many service industries, in contrast, competition is based as much on the quality and range of services delivered as on price. Thus, automating an existing task is less important than enabling new tasks—services—to be performed. The (initial) cost of providing a product or service may actually increase as technology is added to production or service delivery processes; rather than displacing workers, technology largely complements their activities. Labour-cost savings, already only one of several rationales for the purchase of OA equipment, will decline in importance as managerial, technical and sales jobs become increasingly automated. Under these conditions, a firm that automates may force rapid adoption of OA equipment among its competitors even if the new equipment generates net losses. A case in point is automated teller machines, which initially did not prove profitable for banks but which have spread rapidly. Another is American Hospital Supply Inc., which increased its market share dramatically by installing terminals in hospitals linked to its sales organisation. In sum, because the pay-off to the firm from OA will not be manifested entirely, or perhaps even primarily, in cost savings from reduced labour requirements, market penetration should occur more rapidly in industries that compete on the basis of product quality than in industries that compete on the basis of product or service price. Clearly, analysis is needed of the role that OA plays relative to other factors in enhancing a firm's competitive position.

3. *Industry Structure.* All other things equal, the reduced competitiveness of monopolistic or oligopolistic industries, and of industries comprised of widely dispersed, small firms that do not compete directly, would yield slower response to the promise of OA. Regulated industries probably will vary widely in their response. Early experimentation with OA is likely to occur even in monopolistic industries such as utilities, provided slack (uncommitted) resources are available within the firm. But in the absence of competitive pressures, this experimental stage could continue well beyond the time when more competitive industries become committed to office automation. When air fares were regulated in the US, competition based on service delivery encouraged the introduction of computerised reservations; more recently, computerisation of frequent-flyer records occurred for similar reasons. In both cases, automation was introduced under conditions of con-

strained competition when a very specific application was evident. Additionally, industries dominated by small firms are likely to move more slowly because of the initial costs and uncertain payoffs associated with OA. Anecdotal evidence suggests that software costs, training costs and the time required to become informed about OA appear to be particularly burdensome to small firms (*Business Week*, 8 October 1984, pp. 126, 130). These inhibiting factors may be offset by the availability of new, small, multitasking systems designed for small businesses.

Behavioural Factors

Considerable anecdotal material describes a variety of behavioural factors that have limited or even halted the introduction of office automation. Some managers fear or are ignorant of the computer and do not want to reveal their inadequacies; they cannot or do not want to learn to use a keyboard; they do not want to lose the status symbol of a personal secretary. Middle managers fear that many of their functions such as data manipulation and forecasting will be automated. Clerical workers are concerned about electronic sweatshops in which a piece-work situation is closely monitored by machines. To my knowledge, little work has been done to document the pervasiveness or intensity of these situations. One exception is an ongoing study of the implementation of OA equipment and procedures in fifty-five offices representing a range of industries and staffing patterns (Bikson and Gutek, 1983). The first round of interviews with 530 employees suggested that the amount of resistance to OA amongst office workers may be less than is often assumed. The researchers found that problems with implementing computer systems in white collar settings are not technological in nature. Users think the technology itself is fine. What affects how much and how well it is used are organisational, environmental and training matters that organisations should be able to address.

Key management factors that were important to productive use of the technology were the organisation's orientation to change and the variety of tasks included in a single job. Many anecdotal sources and case studies suggest, on the contrary, that numerous behavioural factors will impede the spread of OA. For example, face-to-face communication, an integral element of job satisfaction, could be threatened by electronic communication made possible by OA. Changes in direct control over employees can be manifested, negatively, at several levels. The geographic decentralisation of work means that some managers will have fewer opportunities to observe their employees directly, possibly leaving them frustrated by a sense of lost control. At the other extreme, some clerical workers now can be monitored electronically, their output and error rates recorded and used as the basis for pay and promotion (National Science Foundation, 1983; Gregory and Nussbaum, 1982). Some managers simply fail to appreciate the relatively intangible advantages offered by office automation. In the absence of clear labour savings, managers are reluctant. Some comments from managers will illustrate (from Curley and Pyburn, 1982, p. 32):

> We feel that we've conquered the secretarial problem, but we're not sure if office automation will really improve managerial productivity or just give them more time to play golf.
>
> If I can buy a metal stamping machine that turns out 100 parts per day, I know what I'm getting. If I can increase its output by 10 per cent, I know I'll get 110 parts per day out of it. I can calculate the cost of the machine, figure out the return it will give me, and tell you whether it's a good investment or not. I don't know what buying electronic mail will do for me.Finally, the power of habit, familiarity, and inertia should not be underestimated. In those service industries where the competitive advantages to be gained from automating are unclear, the pace of change will be slow.

Organisational Factors

If Strassmann (1980) is correct, the productivity gains of office automation will not be fully realised until office work at the most fundamental levels is reorganised. The first wave of OA targeted clerical functions. OA simply replaced mechanical devices such as typewriters and calculators with their microprocessor-based counterparts. Its implementation required only superficial reorganisation of work. While some gains in efficiency have resulted, major gains in office productivity must await the second wave of automation, which will involve managers and technical-professional workers. Workflow and jobs will be linked to the introduction of local area networks (LANs) that permit communication among work stations. This restructuring of non-clerical office functions, workflow, and jobs will be costly and will occur over an extended period. Once a firm carries out such a restructuring, the resulting gains in competitive advantage may drive other firms in the industry to follow suit rapidly. These changes will occur at widely varying times in different industries over the next several decades. Strassmann estimates that the major changes envisioned will be occurring well into the next century. Moreover, the process by which this restructuring occurs is unlikely to follow the rational planning approach in which the productivity gains of OA under different organisational configurations are analysed in advance of purchasing decisions. Restructuring is likely to be evolutionary and reactive in character. In keeping with past experience with strategic planning and large-scale organisational change, companies will back into strategic planning for OA only after it has already become a significant, visible and costly element of a firm's activities.

Macro-Economic Conditions

The overall strength of the economy assuredly will affect the rate of office automation's penetration. In general, high rates of economic growth will mean increased rates of investment in office automation, but very rapid rates of market penetration could produce the threat or actuality of structural unemployment. Reactions to such a threat might include white-collar unions, stronger social pressure on firms to retain surplus workers, increased retraining expenditures, or even a shortened work week. Generating forecasts using different macro-economic assumptions is one accepted way to deal with the problem. While we do not know the penetration rate above which structural unemployment will occur, we know that the rate with which mainframe computers penetrated industrial markets in the 1960s and 1970s did not produce large-scale unemployment of clerical workers. But there are several reasons for believing that the past may be an unreliable guide to the future in the case of office automation (see, for example, Roessner, 1985a).

Forecasts of the Penetration of Office Automation Equipment

Forecasts of OA market penetration take two forms: (1) the dollar value of manufacturer shipments and/or sales of particular products; and (2) the number of units of products shipped or in place (installed base) in a given year. Each is useful for different purposes, and different data can be brought to bear to assess the credibility of forecasts or to help set the context for additional forecasts. In particular, sales forecasts obviously are useful for estimating the implications of OA for the national economy, for vendor-dealer revenues, and for end user expenditures. For these forecasts, trend data and extrapolations of annual expenditures by industry for new equipment offer one means of bounding estimates of future expenditures by users for OA equipment. Forecasts of unit penetration, especially installed base, can be compared with the number of likely users to gain some idea of current and projected market penetration as a proportion of the total eventual market. As I have noted elsewhere, publicly available sales forecasts are difficult to interpret because of differing definitions of the technology involved, the effects of inflation, and the difficulty of obtaining the full reports produced by firms that sell forecasts (Roessner, 1985b). Here I discuss only forecasts of the installed base of OA equipment in the US. With radically new technology such as OA equipment that does not simply replace older technology, it is very difficult to specify in advance what the size of the eventual market will be. (In the 1950s IBM thought they would sell only a few mainframe computers, primarily to government.)

Thus, during the initial stages of the diffusion of an innovation, it is difficult to specify what level of penetration (as a proportion of the total potential market) has been achieved, or what level of penetration a particular forecast (measured in sales or number of units) represents. In the case of OA technology, however, we do know how many of what types of office (white-collar) workers are in the US labourforce. The Bureau of Labor Studies prepares projections of employment by major occupational category (managers, professional-technical, clerical, sales, etc.). Table 10.3 shows 1982 levels and projections for 1990 and 1995 for the four categories of white-collar workers.

Table 10.3: US Civilian Employment by Occupation, 1982, and Projected 1995 (in millions of workers)

	Total Employment			
Occupation	1982	1995 (low)	1995 (med)	1995 (high)
All Occupations	101.5	124.8	127.1	129.9
Professional, technical and related	16.6	21.5	21.8	22.3
Managers, officials and proprietors	9.5	12.0	12.2	12.5
Salesworkers	7.0	8.5	8.8	8.9
Clerical workers	19.0	23.5	24.0	24.5
Craft and related	11.6	14.5	14.8	15.1
Operatives	13.0	15.0	15.4	15.8
Service workers	16.2	20.4	20.7	21.1
Labourers, except farm	5.9	6.9	7.1	7.2
Farmers and farmworkers	2.7	2.4	2.4	2.4

Source: Silvestri, Lukasiewicz and Einstein (1983).

The IDC states that by the end of 1983 there were about 18.5 million electronic-keyboard devices (PCs, other computers, terminals and word processors) in use in the US, or about one for every three of the 55 million white-collar workers. By 1987, they forecast, there will be 54 million electronic-keyboard devices—or one for every white-collar worker (IDC, 1983a, p. 94). They further forecast an installed base of about 19 million PCs in the US business-professional market by 1987, or about two PCs for every three managers and professional workers, assuming these occupations are the first to enjoy the benefits of the PC (IDC, 1983b, p. 2). At this rate of market penetration, by 1990 virtually all white collar workers in the US will be working with electronic keyboards, and most, if not all, managers, professionals and sales workers will be working regularly with a PC or electronic workstation (see Figure 10.5).

Figure 10.5 Forecasts of Installed Base of OA Equipment and White-Collar Employment. **Source:** International Data Corporation (1983b, p.2).

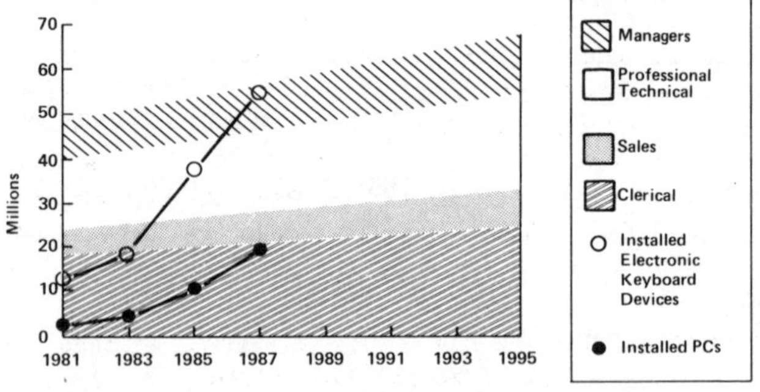

The IDC forecast is consistent with that of the firm Future Computing, which calls for an installed base of 9-10 million PCs in 1985, 32-33 million by 1990, and 53-54 million by 1995. Future Computing used BLS forecasts of occupational employment to generate its forecasts. In their view, by 1995 50-60 per cent of office workers will have personal computers, with 25-30 per cent having more than one (Biagiotti and Ablondi, 1984, p. 59). By about 1990, analyses of the impact of OA on office work and workers would shift from consideration of the shock of—and resistance to—initial exposure to OA equipment, to how, and how rapidly, office work and occupations will be restructured to take advantage of the continuously-improving office technologies that will be available in the marketplace. Stated differently, future OA sales will be a function of the replacement rates for in-place, automated equipment rather than of the rate of acceptance of OA equipment among the uninitiated. If these forecasts are prescient, this transition will occur in the US about 1990 for OA equipment intended to perform or complement managerial and professional-technical functions.

CONCLUSIONS

Office-intensive industries such as banking and insurance are rapidly increasing their capital:labour ratios. Office automation equipment represents a significant proportion, now 40 per cent in some service industries, of total annual expenditures for new equipment. Although equipment investment levels in wholesale and retail trade are not changing dramatically, they are in finance, insurance and real estate, where investment in equipment as a proportion of industry product is increasing at a significantly faster rate than in manufacturing, trade or the total economy.

The competitive edge that office automation offers through new services and higher quality output is a powerful force working to increase investment in OA equipment in service-producing industries. These forces will be offset to some extent by the shortage and high cost of skilled labour to operate the new machines and train others to use them. As the next wave of automation gains momentum toward the end of the century, the need to restructure work in fundamental ways may raise considerably the cost of moving to the next stage of automation that involves all levels of the office work force. In the meanwhile, managers' fears and inertia will present significant but relatively short-term constraints on the rapid spread of OA equipment.

The cost and availability of capital do not appear to have constrained industry's (short-term) rate of investment in OA equipment in the past. Training and implementation costs will assume greater importance as managers and professionals must learn to use the new equipment. These costs may greatly exceed hardware costs and limit the pace of penetration significantly. Very limited evidence suggests that, while managers will have some difficulty justifying the purchase of OA equipment on the basis of cost savings, they will have even greater difficulty justifying the larger, yet essential, costs of implementing and supporting it.

Several factors I have mentioned will have relatively little effect on the pace and pattern of market penetration. There is little doubt that cost-effective (measured in traditional fashion) hardware that promises major productivity gains will be available in the marketplace. The cost of this hardware will not prove to be a major barrier. Estimates of savings in labour costs, and traditional comparisons of labour costs against the cost of capital equipment to replace that labour, will not loom large in future decisions to purchase OA equipment. Forecasts of the installed base of OA equipment suggest that by about 1990 virtually every white-collar worker in the US will be working with some form of microprocessor-based keyboard (or other input-technology) device. At this rate of penetration, by 1995 every manager, professional-technical worker and salesperson will be working with a PC or work station. This represents an installed base of more than 40 million units. If market penetration of OA equipment in the US is this rapid, it will pose significant challenges to managers in white-collar industries, human-resource professionals and government policy-makers. Now is the time to anticipate these challenges and begin to address them. Specific actions should include research and analysis that would yield more valid forecasts of the spread of OA technology and greater understanding of the factors shaping its penetration of various markets. In industry and government, steps should be taken to strengthen and, if necessary, redesign training and education programs to accommodate the new requirements of rapidly changing labour markets.

REFERENCES

Biagiotti, S. J. and Ablondi, W. F. (1984) 'Micros by the Millions: Future Computing Sizes Up the Desktop User Scene', *Management Technology*, December, 58-63.

Bikson, T. K. and Gutek, B. A. (1983) *Advanced Office Systems: An Empirical Look at Utilisation and Satisfaction*, The Rand Corporation, Santa Monica, California.

——— and ——— (1984) 'Training in Automated Offices: An Empirical Study of Design and Methods', *Training for Tomorrow*, Pergamon Press, Oxford.

Curley, K. F. and Pyburn, P. J. (1982) 'Intellectual Technologies: The Key to Improving White-Collar Productivity', *Sloan Management Review*, Fall, 31-9.

Friedman, S. (1982) 'Tools for Training Users of Automated Equipment', *Administrative Management*, July, 38-57.

Gregory, J. and Nussbaum, K. (1982) 'Race Against Time: Automation of the Office', *Office: Technology and People*, 1, 197-236.

Hurter, A. P. and Rubenstein, A. H. (1978) 'Market Penetration by New Innovations: The Technological Literature', *Technological Forecasting and Social Change*, 11, 197-221.

International Data Corporation (IDC) (1983a) 'Office Systems for the Eighties: Automation and the Bottom Line', *Fortune*, 3 October, 89-162.

——— (1983b) *EDP Industry Report*, Framingham, Mass.

Kettinger, W. J. (1983) 'Models of Office Productivity: What Really Can Be Expected?', *Office Automation Conference Digest*, American Federation of Information Processing Societies.

Malecki, E. J. (1977) 'Firms and Innovation Diffusion: Examples from Banking', *Environment and Planning A*, 9, 1291-1305.

Mansfield, E. (1968) *The Economics of Technological Change*, Norton, New York.

Mansfield, E., Rapoport, J., Romeo, A., Villani, E., Wagner, S. and Husic, F. (1977) *The Production and Application of New Industrial Technology*, Norton, New York.

Nabseth, L. and Ray, G. F. (1974) *The Diffusion of New Industrial Processes*, Cambridge University Press, Cambridge, England.

National Science Foundation (1983) 'Trends in Computers and Communication: The Office of the Future', in *Emerging Issues in Science and Technology, 1982*, National Science Foundation, Washington, DC.

Roessner, J. D. (1980) 'Technological Diffusion Research and National Policy Issues', *Knowledge: Creation, Diffusion, Utilization*, 2, 179-201.

——— (1985a) 'Implications of Office Automation for Government Employment and Training Programs', prepared for the National Conference on Employment Policy, Washington, DC, July.

——— (1985b) 'Market Penetration of Office Automation Equipment: Trends and Forecasts', unpublished paper, Georgia Institute of Technology, Atlanta.

———, Mason, R., Porter, A. L., Rossini, F. A., Schwartz, A. P. and Nelms, K. R. (1985) *The Impact of Office Automation on Clerical Employment, 1985-2000*, Quorum Books, Westport, Connecticut.

Silvestri, G. T. Lukasiewicz, J. M. and Einstein, M. E. (1983) 'Occupational Employment Projections through 1995', *Monthly Labor Review*, November, 24-35.

Strassmann, P. (1980) 'The Office of the Future: Information Management for the New Age', *Technology Review*, 82 (December-January), 54-65.

US Department of Labor, Bureau of Labor Statistics (1984) Employment Projections for 1995, *Bulletin 2197*, Washington, DC.

Uttal, B. (1982) 'What's Detaining the Office of the Future?' *Fortune*, 3 May, 176-96.

Utterback, J. M. (1974) 'Innovation in Industry and the Diffusion of Technology', *Science*, 183 (15 February), 620-6.

Note: Much of the available forecast data are generated by vendors or consultants who sell information to vendors. This could introduce a bias in forecasts toward more rapid penetration.

PART 3.
NEW TECHNOLOGY AND SPACE

Chapter 11

THE GEOGRAPHY OF HIGH TECHNOLOGY: AN ANGLO-AMERICAN COMPARISON

P. Hall

In the mid 1980s, high technology has become a kind of instant solution to the economic ills of the Western world. It is hailed as the remedy for the decline of the smokestack industries and the source of the new jobs that will cut the unemployment queues. More than that, it is seen as the source of a new industrial revolution that will transform the economies of the advanced industrial nations, allowing them to retain their economic pre-eminence as newly industrialising countries crowd into the production of goods that were once the monopoly. Each nation, each region, each city is acutely anxious to enter the high-technology game.

Yet in all this euphoria, there is a general lack of hard evidence. Most of the journalistic reporting is just that. Most of the academic contributions are partial, dealing with single aspects of the phenomenon. General accounts have been conspicuously lacking. Now, however, we have the results of two parallel studies of the high-tech phenomenon—one for Great Britain, the other for the United States (Breheny and McQuaid, 1985; Breheny, Hall, Hart and McQuaid, 1986; Markusen, Hall and Glasmeier, 1986). They allow us for the first time to draw some balanced general conclusions. In particular, they begin to answer some specific questions.

First, a question that might seem superfluous, but is very much not so. What exactly is high-tech industry? Secondly, how many jobs does it create? How many high-tech jobs are there now, and how many new high-tech jobs are being added? Thirdly, where does high-tech locate? Does it, as popularly supposed, cluster in a few locations—Silicon Valley, Silicon Prairie, Silicon Glen and the other select silicon landscapes (Hall and Markusen, 1985)? Or do its benefits spread more widely? How far does it locate in older industrial regions; how far does it avoid them? And in what places is it growing most rapidly? Is it continuing to concentrate in its original seedbeds, or is it beginning to decentralise into other regions and cities? Finally, and most importantly, why is there a convincing general theory of high-tech location? And how well does it explain the observed patterns of location and of locational change?

WHAT? PROBLEMS OF HIGH TECH DEFINITION

It might seem easy to define high-tech: it is the group of industries that embraces the new information-technologies, biotechnology and a number of other technologies at the forefront of scientific and technological advance. The problem is to define exactly what these industries have in common, and how to measure it. In fact there have been relatively few serious attempts to define high-tech, and they have produced remarkably different results.

One obvious approach tries to measure the perceived degree of technical sophistication of an industry. The problem is that the resulting list is arbitrary; it offers no criterion for selection. An alternative is to measure growth in employment. The problem here is that many fast-growing industries are not high-tech by any reasonable criterion, while some clearly high-tech industries (such as biotech) are too new to show much growth, and others (such as defence-related industries) show very volatile patterns of alternating growth and decline. In fact, the sources of job growth are many and varied, and high-technology content is only one among them. A third possibility is to use research and development intensity. The problem here is different: if R & D spending is measured against sales, in larger and better established industries (such as chemicals or petroleum) it actually appears quite insignificant.

Therefore, both the American and the British studies were driven to rely on a fourth criterion: occupational profile. High-tech industries are those in which the percentage of scientific and technological occupations is greater than the industrial average. This has the advantage that good, standardised data, based on firm criteria, exist. And it clearly implies that innovation is occurring.

In the American study, use of this criterion resulted in twenty-nine high-tech industrial sectors embracing exactly 100 high-tech separate industries. They ranged from industries that everyone would popularly regard as high-tech (electronic computing equipment, semiconductors and related devices, aircraft and parts, guided missiles and space vehicles and parts, surgical and medical equipment, etc.) to others that are perhaps less obvious (soap and other detergents, nitrogenous fertilisers, printing ink, carbon black). The parallel British definition consists of only seven 3-digit sectors (as defined in the 1968 version of the Standard Industrial Classification): pharmaceutical chemicals and preparations, telegraph and telephone apparatus and equipment, radio and electronic components, broadcast receiving and some reproducing equipment, electronic computers, radio, radar and electronic capital goods, aerospace equipment manufacture and repair.

Table 11.1: Great Britain and United States: High Tech Employment

	Great Britain				United States		
	1981 000s	1971-1981 000s	%		1981 000s	1972-1981 000s	%
Pharmaceutical chemicals, etc.	69.3	-3.3	-4.6	Pharmaceutical preparations	130.5	+18.5	+16.5
Telegraph and telephone	64.5	-19.8	-23.5	Telephone, telegraph app.	147.4	+13.1	+9.8
Radio and electronic etc.	103.1	-25.2	-19.6	Radio, TV transmission, etc.	426.9	+18.8	+34.2
Broadcast receiving etc.	37.5	-10.8	-22.4	Electronic computing eqmt.	320.7	+176.1	+121.7
Electronic computers	42.1	-8.0	-16.0	Aircraft	301.1	+69.2	+29.8
Radio, radar, etc. capital goods	100.1	+6.0	+6.4	Electronic components, NEC	190.0	+100.6	+88.9
Aerospace equipment etc.	197.3	-14.1	-6.7	Semiconductors related devices	169.5	+71.9	+73.7
				Construction machinery eqmt.	145.9	+12.3	+9.2
				Aircraft parts, etc.	140.3	+38.2	+37.4
				Aircraft engines, parts	140.0	+35.2	+33.6
TOTAL (7)	613.9	-75.2	-10.9	TOTAL (10)	2112.3	+632.7	+42.8
TOTAL (43)	3253.2	-366.6	-10.1	TOTAL (100)	5475.2	+1090.9	+24.9

Source: Breheny and McQuaid (1985), based on GB Annual Census of Employment; Markusen, Hall and Glasmeier (1986), based on US Annual Survey of Manufacturers (1981), and Census of Manufacturers (1972 and 1977).

HOW MANY? HIGH TECH AND JOB CREATION

Just as high tech proves unexpectedly difficult to define by constant criteria, so it proves curiously intractable to chart in constant statistics.

The reason is that the industries that constitute it are, by definition, constantly being created and subdivided. In 1950, the statistics did not recognise a computer industry; in 1920 they did not identify a radio industry. The farther back in time we try to take a comparison the more we shall find today's high-tech industries buried in large aggregates.

In practice, therefore, we are limited to fairly short-term comparisons. Table 11.1 shows the pattern of high-tech change in Britain from 1971-1981 and the equivalent pattern (so far as available figures allow) for the United States from 1972-1981.

The contrast is startling. Whereas the United States added 1,091,000 high-tech jobs during the 1970s and 728,000 in the second half of that decade, the British economy actually managed to record losses of 48,000 and 42,000 in the equivalent periods. For individual industries, though the comparisons are complicated by differences of classification, the comparisons are equally telling; for electronic computers, a gain of 176,000 jobs against a gain of only 5,000; for aircraft, a gain of 69,000 against a loss of over 14,000; for electronic components, a gain of 89,000 against a loss of nearly 20,000.

The difference, in essence, is this: in the United States, high tech contributed 87 per cent of an impressive 1,235,000 gain in manufacturing jobs over the decade; in Britain, high tech actually contributed a small amount to a catastrophic loss of 1,963,000 manufacturing jobs—almost exactly one-quarter of the 1971 total. The best that can be said is that high tech helped a little to reduce the overall rate of decline. Further, there was no change in this position in the early 1980s; high-tech industries (by now differently defined) shed a further 13,200 jobs from 1981-1984 (Breheny and McQuaid, 1985, p.7).

One point not demonstrated in Table 11.1 does need underlining: that even in the United States, though the great majority of individual high-tech industries registered employment gains, nevertheless a significant minority—almost one-third of the total, over the 1972-81 period—recorded job losses. Further, some of these losses were large: 33,000 in large-arms ammunition, 26,000 in consumer radio and TV, nearly 16,000 in synthetic fibres. Some of these declining industries were suffering from a saturated market (alkalis and chlorine, reclaimed rubber), others from foreign competition (consumer radio and television). Even in America, then, high tech is no guarantee of job growth; it has to be the right kind of high tech.

Nevertheless, no fewer than seventy-two out of the individual 100 high-tech industries did record job gains. True, only six of these—computing, radio and television communications, electronic components, semiconductors, aircraft and oilfield machinery—accounted for no less than 53 per cent of all high-tech job growth during the period. But when we turn from absolute to percentage gains, a different list emerges, one dominated by smaller industries, principally energy-related in the middle 1970s, principally defence-related in the late 1970s. This marks an important point: that high-tech industries generally, and defence-related ones especially, are highly responsive to shifts in demand, in particular to government demand. They tend to have volatile growth patterns.

Table 11.2 : Great Britain: Concentrations of High Tech by County Groupings, 1981

	Employment	LQ
1. London Western Crescent		
Greater London	91,400	0.85
Hertfordshire	45,100	3.60
Hampshire	33,800	2.00
Berkshire	19,700	2.04
Buckinghamshire	6,900	1.15
Surrey	18,200	1.79
TOTAL	215,100	
2. North-west England		
Greater Manchester	27,300	0.86
Lancashire	30,000	2.01
TOTAL	57,300	
3. 'Silicon Glen' (Central Scotland)		
Strathclyde	23,600	0.88
Lothians	9,000	0.93
Fife	8,100	2.24
TOTAL	40,700	
GREAT BRITAIN	640,900	

Source: Breheny and McQuaid (1985), Table 3.3; based on GB Annual Census of Employment

WHERE? THE GEOGRAPHY OF HIGH TECH

In Great Britain, Table 11.2 shows that high technology jobs are heavily concentrated in two main core areas: one, by far the more important, consisting of London and the counties immediately north and west of it; the other in north-west England. In 1981 Greater London recorded over 91,000 high-tech jobs; Hertfordshire, immediately to the north, another 45,000; Buckinghamshire, to the north-west, nearly 7,000; Berkshire, to the west, close on 20,000; Surrey, to the south-west, over 18,000; Hampshire, also the south-west, nearly 34,000. In the north-west, Greater Manchester had over 27,000 jobs, while Lancashire, immediately to the north of it, recorded 30,000. A third area, the 'Silicon Glen' of Central Scotland, had—despite its media reputation—relatively few jobs in comparison: 24,000 in Strathclyde, 9,000 in Lothians region, 8,000 in Fife.

Table 11.3: Great Britain: Major Concentrations of High Tech by County, 1981

Ranked by Employment	Employm.	LQ	Ranked by Location Quotient	LQ	Employm.
1. Greater London	91,400	0.85	1. Hertfordshire	3.60	45,100
2. Hertfordshire	45,100	3.60	2. Avon	2.32	24,200
3. West Midlands	36,500	1.02	3. Fife	2.24	8,100
4. Somerset	33,800	2.11	4. Somerset	2.11	33,800
5. Hampshire	33,800	2.00	5. Berkshire	2.04	19,700
6. Lancashire	30,000	2.01	6. West Sussex	2.03	14,700
7. Gtr Manchester	27,300	0.86	7. Lancashire	2.01	30,000

Source: Breheny and McQuaid (1985), Table 3.3: based on GB Annual Census of Employment

Table 11.4: United States: Major Concentrations of High Tech by State, 1977

Ranked by Employment	Employm.	LQ	Ranked by Location Quotient	LQ	Employm.
1. California	641,300	1.49	1. Arizona	1.80	45,900
2. Illinois	360,300	1.15	2. Connecticut	1.65	160,100
3. New York	336,800	0.98	3. Kansas	1.63	62,800
4. Pennsylvania	314,300	0.96	4. Maryland	1.60	48,500
5. Ohio	295,100	0.91	5. Colorado	1.57	52,900
6. Texas	285,700	1.33	6. California	1.49	641,300
7. New Jersey	232,300	1.23	7. Massachusetts	1.43	204,600

Source: Markusen, Hall and Glasmeier (1986), Table 7.1

Table 11.5: United States: Major Concentrations of High Tech by State Groupings, 1977

	Employment	LQ
1. Pacific South-west		
California	641,300	1.49
Arizona	45,900	1.80
TOTAL	687,200	
2. Western Gulf		
Texas	285,700	1.33
Louisiana	56,400	1.22
Oklahoma	53,900	1.38
TOTAL	396,000	
3. Chesapeake/Delaware River		
New Jersey	232,300	1.23
Maryland	48,400	1.60
TOTAL	280,700	
4. Old New England		
Massachusetts	204,600	1.43
Connecticut	160,000	1.85
TOTAL	364,600	
5. Lower Great Lakes		
Illinois	360,600	1.15
TOTAL	360,600	
GRAND TOTAL (5 areas)	2,099,200	
USA	4,760,100	

Source: Markusen, Hall and Glasmeier (1986), Table 7.2

Thus, of a total of 641,000 high-tech jobs in Great Britain, the five south-eastern counties had over 32 per cent; the two north-western counties had close on 9 per cent; Central Scotland a mere 6 per cent.

These concentrations of course partly reflect the general concentration of employment in some of Britain's most populous counties. Relative concentrations of high-tech industry, as measured by location quotients, give a rather different picture: as Table 11.3 shows, Hertfordshire, with an LQ of 3.60, comfortably leads the list, followed by Avon with 2.32, Fife with 2.24 and Somerset with 2.11 (Breheny and McQuaid, 1985, p.11).

The American figures, presented in Tables 11.4 and 11.5, give a similar picture of concentration. No less than 42 per cent of all high-tech employment in 1977 was found in five major areas including a total of just ten states. (An additional 7 per cent was found in five other states—Florida, Minnesota, Kansas, Colorado and Utah—bringing the total share to very nearly one-half.) Here, too, however, absolute employment totals tend to be highest in populous and highly industrialised states, many of which prove to have weak, relative high-tech concentrations as measured by the location quotient; thus States like New York, Ohio and Pennsylvania appear in the left-hand column of Table 11.4 but prove to have LQs of less than one, so that they are excluded from the groupings in Table 11.5.

The places where high tech is most concentrated are not necessarily the places where it is growing the fastest. Of the five major American groupings in Table 11.5, two—those located in the older manufacturing belt—grew more slowly than the national average. New Jersey, with 4.9 per cent of national high-tech jobs in 1977, had actually suffered a fall in jobs since 1972; Illinois, with 7.6 per cent, had reported a mere 1 per cent of new high-tech jobs in the preceding period. Conversely, some Sunbelt States showed spectacular gains: Utah, with a mere 0.5 per cent of jobs, had registered 1.5 per cent of new jobs; Texas, with 6.0 per cent, had posted a 12.9 per cent of gains in employment.

Looking more closely at this pattern, it is clear that a fairly systematic shift was taking place: the five major core areas, with the notable exception of the Chesapeake-Delaware River group, were decentralising locally outwards into neighbouring states (Table 11.6). Thus California grew less rapidly than Arizona or Nevada; Texas less rapidly than Oklahoma and Arkansas; Massachusetts less rapidly than Vermont and New Hampshire. This does not necessarily mean that plants were moving from California to Arizona, or from Massachusetts to New Hampshire; the growth in these fringe States may well have come from elsewhere.

In Britain a similar pattern of outward shift is evident, though it is muted because of the generally poor performance of the high-tech industries there (Table 11.7). Between 1975 and 1981 Greater London lost 15.7 per cent of its high-tech jobs, while the counties to the north and west made corresponding gains: Hampshire grew by 6.8 per cent, Surrey by 11.1 per cent, Hertfordshire by 15.1 per cent and Berkshire by a staggering 62.6 per cent (Breheny and McQuaid, 1985, p.10). The same process is observable in the north-west, where Greater Manchester lost jobs while neighbouring Lancashire gained; and in Central Scotland, where Strathclyde lost and Lothians gained. Again, the reservation is necessary that these gains did not necessarily mean actual shifts of high tech from London to the Home Counties.

Table 11.6: United States: High Tech Employment Change, Selected States, 1972-77

		Employment 1977	Employment Change 1972-77	
			Abs.	%
1.	California	641,300	97,200	17.9
	Arizona	45,900	9,500	26.1
	Nevada	3,500	1,700	94.4
2.	Texas	285,700	49,300	20.9
	Oklahoma	53,900	13,800	34.4
	Arkansas	38,000	12,300	47.9
3.	Massachusetts	204,600	22,200	12.1
	New Hampshire	29,500	4,300	32.9
	Vermont	17,200	4,400	34.3

Source: Markusen, Hall and Glasmeier (1986), Table 7.4

On a finer-grained geographical scale, too, the same tendencies to deconcentration are evident. In the United States, we can identify a number of metropolitan high-tech clusters, each composed of a large central Standard Metropolitan Statistical Area (SMSA) and several neighbouring smaller SMSAs, within which the smaller peripheral units demonstrated faster growth than the big central one. This is neatly illustrated by the Los Angeles Basin, where the Los Angeles SMSA—the top metropolitan area in terms of high-tech jobs and plants in 1977—was nevertheless one of the poorest growth performers in the 1972-77 period, with a marginal contraction in high-tech jobs, while the adjacent Anaheim-Santa Ana-Garden Grove SMSA posted a 49 per cent job increase; other adjacent SMSAs (Oxnard-Ventura, Santa Barbara, Bakersfield, Fresno) posted even higher rates of growth. To the north, in the cluster of SMSAs that make up the San Francisco Bay Area, San Francisco recorded a 17 per cent increase, but neighbouring San Jose (Silicon Valley) posted four times as many new jobs, while Santa Cruz, even farther out, more than doubled its high-tech employment.

Table 11.7: Great Britain: High Tech Employment Change, Selected Counties, 1975-81

		Employment 1981	Employment Change 1975-81	
			Abs.	%
1.	Greater London	91,448	-17,012	-15.7
	Hertfordshire	45,060	5,914	15.1
	Berkshire	19,732	7,600	62.6
	Surrey	18,191	1,811	11.1
	Hampshire	33,807	2,179	6.8
2.	Greater Manchester	27,300	-2,980	-9.8
	Lancashire	30,003	1,392	4.8
3.	Strathclyde	23,575	-3,847	-14.0
	Lothians	9,042	1,461	19.3

Source: Breheny and McQuaid (1985), based on GB Annual Census of Employment

The same kind of outward movement is evident in south-east England. The Functional Urban Regions (FURs) of Hemel Hempstead (Herts), Bracknell (Berks), Reading (Berks) and Portsmouth (Hants)—all lying in the crescent west of London, between the M1 motorway and the English Channel—showed the fastest rates of high-tech employment growth between 1975 and 1981. London's FUR, predictably, recorded a decline. So, more surprisingly, did the FURs of Harlow and Cambridge along the so-called M11 Corridor north-east of London—often supposed to be a line of high-tech growth—and, farther out, Swindon. Very probably both Cambridge and Swindon have seen a dramatic upswing since 1981 when this series of figures stops; we have detailed figures only for Berkshire in 1984, which show that the new town of Bracknell is by far the leading centre with an estimated 9,000 high-tech jobs, followed by Reading and Slough with an estimated 6,000 each (Breheny and McQuaid, 1985, p.20).

The picture that emerges of British high-tech, then, is one of concentration not so much in the western corridor so emphasised by the media, as in a much broader western crescent that sweeps round London from the north-west to the south-west, within which employment is deconcentrating out of London into towns some 25-40 miles distant such as Reading, Bracknell and Hemel Hempstead. Within it—or at least within the Berkshire portion of it—a notable feature is the relatively low proportion of actual manufacturing workers (about 44 per cent) and, correspondingly, the relatively high proportion of office workers (33 per cent). This, which is clearly related to the large number of small indigenous firms that have no other facility, may make the area relatively resilient to recession in comparison with other areas, such as Central Scotland, that concentrate on volume production for companies headquartered elsewhere.

WHY? EXPLAINING HIGH TECH GENESIS AND GROWTH

Conventional neoclassical location theory, as developed in such classics as Weber (1929, 1909) and Hoover (1948), does a very mixed job in explaining why high-tech industries are where they are. Two of the corners of Weber's celebrated locational triangle—materials and market pulls—seem almost irrelevant for a group of industries with such light, valuable inputs and products. A third, labour, is clearly more important but is highly segmented, with a minority of very skilled and very mobile scientists and engineers, and a majority of relatively low-paid process and assembly workers, the first leading the industry to high-amenity locations (Berry, 1970), the second drawing it to low-wage areas.

The other major element of Weberian theory, agglomeration economies, may be highly significant for small firms in the early stage of an industry's development. Such firms may be drawn together by their mutual dependence on shared services, shared information and shared skilled labour, forming an industrial quarter such as was once typical of inner-city locations (Marshall, 1920 (1890)). This, however, still leaves the critical question: why, apparently, does high-tech industry now reject older central cities, thus denying the seedbed function once thought critical for such places?

This suggests a central weakness of traditional theory: that, being essentially static, it cannot handle the characteristically dynamic nature of innovative capitalist industry. A vital clue here may be provided by the product cycle theory of Raymond Vernon (1966) and by the related theory of long-wave cyclical innovation swarming as developed by Kondratieff (1935) and Schumpeter (1939), which has recently received intense critical attention (Mandel, 1975, 1980; Mensch, 1979; Freeman, Clark and Soete, 1982; van Duijn, 1983; Clark, Freeman and Soete, 1984). This would suggest that at intervals, a bunch of product innovations—the commercial application of inventions—in effect create new high-tech industries, dominated by new small firms that agglomerate in a particular location; only later, with maturity and larger firm size, some of these feel free to establish branches or to relocate outside the original core region (Markusen, 1985).

Though reasonably convincing as an account of industrial change over the last 100 years, this still leaves open the elusive question of the original choice of location. Where and why, precisely, does innovation occur? Studies in economic history (Clapham, 1910; Carus-Wilson, 1941; Checkland, 1975) suggest that older-established industrial areas may develop some kind of hardening of the economic arteries, so that they actually discourage innovation. But this fails to take account of the fact that increasingly in the modern economy, innovation comes not from serendipity but from organised research and development (R & D).

Hence, the location of the R & D function may be the critical element. Malecki's work in the United States (1980a,b,c, 1981a,b) shows that there, R & D takes two main forms, industrial and federal. The first is largely conducted in older industrial cities, many of them headquarters of old-established industrial corporations. The second tends to be located in quite different places, many of them in the so-called Sunbelt of the south and west, away from the historic manufacturing belt; a large part of it is defence-related or space-related. Similarly, work in Britain (Buswell and Lewis, 1970; Howells, 1984) shows that R & D is disproportionately concentrated in south-east England, especially in the western crescent, where it is particularly associated with the bunching of government research establishments (GREs), again, largely involved in defence and related research.

The history of these establishments is therefore of particular interest. They are located in the western crescent for a variety of reasons, some going back almost into antiquity. Thus the naval research establishments, which have always kept separate from the rest, are concentrated around Portsmouth (with some outstations westward into Dorset) for the simple reason that Portsmouth has been the main centre of the Royal Navy since Tudor times; the Royal Aircraft Establishment, by far the largest of the establishments and one of the most important for industrial procurement, is located at Farnborough, south-west of London, because the Royal Engineers placed their new Balloon School in nearby Aldershot in 1892; the National Physical Laboratory in Bushy Park, next to Hampton Court in south-west London, located there on its establishment in 1900 because it was an offshoot of the Royal Society and because a royal grace-and-favour residence was available (Breheny, Hall, Hart and McQuaid, 1986).

Generalising, it appears that the research establishments were all located close to London because London, as capital, was the centre of the scientific and military establishments, and because the latter logically placed their training and research establishments in military garrisons close by. Once these critical initial decisions were taken, inertia exerted a powerful influence to keep the centres there, or at very least to ensure that new branches or derivatives were located close by—as in the cluster of Army research establishments on Salisbury Plain, some 50 miles west of the Aldershot complex (ibid.).

The equivalent American research is now in progress. It is too early to draw any conclusions, but it would be surprising if the location of military research were not dominated first by decisions reached at speed in World War II, and second by political factors at critical moments, as, for instance, the location of facilities in Texas during the Johnson presidency.

The strong suspicion, then, is that the development of high tech in new industrial regions is not a random process. It arises from conscious decisions to locate R & D and related procurement facilities at certain times, often under military stress in times of war or defensive preparation, and with a strong strategic component in the locational decision. These decisions tend to be taken at times when, whether due to innovation bunching or to deliberate military promotion of new technology—there is fierce debate among scholars on this point—the resulting product innovations are effectively creating new high-tech industries.

That is the thesis. The empirical evidence, which has emerged from the American work, partly confirms and partly denies it. Surprisingly, when high-tech location and growth is regressed at SMSA level against a series of possible explanatory variables, the intensity of R & D (as measured by the volume of Federal research contracts) does not emerge as significant; but defence spending very definitely does. The suspicion is that Federal research spending covers a wide field, much of which does not have an industrial spin-off, but that defence spending does have such spin-off capacities.

The statistical analysis provided some more surprises. Thus American high tech does not appear to be attracted by features of the labourforce such as wage rates or unionisation; it is much more positively drawn to places with good amenities like climate and a wide range of educational facilities, by access features like freeways and airports, and by the presence of agglomeration economies in the form of major headquarters offices and concentrations of specialised business services. But, as already noted, concentrations of basic research do not figure among these agglomeration features.

To trace why this should be so, while defence spending is significant, would need detailed work at the level of the area and of the firm, which is now under way. The local work so far relates to California's Silicon Valley and to Massachusetts Route 128 (Rogers and Larsen, 1984; Hall, 1985; Saxenian, 1985a,b). It strongly suggests that university research, funded by the military immediately after World War II, was instrumental in helping to create new high-tech firms that relied on their access to research and were in fact commercial spin-offs from it. Further, the very existence of one firm then tended to cause progressive spawning-off through breakaways by innovative individuals; very naturally, therefore, an agglomerated industrial quarter developed.

Such local anecdotal evidence is thus difficult to relate to the macro-level statistical analysis which suggests that research concentrations are relatively insignificant in high-tech generation. Perhaps the clue is that such concentrations are significant only at certain points in time. Or, perhaps, only certain kinds of concentration will prove significant. Teasing out these connections is almost certainly the next major task in explaining the geography of high technology.

REFERENCES

Berry, B. J. L. (1970) 'The Geography of the United States in the Year 2000', *Institute of British Geographers, Transactions*, 51, 21-53.

Breheny, M. and McQuaid, R. (1985) *The M4 Corridor: Patterns and Causes of Growth in High Technology Industries*, Geographical Papers, 87, Department of Geography, University of Reading, Reading.

———, Hall, P., Hart, D. and McQuaid, R. (1986) *Western Sunrise: The Genesis and Growth of Britain's High Tech Corridor*, George Allen and Unwin, London.

Buswell, R. J. and Lewis, E. W. (1970) 'The Geographical Distribution of Industrial Research Activity in the United Kingdom', *Regional Studies*, 4, 297-306.

Carus-Wilson, E. M. (1941) 'An Industrial Revolution of the Thirteenth Century', *Economic History Review*, 11, 39-60.

Checkland, S. (1975) *The Upas Tree*, Glasgow University Press, Glasgow.

Clapham, J. H. (1910) 'The Transference of the Worsted Industry from Norfolk to the West Riding', *Economic Journal*, 20, 195-210.

Clark, J., Freeman, C. and Soete, L. (1984) 'Long Waves, Inventions and Innovations', in C. Freeman (ed.) *Long Waves in the World Economy*, Butterworth, Guildford.

Freeman, C., Clark, C. and Soete, L. (1982) *Unemployment and Technological Innovation: A Study of Long Waves and Economic Development*, Frances Pinter, London.

Hall, P. (1985) 'The Geography of the Fifth Kondratieff', in P. Hall and A. Markusen (eds) *Silicon Landscapes*, Allen and Unwin, Boston.

Hall, P. and Markusen, A. (eds) (1985) *Silicon Landscapes*, Allen and Unwin, Boston.

Howells, J. R. L. (1984) 'The Location of Research and Development: Some Observations and Evidence from Britain', *Regional Studies*, 18, 13-30.

Hoover, E. M. (1948) *The Location of Economic Activity*, McGraw-Hill, New York.

Kondratieff, N. D. (1935) 'The Long Waves in Economic Life', *Review of Economic Statistics*, 17, 105-15.

Malecki, E. J. (1980a) 'Corporate Organization of R & D and the Location of Technological Activities', *Regional Studies*, 14, 219-34.

——— (1980b) 'Science and Technology in the American Urban System', in S. D. Brunn and J. D. Wheeler (eds) *The American Metropolitan System: Past and Future*, Edward Arnold, London.

——— (1980c) 'Technological Change: British and American Research Themes', *Area*, 12, 253-60.

——— (1981a) 'Public and Private Sector Interrelationships, Technological Change, and Regional Development', *Papers of the Regional Science Association*, 47, 121-38.

——— (1981b) 'Government-Funded R & D: Some Regional Economic Implications', *Professional Geographer*, 33, 72-82.

Mandel, E. (1975) *Late Capitalism*, Verso, London.

——— (1980) *Long Waves of Capitalist Development*, Cambridge University Press, Cambridge, England.

Marshall, A. (1890) *Principles of Economics*, Macmillan, London (1920 edn).

Markusen, A. (1985) *Profit Cycles, Oligopoly and Regional Development*, MIT Press, Cambridge, Mass.

———, Hall, P. and Glasmeier, A. (1986) *High-Tech America: The What, How, Where and Why of the Sunrise Industries*, Allen and Unwin, Boston.

Mensch, G. (1979) *Stalemate in Technology: Innovations Overcome the Depression*, Ballinger, Cambridge, Mass.

Rogers, E. M. and Larsen, J. K. (1984) *Silicon Valley Fever: Growth of High-Technology Culture*, Basic Books, New York.

Saxenian, A. (1985a) 'The Genesis of Silicon Valley', in P. Hall and A. Markusen (eds) *Silicon Landscapes*, Allen and Unwin, Boston.

Saxenian, A. (1985b) 'Silicon Valley and Route 128: Regional Prototypes or Historic Exceptions?' in M. Castells (ed.) *High Technology, Space and Society*, Urban Affairs Annual Reviews, Volume 28, Sage, Beverly Hills.

Schumpeter, J. A. (1939) *Business Cycles: A Theoretical, Historical and Statistical Account of the Capitalist Process*, (2 Volumes), McGraw-Hill, New York (reprinted 1982, Porcupine Press, Philadelphia).

van Duijn, J. J. (1983) *The Long Wave in Economic Life*, George Allen and Unwin, London.

Vernon, R. (1966) 'International Investment and International Trade in the Product Cycle', *Quarterly Journal of Economics*, 80, 190-207.
Weber, A. (1909) *Theory of the Location of Industry*, trans. C. J. Friedrich, University of Chicago Press, Chicago (1929 edn).

Chapter 12

THE IMPACT OF INDUSTRIAL STRUCTURE ON GLOBAL HIGH TECHNOLOGY LOCATION

R. Gordon and L. Kimball

The new flexibility of modern business practice, particularly in the high-technology sector, has wrought confusion within industrial location theory. Theories of location traditionally have been preoccupied with the impact of spatial constraint on regional industrial development: locational decisions concerned the possibilities for establishing economically efficient (cost-minimising or revenue-maximising) linkages inherent in the prevailing geographical distribution of production factors and markets. In contrast, the contemporary global re-organisation of industrial production transcends spatial barriers and appears to negate the very idea of locational constraint. Mesmerised by this new form of capital mobility, location theory has reacted by exaggerating pre-existing tendencies towards abstraction and reductionism: the determinate basis and content of location decisions is sought in movement rather than structure, in the isolated self-determination of firms rather than in the linkages between them, and in total freedom rather than constraint. An adequate location theory for high-technology industry, and a corresponding improvement in policy prescription, demands a reversal of these priorities.

HIGH TECHNOLOGY INDUSTRY AND LOCATION THEORY

The locational decisions of high-technology firms are not determined by costs incurred in the movement of materials, products or people across space. The relatively light weight, small size and high value-added nature of high-technology inputs and outputs facilitates low-cost global transportation of raw materials, components and products. Since high-technology inputs typically derive from a variety of sources, and because markets are global, siting decisions designed to concentrate production in geographic locations which minimise costs or maximise market penetration and revenues have been more difficult to calculate than in traditional industries far more dependent upon specific sets of suppliers or customers (Rees and Stafford, 1983). The ability to segment the production process facilitates component standardisation and the re-aggregation of production in different locations. Sophisticated information and communication technology provides a new material basis for maintaining centralised administrative control over geographically distant plants and operations. Declining costs of transport and communication highlight the availability of globally differentiated wage regimes. The relative insensitivity of high technology firms to natural or material factor supply differentials, as well as their ability to deploy malleable technologies and labour processes, have enabled the electronics industry to create an integrated global network of industrial production which apparently reduces or eliminates the need for spatial propinquity (Castells, 1984).

Traditional location theory concerned itself primarily with the cost of physical transactions between discrete industrial units which assessed their locational strategies on the basis of an existing distribution of factors and markets (Weber, 1909; Lösch, 1954; Isard, 1956). Locational decision making was presumed to be a rational choice standing apart from the complex internal organisational structure of the firm or the existence of those linkages that transcend cost-efficient commodified material input-output linkages. Thus location theory initially highlighted a specifically quantitative conception of interfirm material linkage within a formalistic framework of spatial constraint (the given distribution of factors and markets) that was unconcerned, at least in any systematic manner, with the production, organisation or industrial structure of the firms evaluating their locational priorities.

Mirroring high-technology industry's apparent escape from pressures for territorial confinement, contemporary location theory has shifted its emphasis from frictional variables associated with cost-distance ratios to area variables, that is, to the specific attributes of areas presumed favourable to the growth of high-technology production (Rees and Stafford, 1983). Analysis has delineated a wide range of decisive area characteristics. Thus, the close linkage between basic and applied research ostensibly renders proximity to university research locations more important to high-technology firms than relations of supply and distribution (Joint Economic Committee, 1982; Malecki, 1984a). Venture capital's apparent contribution to the very emergence of the electronics industry has highlighted the significance of new forms of financial infrastructure and local capital availability for high technology development (Rogers and Larsen,

1984). The rebirth of classic free-enterprise values and anti-unionism in the high-technology sector has made a community's general business climate an important locational consideration, even generating projects for domestic processing zones emulating the business advantages found in offshore production platforms (Bluestone and Harrison, 1982). The industry's dependence upon scarce scientific and engineering talent renders it sensitive to 'quality-of-life' considerations which sustain an attractive living and working environment for highly-paid scientific and technical professionals (Malecki, 1980; Rees and Stafford, 1983). Others have emphasised conversely that industrial restructuring is a product of the search for control of cheap production labour as freedom from other locational constraints focuses comparative advantage on the problem of labour supply (Fröbel, Heinrichs and Kreye, 1980). In somewhat more sophisticated versions of this same thesis, it is the search for variable labour supply that is paramount in locational decisions, firms co-ordinating segmented labour markets in globally dispersed locations (Sassen-Koob, 1982; Malecki, 1984b).

Each of these factors may be more or less significant in particular instances but this form of locational analysis cannot provide an adequate or coherent explanation of the global organisation of high technology industry as a whole. Exaggerating the importance of basic research in comparison with immediately commercial technical information, and underestimating the reverse information flows from industry to academia, most university-based research parks have failed to achieve economic success or generate substantial industrial development.[1] State demand and financial support historically has been as significant as venture capital to the creation of high technology formations; indeed, the existence of these agglomerations (Silicon Valley, Route 128) was as much the precondition for venture-capital operations as the latter were responsible for the industry's advance. Disappointed policy-makers in many communities have discovered that high technology is far more geographically restricted than theory suggests and that firms are resistant to traditional economic incentive packages, favourable business climate or rich cultural and educational amenities.[2]

Similarly, while labour costs undoubtedly remain important locational considerations, particularly in site selection for routine manufacturing and assembly work, the overall global organisation of the high technology production process cannot be reduced to decisions about one of its components.[3] Direct overseas investment in high-technology industry still concentrates disproportionately within other advanced industrial societies as opposed to the peripheral labour markets of the Third World (Sciberras, 1977; Hood and Young, 1979; Flamm, 1985a). Subtle, non-economic distinctions are made within Third World labour markets (Fernandez-Kelly, 1983). As Ernst (1981) has argued, with capital intensification increasing rapidly in electronics, rationalising constant capital and reducing organisational vulnerability assume at least as much importance in locational patterns as lowering labour costs. The quality of labour also is becoming an increasingly significant consideration. Rapid product innovation presupposes constant and immediate access to a broad range of technical skills and equipment (Malecki, 1984a; Oakey, 1984). Frequent product and process changes necessitate integrated working relations between different segments of the labourforce (between engineers and production workers, for example, machine operators and programmers, or R & D professionals and marketing personnel). Growing marketing budgets and sales organisations as well as the crisis-prone advance of the industry in the last decade reveals the importance of expanding demand as a means of offsetting declining profits: marketing strategies are becoming more differentiated and some offshore locations in Europe and the Third World are now being employed not simply as sources of cheaper labour but as platforms for the construction of regional markets and high-tech economies (Lim, 1986; O'Connor, 1983; Goldstein, 1985).

Finally, the geographic separation of control and R & D functions from routine manufacture is in fact merely one strategy for the spatial division of labour: it is exemplary rather than definitive of locational strategy *per se*.[4] This particular division of labour will probably continue to be replicated. But skills are not functionally distributed on a geographical basis: two significant tendencies within contemporary high-technology production and location patterns concern the upgrading of certain peripheral sites (e.g. Singapore, Taiwan, Scotland) to incorporate more advanced technical processes (automation, final test) and engineering functions utilising indigenous technical skills and, conversely, the employment of large quantities of unskilled, peripheral, immigrant labour for basic manufacturing and assembly work in domestic core cities (Los Angeles, San Diego, New York, Silicon Valley) (Ernst, 1986; Fernandez-Kelly and Garcia, 1986; Sassen-Koob, 1982).

Reformulated location theory, therefore, undertakes a search for the decisive determinants of 'regional innovation potential' (Boeckhout and Molle, 1982) whose introduction, either singly or in aggregate, will autonomously incubate and promote local high-technology capabilities. These synergies ostensibly replace traditional linkage structures. Contemporary location theory ironically then emphasises production requirements over exchange while discarding earlier emphases upon intrafirm connections. In contrast with traditional location theory, current theory (and policy) maintain a narrow concern for production requirements within a formalistic geographical framework (areal attributes) that obscures the independent significance of linkage and industrial structure. Revisionist marxism similarly emphasises abstract organisational changes situated within a formalistic understanding of production (capitalist accumulation) that vitiates the critical importance of intra-sectoral linkages and socio-spatial constraint.[5] The positive contribution of each approach—recognition of the continuing importance of interfirm relations, concern for the unique requirements of high-technology production, and the stipulation of historical changes in the global organisational capacity of capitalist enterprise—is immediately denied by the others, a shared myopia that renders problematic any commitment to any single perspective and imposes the need for a more synthetic approach.

The limitations of each perspective are also mutually reinforcing, the weaknesses of one argument in part revealed by the strengths inherent in the others. Traditional linkage analysis, for example, fails to account for corporate internalisation of transactions and the increasing centrality of non-material and non-market relations between firms, obscuring the significance of linkages relating to the mutual interdependency of production relations as opposed to relations of exchange between independent entities. Areal analysis underestimates the subordination of local resource environments to the changing historical dynamics of industrial organisation within high technology. The image of the omnipotent and footloose multinational firm on the other hand tends to mirror, at a higher level of organisation, the mythology of individual entrepreneurial genius and similarly denigrate the specific social conditions of high-technology innovation and production. Geographic formalism obscures the fact that locational choices are neither prior to, nor independent of, investment decisions and production process organisation (Walker and Storper, 1981; Walker, 1985, p. 250), while production formalism neglects the expanded importance of a range of specific production conditions and forms of social organisation within different locational clusters as firms extend their spatial ambit.

The essential reductionism of these arguments precludes a more inclusive or integrated understanding of business organisation, production processes and links between firms and sectors. The characteristic abstraction of perspectives which individualise the issue of locational choice or derive it mechanistically from law-like processes prevents a more historically determinate appreciation of the relationship between industrial structure and spatial organisation.

SILICON VALLEY AND CHANGING INDUSTRIAL STRUCTURE

What are the locational options and constraints inherent in the corporate strategic appropriation of production and linkage structures within contemporary high-technology industry? Any treatment of locational strategies and changing industrial structure in electronics inevitably must assess developments along both dimensions within Silicon Valley itself, the seedbed of global high technology.

Oft-repeated mythology stresses the contingent and highly localised emergence of the electronics industry in post-World War II Santa Clara County. The stories are legendary: Frederick Terman's plans for the commercialisation of electrical engineering at Stanford University; the fortuitous residence of William Shockley's mother in Palo Alto; the garages of Hewlett and Packard, Wozniak and Jobs (Osborne, 1979; Rogers and Larsen, 1984). Contingent personal events, unforeseeable technological breakthroughs, and accidents of geography undoubtedly contributed to the particular pace and texture of Silicon Valley's evolution (Hanson, 1982). But too much attention has been accorded these heroic minutiae for, despite the genuine significance of individual contributions and chance occurrence, the real presuppositions of the post-war micro-electronics industry in Silicon Valley (as well as in Massachusetts) were global from the outset.[6]

Even the local educational, industrial and financial complex which provided the material infrastructure for high-technology development in Santa Clara County was sustained by national and international linkages. Stanford University's critical contribution to the evolution of Silicon Valley's science-based industry rested less on the university's scientific and engineering research strengths—a feature that has (mistakenly) appeared duplicable to academic institutions around the globe attempting to create the nuclei of new regional centres for high technology innovation—than on the specific social form in which its expertise was organised. Along with MIT, the other academic institution to engender a major concentration of high-technology industry, Stanford University, exercised national, even world research leadership in electrical engineering and, in conjunction with UC, Berkeley, produced the largest concentration of high-quality, advanced electrical-engineering graduates in the United States (Malecki, 1984a). The university's primacy was reflected in its receipt of the largest share of Federal Government electronics-related research funding, initiating a virtuous circle that was closed with the university's provision of an unusually large supply of trained scientific and technical manpower for local industry (Saxenian, 1981, pp. 50-58). This interchange with industry was facilitated by the university's historical rejection of traditional academic insularity and its long-standing encouragement of interpenetration with commercial enterprise. These forces crystallised in the formation of a pool of scientific entrepreneurial initiative, perhaps the university's most decisive contribution to the growth of Silicon Valley.

The existing industrial milieu, which provides such an unusually favourable demand structure for continuous innovation in electronics, was not solely an indigenous development either. California's significance as the Pacific gateway during World War II focused an enormous concentration of US military production in the State, an influx which made it pre-eminent in aircraft, radar, atomic energy and, subsequently, in aerospace, missiles and advanced electronics. Vast private investment was reinforced by extensive government funding (between WW II and the 1970s California received the largest share of military prime contract awards as well as the overwhelming proportion of Department of Defence (DoD) and NASA research and development expenditures), and the extensive collaboration required between contractors and manufacturers placed a premium on close proximity between defence industries and electronics firms. The existence of this support for industrial development in electronics proceeded from an extraordinary commitment of public resources over several decades (SRI International, 1980; Saxenian, 1981, pp. 55-57; Markusen *et al.*, 1984; Tiebout, 1966; Markusen, 1984).[7]

Capital availability was the third precondition necessary for the formation of Silicon Valley's high-technology industry and these resources too were ultimately global in character. San Francisco (along with New York and Boston) is a pre-eminent US financial centre, a legacy of the area's rich agricultural heritage, its links with national railroad development in the nineteenth century, its position as an entrepôt for military and trade flows throughout the Pacific region during and after World War II and, more recently, its attraction to transnational corporate and banking interests. Technological innovation, industrial development and economic expansion in Silicon Valley were able to call upon local financing resources that derived ultimately from the region's location at the core of a network of international financial activities and economic exchange (Mollenkopf, 1983; Noyelle and Stanback, 1984, pp. 144-158).

Despite their mutual reinforcement, these factors alone were not sufficient to engender an electronics agglomeration of the size or sophistication of Silicon Valley. In the first phase of its evolution up to the mid 1960s, the relatively small-scale micro-electronics industry was more directly dependent upon the unique linkage established between an incipient autonomous capacity for rapid technological innovation and the driving force of highly centralised and technically sophisticated state demand. Far from being created by ineffable scientific genius and heroic entrepreneurialism, the electronics industry in its infancy proceeded ultimately from extensive state intervention.[8] The military's role as first-user can hardly be overestimated for it not only propelled the new industry beyond the point of take-off but also, and not without a fair measure of foresight, established the conditions for US high-technology leadership in subsequent decades.[9]

Military and aerospace demand provided the principal market for initial production of innovative devices, the military's share declining as product lines were consolidated and costs fell to a level where commercial or industrial sales became viable. Government support weighed heavily in the establishment of scientific and technical priorities, re-orienting research from germanium to silicon semiconductors, for example, and consolidating the breakthrough of integrated circuit design. Defence commitments extended to process technology as well as product innovation promoting component miniaturisation, facilitating refinements of fabrication methods and establishing semiconductor quality control criteria. Quite apart from an impressive variety of specific contributions, defence demand more generally provided a critical incentive for investment in this particular area of research and manufacture and accelerated the transformation of new ideas into products.

In a period when military and commercial technical requirements were generally compatible, defence research programs in university and private laboratories formed a cadre of skilled scientists, engineers and technical workers whose movements between public and private sectors, and from firm to firm, promoted the diffusion of new ideas, techniques and products. Government support for technical and scientific education was vital to the careers of nearly all the major innovators in semiconductor development. Commercial spin-offs from military-related products were less important for the industry's evolution than the transferability of basic knowledge, skills and techniques acquired in military research and development projects. Established electronics-industry leaders received substantial military funding for R & D in micro-electronics but defence contracts also were awarded to smaller, non-established start-up companies, a policy that helped create the diversified structure of the emergent modern semiconductor industry (Tilton, 1971).

The total volume of military and aerospace procurement was comparatively small by contemporary industry standards. However, premium prices and guaranteed state demand stabilised high profits and underwrote the costs and risks of product development in the private sector, effectively subsidising the expansion of private productive capacity and new investment, stabilising production organisation and focusing efforts on technical improvements. Government contracts also encouraged private investment in the fledgling industry (Okimoto *et al.*, 1984, p. 87). High profits from military sales permitted firms to lower prices in commercial markets, thereby expanding civilian applications of the new technology. Indeed, defence support played an important role in establishing the structure of learning economies which governed the subsequent evolution of the industry: since market growth and expanded production volumes resulting from military sales accelerated cost reductions, US firms were able to move faster and earlier than their foreign competitors down the learning curve towards more consistent quality, higher yields and lower prices. In this way military demand was instrumental in creating and maintaining US market leadership in microelectronics.[10]

Still more directly, the performance specifications of the world's most sophisticated military power pushed research beyond existing technological frontiers. Similar types of European and Japanese government economic support could not promote the same level of technical supremacy. Military requirements regarding miniaturisation, functional complexity and reliability helped to determine specific paths of technical development. Moreover, government standards—oriented to performance requirements rather than detailed product specifications—meshed broadly in this period with civilian trajectories of technological development.[11]

Military demand therefore decisively affected the structure of domestic electronics industrial organisation and the emerging framework of international competition. Private companies still largely determined specific research priorities—none of the major technological breakthroughs in semiconductors resulted from direct government funding—but military development priorities were frequently instrumental in determining choices between competing technological alternatives and generally established US primacy in technological innovation. Their permanently advanced competitive position on the learning curve permitted US firms driven by military demand to translate their technical superiority into commercial dominance within world markets.

The second phase of Silicon Valley's growth in the decade between the mid 1960s and the mid 1970s involved a substantial demilitarisation of the industry and the predominance of entrepreneurial capitalism. The electronics industry was characterised by the phenomenally rapid growth of small and medium-sized firms oriented to technological innovation and market penetration, highly competitive and supported by a host of independent ancillary and service organisations. Between 1963 and 1972 a new firm entered the semiconductor industry in Silicon Valley every two weeks on average and substantial technical breakthroughs (MOS, the microprocessor, VLSI, microcomputers) seemed to occur almost as rapidly (Wilson, Ashton and Egan, 1980; Hanson, 1982; Saxenian, 1981).

The existing technological leadership of US firms, a legacy of prior military support, positioned the industry perfectly for dominion over emerging mass markets. The extraordinary expansion of relatively small-scale entrepreneurial firms was based upon the synergistic integration of semiconductor manufacture into global consumer, computer and industrial markets (Borrus, Millstein and Zysman, 1982; Linvill, 1984; Okimoto *et al.*, 1984; Alic, Harris and Miller, 1986). The dramatic increase in circuit complexity and corresponding decline in functional cost—the provision of increased performance at reduced prices—fuelled an applications explosion while expanding product demand in turn promoted further technological innovation in micro-electronics. In this way, the conditions for continued technical development and industrial growth in electronics became less dependent upon the state and instead became internalised within the competitive structure of the industry itself. Technical dominance and expansive markets subject to minimal foreign competition provided a higher rate of potentially marketable innovations in conjunction with a favourable structure of incentives for economic risk-taking and technical experimentation. Entry costs remained low for most of this period, permitting new market entrants.[12] Cost minimisation became the critical means of economic survival as firms lowered fixed costs and attempted to profit from early installation of learning economies: advanced movement down the learning curve provided a competitive edge by enabling firms to establish product standards, expand production volumes through standardised production processes, lower prices, and gain valuable experience towards the next technological breakthrough (Tilton, 1971; Noyce, 1981; Ernst, 1981).[13] The massive expansion of diverse electronics-related markets encouraged large numbers of new entrants, particularly as entry costs remained relatively restrained. The sophistication of US final systems markets (consumer products being less significant than computer and industrial equipment markets) consolidated the position of US firms at the leading edge of technological innovation (Borrus, 1985; Okimoto *et al*., 1984, pp. 179-180). Though hardly sustainable without scientific commercialism, academic entrepreneurship, or venture capital resources—all factors considered the unique psychological or material edge of the US electronics industry and essential prerequisites for high-technology development elsewhere—entrepreneurial capitalism derived directly in fact from undisputed US hegemony over mass markets in electronic components and industrial products. Far from being intrinsic to the growth of high-technology industry in general, entrepreneurialism derived principally from the conditions unique to a specific historical phase in the evolution of electronics manufacture, essentially unrepeatable conditions which already have been surpassed.

In the third phase of the industry's development (broadly coincident with the advent of VLSI technology) the conditions for successful business advance and technical innovation have been domesticated by large-scale, transnational organisations. Silicon Valley (along with Route 128) has been transformed from a locus of small-scale, competitive capitalism and entrepreneurial initiative into the global corporate headquarters of high-technology growth (Ernst, 1981; O'Connor, 1983, 1986; Alic, Harris and Miller, 1986). New forms of industrial restructuring have emerged in electronics during the past decade. Extensive changes in process and product technology have dramatically increased the costs of circuit development and, as chips have become smaller, companies have increased in size. Entry costs are prohibitive: expenditures for establishing a state-of-the-art wafer-fabrication facility have increased from $1-2 million in 1968 to $75-100 million in 1980; capital expenditures as a percentage of sales in the semiconductor industry have increased from 8 per cent in 1978 to 20 per cent by 1981. Escalating capital requirements, frequent demand slowdowns and crisis susceptibility (1974-75, 1980-82, 1984-85), heightened international competition in key revenue-producing markets, abbreviated product and equipment life-cycles and declining profits have generated substantial movement towards vertical integration in the industry. Semiconductor producers are integrating forward into equipment systems and consumer products in order to capture increased value-added and to protect proprietary designs. Systems producers, as well as companies outside the electronics industry, are absorbing or establishing semiconductor firms as captive manufacturers in order to secure reliable quality supplies. The convergence of circuit and system development reinforces these tendencies, engendering new forms of co-operative or integrative strategies between component manufacturers and system producers, and between equipment makers and a diverse range of end-users. Concentration is still more advanced in computers and telecommunications (Fishman, 1981; Linvill, 1984; OECD, 1983). These new forms of organisation cut across traditional lines of segmentation within the industry. Sectoral boundaries within electronics production, and between electronics and other industries, collapse. Intercompany technical exchanges and co-operative arrangements are expanding. The larger electronics complexes are oriented to a global field of operation. The industry's ownership structure increasingly is internationalised, while the explosion of US-Japanese (and US-European) joint ventures in the past few years presage the emergence of an even more innovative strategy whereby giant transnational combines surmount crisis tendencies by establishing permanent hegemony over the technical standards for future product development in all phases of integrated electronics-based production (Ernst, 1981; Degeorge, 1983; Dataquest, 1985). Capital availability, control of base information technologies and global organisation are fundamental to new industrial configurations in high technology.

Current transmutations in electronics signify not simply a new process of geographical relocation of existing plants and operations but a shift in the structure of industrial production. The geographical shift of production activities is subordinated to the structural transformation from one form of industrial organisation to another, a change propelled by innovations emanating from the leading centres of high technology. In the post-war period international trade regimes, multinational business organisation and competitive state planning and industrial policies objectively internationalised the conditions of production within all advanced industrialised societies. Firms were increasingly compelled by the structure of international competition to meet global standards in costs, technology, product organisation and industrial strategy. In the current epoch, objective interdependence is being replaced by the construction of integrated global networks as the conditions of production are consciously organised by transnational firms capable of permanently defining and controlling the material bases of innovation, finance, communication and production for the world economy as a whole (Degeorge, 1983). Far from emphasising abstract or universal conditions, this structural transformation privileges specificity: the global spatial organisation of production is tied neither to the maximum exploitation of single production factors nor to their aggregate optimisation but, on the contrary, to specific, qualitative combinations of factors and relations of production. Locational decisions are no longer a matter of optimising an exogenous distribution of production factors but involve the strategic creation of linked production complexes, each appropriate to a very diverse mix of specific regional and social endowments and distributed as an inter-regional network in accordance with corporate headquarters' strategic conceptions.[14]

The new global structure of production is characterised by the qualitative ability of contemporary high-technology conglomerates to specify the social, cultural and economic conditions of production on a varied and universal scale. Far from demonstrating their insensitivity to all objective constraints (save the relative intractability of labour) the various production strategies adopted by the modern high-tech firm reveal their appropriation at a higher level of historical development. The freedom of firms to command, manipulate and combine the entire range of social and economic preconditions for profitable production is not absolute (Sayer, 1985). But, operating through control of the material infrastructure defining the universal conditions of possibility of economic activity in the information age, this freedom reveals a new historical advance in the private (and social) appropriation of both nature and cultural organisation. The central question for location theory no longer concerns discovery of those objective factors likely to promote independent generation of indigenous regional high-tech agglomerations but comprehension of a region's social and economic position within the corporately organised integrated circuit of global production.

A TYPOLOGY OF HIGH TECHNOLOGY REGIONS

Global high-technology headquarters such as Silicon Valley, Route 128 and their Japanese equivalents are the dominant spatial configurations of industrial development in micro-electronics. Quite apart from their industrial and technological leadership, these high-technology centres are remarkable for their concentration of manifold and global resources within a localised regional framework. Each centre is constructed around a rare fusion of technical, economic and organisational advantages: world-class research institutions; extensive state funding and support; abundant capital; scientific commercialisation and university-industry entrepreneurial linkages and transnational administrative, financial and scientific corporate organisation. Whatever the initial provenance of electronics production in the region, global headquarters subsequently display an ability to foster and absorb significant concentrations of firms in diverse stages, types and sectors of electronics production. Often forgotten in assessing their importance is the massive infrastructure of support institutions providing the talent, skills, equipment, materials and other specialised inputs which comprise an essential component of the social conditions of production in global centres of innovation and corporate control (Braun and Macdonald, 1982).

Global high-technology headquarters, once established, are not replicable. Firstly, they were formed under specific historical conditions and circumstances that no longer pertain. Their emergence, far from being reducible to a single, easily replicated, origin was dependent precisely upon the favourable confluence of diverse and complex sources of growth whose simulation, particularly in competition with established headquarters locations, would require new and impossibly risky governmental and business strategies as well as the concentration of extraordinarily vast resources. Secondly, rather than other areas creating anew the conditions which made headquarter locations successful, the latter's infrastructural advantages act as a magnet for new businesses and skills which continue to be attracted to such areas rather than locating elsewhere.[15] Thirdly, the conversion of competitive entrepreneurial places into dominant high-technology headquarters renders their duplication unnecessary.[16]

The dominion of the global high-technology headquarters tends to subordinate other forms of high-technology industrial development: distance from the central global location implies either relative insignificance or dependence. The most prevalent form of headquarters extension are illusory Silicon Valleys, high-technology branch-plant economies in which routinised, low-skill, low-wage manufacturing and assembly labour predominates (Watts, 1981; Malecki, 1984a; Luger, 1984; Macdonald, 1983; Bradshaw, 1986). Branch plant economies tend to cluster either around particular phases of high-technology production or around specific sectors of high-technology industry. In areas concentrating diverse firms engaged in similarly standardised phases of the production process, expansion will most likely occur through attraction of additional branch facilities rather than research and development, or other non-routinised, operations.[17] Sectorally concentrated regions tend, on the other hand, to be dominated by a small number of firms. The internal operations of the companies themselves are often more diversified (sometimes embodying more sophisticated technical processes though usually few, if any, R & D functions) but regional expansion is limited by self-sustaining internal corporate production organisation and the correspondingly narrow market for potential external suppliers and infrastructural support firms.[18] Dependent upon external parent-company control, these economies generate neither innovations nor spin-offs and develop meager interconnections with the surrounding local economy. Their long-term employment and development prospects are fundamentally unstable, acutely sensitive to changing corporate policies, technological innovation or business cycle fluctuations. As vertical integration proceeds and a truly integrated global circuit of production is consolidated, a few such regions may be upgraded by the incorporation of more sophisticated technological processes, the utilisation of skilled labour and the elaboration of administrative, sales and other functions. Created principally to serve diverse application needs within critical market growth areas these subordinate production and marketing economies provide somewhat more differentiated, but still narrow and dependent, forms of high-technology industrialisation. They function as localised mediators between corporate headquarters, branch facilities and regional market sites within the world economy.[19]

Attempts to counter, or to duplicate, global headquarters have generally relied upon state-induced high-technology complexes in which exogenous state investment creates essentially distorted imitations of Silicon Valley.[20] These high-technology formations are created by substantial injections of state government industrial support (North Carolina, Austin) or national, usually military-related, regional investment (Los Angeles). As Hall (this volume) and Markusen (1984) observe, state defence spending is the most important single determinant of high-technology geography. State-induced complexes are frequently maintained in a relatively static state at the cost of wasteful protective expenditures and ignorance of alternative and potentially more fruitful local economic growth paths. In marked contrast to global headquarter regions, even more successful state-induced high-technology economies tend to remain permanently indebted to state financing, substituting sectoral concentration for industrial diversity and defence dependence for competitive commercial development. Growth in such areas is neither diversifying nor self-sustaining, regional economies remain vulnerable to cycles in military spending and local communities lack control over economic development decisions. Under conditions of globally integrated production, such centres will tend to remain comparatively insignificant secondary high-tech locales or, on the contrary, will stimulate external corporate infiltration. State investment in the latter instance serves merely to subsidise the area's subsequent integration into a division of labour organised by the corporate headquarters.

The pre-occupation of location theory and local economic development policy with the international dispersion of high-technology employment, however, has also obscured more recent developments involving the expansion of the global headquarters region itself and the increasing dependence of headquarter firms upon clustered, rather than scattered, activities. As we have seen, the objective conditions incubating and sustaining the independent entrepreneurial firms of the preceding epoch have been eradicated by industrial concentration and the globalisation of production. Nevertheless, smaller producers continue to exist, even to proliferate, for though they are unable to surmount entry barriers to full-scale integrated production, they are favoured by tendencies which, in tandem with forces promoting vertical integration, simultaneously encourage organisational decentralisation.[21] The dramatic expansion in the variety of micro-electronics applications provides market niches for small specialised producers, while new technologies enlarge the possibilities for efficient small batch and customised production. Contemporary market expansion in electronics faces an increasingly complex demand structure and requires close working relations between circuit designers and systems producers, between equipment manufacturers and a diverse range of end-users. The erosion of mass markets, the manufacture of diverse and frequently-changing product lines and computer-based production methods introduce new forms of flexibility to large-scale production. New production strategies, such as those enabling silicon foundries to shift circuit-design costs to systems producers, provides a basis for small-firm adaptation to structural change. Ownership and control are increasingly centralised while production processes are downscaled and decentralised: in this way, the resources of large capital are combined with the malleability of small business. Supported by trends towards market diversification and flexible production methods, smaller firms tend to emerge as subordinate components of a division of labour governed by giant conglomerate firms and to be linked directly to their large corporate sponsors or clients rather than acting as the autonomous driving force of electronics industry growth.

The recent restructuring phase of the electronics industry in Silicon Valley, for example, has given rise to an autonomous, innovative, high-technology complex in adjacent Santa Cruz County. This spatial configuration follows an entirely different logic from other high-technology centres. In less than a decade, an extensive network of innovative, indigenous, electronics firms has transformed Santa Cruz County, previously an isolated hinterland of Silicon Valley, into a distinct but interdependent component of a broader regional economy through its integration with the global headquarters of high technology.[22]

The roots of high-technology industry in Santa Cruz County reach back to the late 1950s and 1960s with the sporadic and isolated opening of a few electronics plants. This limited involvement with the electronics industry, however, basically left the area's employment and social structure unchanged despite the consolidation of a booming high-technology economy in adjacent Santa Clara County. Coinciding with the beginning of the most recent structural phase of electronics industry reorganisation in Silicon Valley, Santa Cruz began a period of unanticipated growth and rapid transformation. Between 1976 and 1984 more than 100 of Santa

Cruz's existing high-technology firms commenced business, 41 per cent of them between 1981 and 1984 alone. During this same period the Santa Cruz area experienced sustained population and employment growth, far surpassing growth rates in California as a whole and even in Santa Clara County itself. Electronics production has replaced agriculture-related industry as the principal source of manufacturing employment, and generates the majority of all new jobs in the County. A working-class, farming and tourist community has become a predominantly professional, technical and service middle-class industrialised society.

As in Silicon Valley itself, the Santa Cruz electronics industry is composed mainly of small and medium-sized firms with an important complement of large and very large firms.[23] The majority of firms are particularly R & D intensive with production geared primarily towards small-batch runs, prototype manufacture and custom designs. Unlike more specialised areas, Santa Cruz duplicates the diverse sectoral distribution of electronics production in Silicon Valley. Through the use of sophisticated capital equipment, high levels of productivity, concentrations of skilled labour, and a complicated network of contract arrangements with Silicon Valley operations,[24] Santa Cruz high-technology firms have captured a diverse and constantly changing range of market niches, particularly in the area of high-technology industrial products, capital goods and producer services. Santa Cruz firms tend to export new industrial products and processes for both high technology and more traditional industrial sectors. This strategy has connected Santa Cruz firms with a wide spectrum of local, national and international customers either indirectly, through linkages with Silicon Valley firms, or directly through independent market relations.

The formation of the electronics industry in Santa Cruz does not derive simply from its proximity to Santa Clara—a simple result of external financial investment or branch plant location—nor is it due to specific locational determinants.[25] Instead, this new autonomous complex is a product of the County's involvement in the most recent round of fundamental structural changes in the organisation of the electronics industry as a whole. Based upon the manifold qualitative advantages made possible through the close proximity of firms engaged in differentiated yet interconnected stages of non-routine production, this transformation privileges the broad social framework of innovation and economic viability. It is the qualitative structure of social relations and resources which decisively binds together this production complex, not specific economic calculations or locational factors. Though an independent formation it cannot be understood apart from its position in a widening circle of regional interdependence. The regional extension and diversification of the global headquarter electronics complex itself enhances its cumulative supremacy over other high-technology areas—acting as a magnet for the continued arrival of new firms—while integrating formerly marginal hinterlands directly into the core of the new world economy. If dispersion tends to imply dependence, the dynamics of independent high-technology production appear to presuppose an area's involvement in new social relations of production and its spatial proximity to the global headquarters.

POLICY ALTERNATIVES FOR REGIONAL DEVELOPMENT

The policy implications of this analysis appear bleak for areas not already favourably situated on the terrain of international high-technology competition. The dynamics of high-technology spatial organisation are now a reflection of global high-technology industrial structure, that is dependent on the prevailing distribution of corporate control, the design of linkages within and between regions, and the changing requirements of production. Consolidation of this framework of high-technology location and economic growth is transparently inimical to new efforts aiming to establish independent, high-technology, regional development or commercial supremacy. Inspired initially by the example of Silicon Valley, economic policymakers everywhere hope to create autonomous, high-technology complexes minimally similar to the Santa Cruz prototype. But neither Silicon Valley nor Santa Cruz in fact provides a viable model for other communities or states to emulate, since neither model can be abstracted from its determinate position within the historical structure of high-technology industry as a whole. The intimate connection of autonomous high-technology regions with the social conditions of production in the global headquarters suggests that independence is precisely not geographically transferable while the overall dominion of the global headquarters indicates that the geographical mobility of high technology operations will most likely subject other regions to economic dependence.

Moreover, contrary to common assumption, contemporary trends appear to enhance further the leadership position of US high-technology industry, even against its primary Japanese rivals (Gordon and Kimball, 1985). US design expertise in more complex integrated circuits positions American firms well in emerging upgraded commodity markets. Increasing chip complexity resulting from VLSI technology and the consequent transformation of components into systems open up new markets in application-specific integrated circuits. The close co-operation between semiconductor manufacturers and systems producers required for the production of semi-custom and custom devices promotes stronger relations between domestic electronics firms while the market fragmentation accompanying customisation militates against Japanese volume sales strategies and supports the evolution of myriad small specialised producers occupying market niches under the protective wing of larger diversified firms. The US market, only 15 per cent of which is geared to consumer products (compared with 55 per cent of Japanese semiconductor sales), is driven by demand for more sophisticated and specialised designs (Borrus, 1985), a core requirement for future economic leadership in high technology. Software has become the critical ingredient in the future diffusion of electronics applications: up to 80 per cent or more of final systems' total cost is now absorbed by software development as the expense of software creation and maintenance has escalated. US firms, both hardware manufacturers and the small companies producing software for particular types of application or specific industrial sectors, maintain strong leadership in software and user-support services even as this branch evolves into fifth-generation capabilities.

As global marketing and production strategies continue to increase in importance, the overseas positioning and market penetration of US electronics firms are superior to their European and Japanese competitors (Alic and Miller, 1985). And, finally, as vertical integration on a global scale with the aim of defining universal standards becomes paramount in an industry increasingly based on the fusion of computing and telecommunications, the battle for market dominance between US superfirms IBM and AT&T and their respective allies threatens to swamp competing efforts.

European economies (or, in different ways, local US communities) structurally disadvantaged by the prevailing global organisation of high technology confront severe dilemmas in their attempts to successfully establish regional centres of technological innovation or competitive information technology industries at the national level. Continued investment in the full-scale construction of leading-edge, high-technology industries is unlikely to succeed. At best, it will continue to finance defensive and imitative policies or, at worst, subsidise the absorption of firms or regions into an exogenously controlled and hierarchically organised international division of labour.[26] Reliance upon purchases of basic technology from the US or Japan is equally problematic: shortening product life-cycles and increasing research secrecy will lengthen the distance between leading areas and those falling behind, while failure to maintain indigenous high-technology production will deprive national economies of crucial practical complementarities with other manufacturing and end-user sectors. Efforts to establish superiority in specialised high-technology market niches should be encouraged, but success in relatively limited spheres, should it eventuate, is unlikely to fuel general economic recovery, restore the international competitiveness of European economies or resolve specific problems of pervasive unemployment. The strategy of a qualitative leap to fifth-generation technology—the kind of breakthroughs that might wrench control of future high-technology development from its existing centres—does not seem propitious given the vast technical problems encountered in both US and Japanese fifth generation programs (Okimoto *et al.*, 1984) and the comparatively low levels of European investment in equivalent projects.

Resolution of these bitter choices requires two radical shifts of perspective. Firstly, high-technology policy must assume more concrete forms. Policy is most commonly formulated in an abstract manner: local communities import external models (Silicon Valley) as replacements for existing concentrations of economic activity; national economies, propelled by the strategic compulsions of international competition, attempt to develop parity in the full complement of high-technology industries out of necessity. In both instances, high-technology policy is governed by putatively universal requirements for economic growth while its relationship to existing economic strengths and potentials remains indeterminate, assumed to follow naturalistically from the mere establishment of high technology itself.

Secondly, mastery of high-technology innovation *per se* is less important than mastery of high-technology applications. Economic revitalisation presupposes not simply the creation of new industries but, even more, the ability to restructure traditional sectors. Increased production efficiency is dependent upon new manufacturing technologies, market expansion upon substantial product innovation, adjustment to constant change requires flexible labour process organisation, coherently integrated transformations in technology, product base and production procedures imposes fundamental changes in business organisation. High-technology applications are critical to change in all these areas and industrial restructuring on so extensive a scale necessitates their diffusion throughout the economy as a whole (see also Rothwell and Zegveld, 1985). Competitive capabilities in high-technology development and successful application of computer-based technologies are obviously interconnected, but they are not equivalent as the stark contrast between US high-technology predominance and its continuing decline in traditional basic manufacturing reveals.

Duplication of the entire range of high-technology expertise available in global headquarter locations is less relevant to the cultivation of application capabilities than the deliberate orientation of basic information technology innovation to restructuring needs elsewhere in the economy.[27] Moreover, effective application involves not only provision of basic electronics technologies but the conscious direction of the complex and little understood practice of domestic technological transfer.[28] Shifting the structure of competition from high-technology industrial development alone to systematic economic restructuring also requires fundamental institutional and organisational changes: social innovation is perhaps ultimately more crucial than associated technological innovation. The conflict between the present self-defeating technologisation of society and a self-conscious social organisation of technology will provide the framework for future economic competition between advanced industrial societies and will likely introduce new global dynamics of location in high-technology industry.

NOTES

1. See Battelle Laboratories, 1982; Macdonald, this volume; Nijkamp, this volume. Segal Quince & Partners, 1985, makes clear that the university's commercial and economic development policies have been far more important to the evolution of the Cambridge phenomenon than formal research links between firms and the university. On North Carolina see Luger, 1984; Whittington, 1986. Giese, this volume, notes that technology centres and R & D parks tend to proliferate because they are local-government projects and their establishment is fostered by anarchic competition among local authorities.
2. Rees and Stafford (1983) have noted that the profusion of state and local high-tech incentive packages essentially cancel each other out as far as industry is concerned.
3. Flamm (1985a, p. 119) suggests more correctly that firms seek an optimal combination of risks and returns among diversified sites.
4. Nor does the more schematic product cycle version of this argument apply to the electronics industry. The Japanese push to mechanisation and standardisation did not presuppose any significant offshore dispersal of production as product development matured. Within the US electronics industry increasing industrial centralisation has been accompanied not only by standardisation but also by continuing dependence on skilled labour and advanced forms of product differentiation and customisation.
5. A fine critical analysis of formalistic Marxist approaches, encountered after this chapter had been written, can be found in Sayer, 1985.
6. Though many of the present arguments apply to the Route 128 complex, Massachusetts high technology will not be analysed in detail here. For a general overview see High Tech Research Group, 1984, and Saxenian, 1985.
7. In fiscal year 1984 California continued to rank first in the US in dollar amounts of prime military contracts. Military contract payments to firms in Silicon Valley increased by 18 per cent over fiscal year 1983 and only five States (excluding California) recei-

ved award totals greater than Santa Clara County in fiscal year 1984.

8. Hall (this volume) stresses the importance of state funding, especially defence-related research and development financing, for the establishment of high-technology areas in both the United States and the United Kingdom.
9. For discussions of the significance of military leadership in the early development of the industry see Tilton, 1971; Utterback and Murray, 1977; Braun and Macdonald, 1982, Borrus, Millstein and Zysman, 1982; Dosi, 1984; Flamm, 1985b.
10. Dosi (1984) in particular emphasises this important point.
11. Dosi (1984) points out that military financing permitted a trial-and-error exploration of different technologies that would not have been financed by private industry; hence, the failures supported by the military were an equally important contribution to the specific evolution of the US industry.
12. Dosi (1984) notes that spin-off firms typically arose to exploit technological innovation initiated in large firms and thus were dependent on, rather than responsible for, US technical leadership and market supremacy.
13. The fact that the introduction of the integrated circuit rendered the path of subsequent technical developments more predictable—a self sustaining logic of increasing circuit density according to Moore's law—intensified the rush towards the next technical plateau for both existing firms and new entrants alike.
14. There exists no singular trend towards core concentration or peripheral dispersion of high-technology industrial development precisely because the increasing corporate appropriation of the scientific and social conditions of production permits a diversity of specific regional combinations. Orthodox arguments tend to mistake mobility (a spatial variable) for flexibility (an organisational variable).
15. The rapid increase of foreign investment in Silicon Valley is one manifestation of this trend; the high concentration of software firms, theoretically able to locate anywhere, around Santa Clara County and Cambridge, Mass., is another. Though reliable data are lacking, the continued growth of employment and construction in Silicon Valley during recent years has been extraordinary.
16. Indeed, as Oakey (1984) points out, shortages of skilled labour, high housing costs and other problems encountered by firms and workers in an area like Silicon Valley constitutes perverse evidence of its continuing centrality. Insofar as orthodox spatial utility theory is correct about the locational attributes for high-technology industrial development, it is in reality making an argument not for

the movement of high technology to new locations but for its commitment to the agglomeration economies available in its present location.

17. Such areas differ little from traditional basic manufacturing activities and have little genuine connection with high technology proper.
18. Some excellent examples of these different types are contained in Glasmeier, 1985 (though Glasmeier's analysis differs from our own).
19. For examples of such complexes, see O'Connor, 1986; Lim, 1986; Goldstein, 1985.
20. Illuminating discussions of the importance of state support for different high-tech regions can be found in Malecki, 1984a; Markusen, 1984; Soja *et al.*, 1983; Hall, this volume; Luger, 1984; Macdonald, 1983.
21. For a consideration of this phenomenon in more general terms, see Piore and Sabel, 1985.
22. This section derives from research undertaken by the authors based upon personal interviews with managers and executives of 129 Santa Cruz County high-tech companies employing 6993 workers (representing between 85-90 per cent of existing companies and of the total high-tech workforce in the County). For more details see Gordon and Kimball, 1986.
23. Our data indicate, contrary to much of the literature concerning the relationship between firm size and employment generation, that neither large nor small firms dominate the job-creation process. Rather, new employment tends to be relatively evenly concentrated among a minority of both types of firm, suggesting that job creation policies must focus on both small and large firms and in a manner capable of making qualitative distinctions within, and links between, each category.
24. In contrast with expectations that might perceive Santa Cruz in an entirely subordinate or marginal position vis-à-vis the global headquarters or the overall division of labour generally, it is important to observe that the circles of subcontracting extend outward from Santa Cruz, rather than inwards from Silicon Valley, highlighting the independent nature of the Santa Cruz complex.
25. None of the variables routinely cited to explain location decisions of high technology firms—land, labour, university or research facilities, environmental and cultural amenities—proved determinant. Even the County's principal 'scarce resource'—its proximity to Silicon Valley—operated less as an explicit locational consideration than as an objective taken for granted. Instead, the overwhelming factor in locational choice was a purely personal en-

trepreneurial or managerial consideration of local Santa Cruz residents. The majority of Santa Cruz high-tech firms are independent start-ups, founded within Santa Cruz itself, and have been financed by personal savings and retained earnings.

26. For detailed analysis of the defensive European position in high technology see, for example, Dosi, 1983; von Gizycki and Schubert, 1984; Malerba, 1985. Even successful establishment of high-technology industry will have a problematic impact upon employment creation, job quality, and economic growth; see Gordon and Kimball, 1985.
27. Santa Cruz provides a specialised high technology economy essentially servicing the larger high-technology agglomeration in Silicon Valley. Similarly specialised complexes might well be established, as a matter of deliberate policy, to service other industrial sectors and develop a fusion of high technology expertise and more traditional skills. In this specific respect, the Santa Cruz complex begins to offer an interesting model for other regions.
28. Indeed it is not inconceivable that traditional industrial areas may begin to absorb substantial segments of high technology industry as this process occurs.

REFERENCES

Alic, J. A., Harris, M. C. and Miller, R. R. (1986) 'Microelectronics: Technological and Structural Change', in R. Gordon (ed.) *Microelectronics in Transition: Industrial Transformation and Social Change*, Ablex Publications, New Jersey.

——— and Miller, R. I. (1985) 'Export Strategies in the Computer Industry: Japan and the United States', Paper presented to Conference on Strategic Computing: History, Politics, Epistemology, Silicon Valley Research Group, University of California, Santa Cruz, March.

Battelle Laboratories (1982) *Development of High Technology Industries in New York State*, New York State Science and Technology Foundation, Albany.

Bluestone, B. and Harrison, B. (1982) *The Deindustrialization of America: Plant Closings, Community Abandonment, and the Dismantling of Basic Industry*, Basic Books, New York.

Boeckhout, I. and Molle, W. (1982) *Technological Change, Location Pat-*

terns and Regional Development, FAST Occasional Paper No. 16, Forecasting and Assessment in Science and Technology (FAST), Brussels.

Borrus, M. (1985) *Reversing Attrition: a Strategic Response to the Erosion of US Leadership in Microelectronics*, BRIE Working Paper, University of California, Berkeley.

———, Millstein, J. and Zysman, J. (1982) *US-Japanese Competitition in the Semiconductor Industry*, Institute of International Studies, University of California, Berkeley.

Bradshaw, T. K. (1986) 'The Future of the Electronics Industry in California: Issues of Regional Growth', in R. Gordon (ed.) *Microelectronics in Transition: Industrial Transformation and Social Change*, Ablex Publications, New Jersey.

Braun, E. and Macdonald, S. (1982) *Revolution in Miniature*, Cambridge University Press, Cambridge, England (2nd edn).

Castells, M. (1984) *Towards the Informational City? High Technology, Economic Change, and Spatial Structure: Some Exploratory Hypothesis*, Working Paper No. 430, Institute of Urban and Regional Development, University of California, Berkeley.

Dataquest (1985) *Japanese Research Newsletter*, 18 December.

Degeorge, F. (1983) 'Politique Industrielles et Stratégies des Enterprises dans le Jeu Concurrentiel International en Electronique: Pour Une Voie Européenne', *Memoire de Recherche*, December.

Dosi, G. (1983) 'Semiconductors: Europe's Precarious Survival in High Technology', in G. Shepherd, F. Duchêne and C. Saunders (eds) *Europe's Industries*, Frances Pinter, London.

——— (1984) *Technical Change and Industrial Transformation*, Macmillan, London.

Ernst, D. (1981) *Restructuring World Industry in a Period of Crisis: The Role of Innovation*, United Nations Industrial Development Organization (UNIDO), Vienna.

——— (1986) 'Automation, Employment and the Third World: The Case of the Electronics Industry', in R. Gordon and L. Kimball (eds) *Automation, Work and Employment* (in preparation).

Fernandez-Kelly, M. P. (1983) 'Mexican Border Industrialization, Female Labor Force Participation and Migration', in J. Nash and M. P. Fernandez-Kelly (eds) *Women, Men and the International Division of Labor*, State University of New York Press, Albany.

——— and Garcia, A. (1986) 'Advanced Technology, Regional Development and Hispanic Women's Employment in Southern California', in R. Gordon (ed.) *Microelectronics in Transition: Industrial Transformation and Social Change*, Ablex Publications, New Jersey.

Fishman, K. D. (1981) *The Computer Establishment*, Harper and Row, New York.

Flamm, K. (1985a) 'Internationalization in the Semiconductor Industry', in J. Grunwald and K. Flamm (eds) *The Global Factory*, The Brookings Institution, Washington, DC.

——— (1985b) 'Federal Support for Computer Research: A Brief History', Paper presented to Conference on Strategic Computing: History, Politics, Epistemology, Silicon Valley Research Group, University of California, Santa Cruz, March.

Fröbel, F., Heinrichs, J. and Kreye, O. (1980) *The New International Division of Labour*, Cambridge University Press, Cambridge, England.

Glasmeier,A. (1985) *Spatial Differentiation of High Technology Industries: Implications for Planning*, Ph.D. Dissertation, University of California, Berkeley.

Goldstein, N. (1985) 'Technological Change and the International Restructuring of the Integrated Circuit Industry: The Threat to Women's Employment in Scotland', Paper presented to Conference on Women, High Technology and Society, Silicon Valley Research Group, University of California, Santa Cruz, June.

Gordon, R. and Kimball, L. M. (1985) *High Technology, Employment and the Challenges to Education*, Silicon Valley Research Group Working Paper No. 1, University of California, Santa Cruz.

——— and ——— (1986) *Industrial Structure and the Changing Dynamics of Location in High Technology Industry*, Silicon Valley Research Group Working Paper No. 3, University of California, Santa Cruz.

Hall, P. and Markusen, A. R. (1985) 'High Technology and Regional-Urban Policy', in P. Hall and A. R. Markusen (eds) *Silicon Landscapes*, Allen and Unwin, Boston.

Hanson, D. (1982) *The New Alchemists*, Little, Brown and Company, Boston.

High Tech Research Group (1984) *Massachusetts High Tech: The Promise and the Reality*, High Tech Research Group, Boston.

Hood, N. and Young, S. (1979) *The Economics of Multinational Enterprise*, Longman, London.

Isard, W. (1956) *Location and Space Economy*, Wiley, New York.

Joint Economic Committee (1982) *Location of High Technology Firms and Regional Economic Development*, US Government Printing Office, Washington, DC.

Lim, L. Y. C. (1986) 'Global Factory in the Global City: Labor and Location in the Evolution of the Electronics Industry in Singapore and Southeast Asia', in R. Gordon (ed.) *Microelectronics in Transition: Industrial Transformation and Social Change*, Ablex Publications, New Jersey.

Linvill, J. G. (1984) *The Competitive Status of the US Electronics Industry*, National Academy Press, Washington, DC.

Lösch, A. (1954) *The Economics of Location*, Yale University Press, New Haven.
Luger, M. (1984) 'Does North Carolina's High Tech Development Work?', *Journal of the American Planning Association*, 50(3), 280-9.
Macdonald, S. (1983) 'High Technology Policy and the Silicon Valley Model: An Australian Perspective', *Prometheus*, 1(2), 330-49.
Malecki, E. J. (1980) 'Dimensions of R & D Location in the United States', *Research Policy*, 9(1), 2-22.
———— (1984a) 'R & D and Regional Development', Paper presented to Symposium on Technology, Policy and Regional Change, Syracuse University, New York, April.
———— (1984b) 'High Technology and Local Economic Development', *Journal of the American Planning Association*, 50(3), 262-9.
Malerba, F. (1985) *The Semiconductor Business*, Frances Pinter, London.
Markusen, A. R. (1984) *Defense Spending and the Geography of High Tech Industries*, Working Paper No. 423, Institute of Urban and Regional Development, University of California, Berkeley.
———— *et al.* (1984) *Military Spending and Urban Development in California*, Working Paper No. 425, Institute of Urban and Regional Development, University of California, Berkeley.
Mollenkopf, J. (1983) *The Contested City*, Princeton University Press, Princeton, New Jersey.
Noyce, R., (1981) 'Microelectronics', in T. Forester (ed.) *The Microelectronics Revolution*, MIT Press, Cambridge, Mass.
Noyelle, T. and Stanback, T. Jnr. (1984) *The Economic Transformation of American Cities*, Rowman and Allanheld, Totowa, New Jersey.
O'Connor, D. (1983) *Transnational Corporations in the International Semiconductor Industry*, United Nations Center on Transnational Corporations, New York.
———— (1986) 'Changing Patterns of International Production in the Semiconductor Industry', in R. Gordon (ed.) *Microelectronics in Transition: Industrial Transformation and Social Change*, Ablex Publications, New Jersey.
Oakey,R. (1984) *High Technology Small Firms: Innovation and Regional Development in Britain and the United States*, Frances Pinter, London.
OECD (1983) *Telecommunications*, OECD, Paris.
Okimoto, D. I., Sugano, T. and Weinstein, F. B. (eds) (1984) *Competitive Edge: The Semiconductor Industry in the US and Japan*, Stanford University Press, Stanford.
Osborne, A. (1979) *Running Wild: The Next Industrial Revolution*, McGraw-Hill, Berkeley.
Piore, M. and Sabel, C. (1985) *The Second Industrial Divide*, Basic Books, New York.
Rees, J. and Stafford, H. (1983) *A Review of Regional Growth and Indus-*

trial Location Theory: Towards Understanding the Development of High Technology Complexes in the United States, Report prepared for Office of Technology Assessment, US Congress, Washington, DC.

Rogers, E. and Larsen, J. (1984) *Silicon Valley Fever*, Basic Books, New York.

Rothwell, R. and Zegveld, W. (1985) *Reindustrialization and Technology*, Longman, Essex.

Sassen-Koob, S. (1982) 'Recomposition and Peripheralization at the Core', *Contemporary Marxism*, 5 (Summer), 88-100.

Saxenian, A. (1981) *Silicon Chips and Spatial Structure: The Industrial Basis of Urbanization in Santa Clara County*, Working Paper No. 346, Institute of Urban and Regional Development, University of California, Berkeley.

——— (1985) 'Silicon Valley and Route 128: Regional Prototypes or Historic Exceptions?', in M. Castells (ed.) *High Technology, Space and Society*, Urban Affairs Annual Reviews, Volume 28, Sage Publications, Beverly Hills.

Sayer, R. A. (1985) 'Industry and Space: A Sympathetic Critique of Radical Research', *Environment and Planning D: Society and Space*, 3, 3-29.

Sciberras, E. (1977) *Multinational Electronics Companies and National Economic Policies*, JAI Press, Greenwich, Conn.

Segal Quince and Partners (1985) *The Cambridge Phenomenon*, Cambridge, England.

Soja, E., Morales, R. and Wolff, G. (1983) 'Urban Restructuring: An Analysis of Social and Spatial Change in Los Angeles', *Economic Geography*, 59(2), 195-229.

SRI International (1980) *The Role of Defense in Santa Clara County's Economy*, Preliminary Report, Menlo Park.

Tiebout, C. (1966) 'The Regional Impact of Defense Expenditures: Its Measurement and Problems of Adjustment', in R. Bolton (ed.) *Defense and Disarmament*, Prentice-Hall, Englewood Cliffs, New Jersey.

Tilton, J. E. (1971) *International Diffusion of Technology: The Case of Semiconductors*, The Brookings Institution, Washington, DC.

Utterback, J. and Murray, A. (1977) *The Influence of Defense Procurement and Sponsorship of Research and Development on the Civilian Electronics Industry*, Center for Policy Alternatives, Massachusetts Institute of Technology, Cambridge, Mass.

von Gizycki, R. and Schubert, I. (1984) *Microelectronics: A Challenge for Europe's Industrial Survival*, Oldenberg Verlag, Munich.

Walker, R. (1985) 'Technological Determination and Determinism: Industrial Growth and Location', in M. Castells (ed.) *High Technology, Space and Society*, Urban Affairs Annual Reviews, Volume 28, Sage Publications, Beverly Hills.

——— and Storper, M. (1981) 'Capital and Industrial Location', *Progress in Human Geography*, 5(4), 473-509.

Watts, H. D. (1981) *The Branch Plant Economy*, Longman, London.

Weber, A. (1909) *Theory of the Location of Industry*, University of Chicago Press, Chicago (1929 edn).

Whittington, D. (1986) 'Planning for the Microelectronics Industry: State Policy in North Carolina', in R. Gordon (ed.) *Microelectronics in Transition: Industrial Transformation and Social Change*, Ablex Publications, New Jersey.

Wilson, R. W., Ashton, P. K. and Egan, T. P. (1980) *Innovation, Competition and Government Policy in the Semiconductor Industry*, D. C. Heath, Lexington, Mass.

Chapter 13

THE SPATIAL IMPACT OF COMPUTER CULTURE: POST-INDUSTRIALISM AND THE SPREAD OF THE PERSONAL COMPUTER IN BRITAIN

M. Batty

> Mr Smith, like almost everybody I met in Wigan, was a very decent person. 'Decency' was a quality Orwell cherished in the English working class; Wigan is still alive with it. Mr Smith took me to watch Wigan Athletic being beaten 1-0 by Newport County in the rain. He took me home to meet his wife. He took me to the site of Orwell's lodgings—a grassy mound, the statutory saplings and the statutory rain. Meanwhile, he recalled interesting things. There were a thousand old mine shafts within five miles of the town centre. The Wigan Coal and Iron Company had once been a larger consortium than Krupps. Occasionally, with a distant smile, he would recall poverty. "The unemployed used to say: 'Let's go to Woolworths and have a warm . . .' I can remember the days when the top of an egg were a luxury." (Sadly, such remarks have been devalued by the Hovis commercials.) But Mr Smith's most interesting revelation concerned the present; sales of magazines stocked by his shop. Here they are (and remember this is in a solid Labour-voting town with hardly a middle class to speak of): *Tribune*, 7 copies; *Labour Weekly*, 12; *New Statesman*, 13; *Economist*, 22; *Investor's Chronicle*, 2; *The Lady*, 36; *Private Eye*, 200; and computer magazines (twenty-five different titles) 2,500 copies a month. Sydney Smith said that it astonished him too, this great Wigan interest in computers and then, as a man who remembered the top of an egg as a luxury, he laughed with bafflement and delight.
>
> Anon, *Sunday Times Colour Magazine*,
> 13 February 1983.

A survey in July 1985 revealed that 23 per cent of British households now own a personal computer (*Retail Business*, 1985). A decade ago such a speculation would have proved unthinkable. Indeed in 1968, Donald Michie (1974), the doyen of artificial intelligence in Britain, declared quite categorically: 'For many years, only the rich will be able to install terminals in their private homes . . .'. Although the emergence of the microprocessor was foreseen in the early 1970s, its impact on society at large was in no way predictable. The speed of change during the last decade has been so great that we have barely begun to record the detail of this impact and only now are conscious efforts being made to understand it.

With respect to the spatial impact of new technology and technological change, we know little about the impact in other than conventional terms. Although a new industrial geography is being forged from studies of location and the spatial divisions of labour (Massey, 1984), the impact of new technology on consumer activities and on leisure has hardly been considered. Where attempts have been made to study spatial impact more comprehensively, these have been entirely anecdotal. It is the purpose of this chapter to extend this speculation and suggest how it might be grounded in a more formal approach.

It is tempting to suggest that the spatial impact of personal computers in particular and micro-electronics more generally follows well-known and well-entrenched spatial patterns such as unemployment and social class. In Britain, for example, this suggests that the traditional north-south divide might represent the frontier between information-poor and information-rich regions on the basis of indices of personal computer ownership, software production and so on. There is, however, much anecdotal evidence to the contrary. On the producer side, although the south has significant high-tech activity in the western corridor following the M4 from London to south Wales (Sunrise Strip) and in East Anglia around Cambridge (Silicon Fen), Scotland can claim Silicon Glen in the central lowlands. Other peripheral areas particularly south Wales and north-east England, have significant investments in consumer electronics industries. The picture is by no means clear but when confronted with observations about what people are interested in, such as Mr Smith's comments on 'computer culture in Wigan' which preface this essay, some doubt emerges as to whether the standard regional divisions and our perceptions of them are of much use in assessing spatial impacts. The pattern may be as surprising as in most other areas of the present technological revolution, thus reinforcing Adam Osborne's dictum that 'the most paralyzing aspect of the microelectronics industrial revolution is the inability of law makers and sociologists to cope with what is occurring because no one knows where to begin.' (Osborne, 1979). But begin we must, for unevenness in the penetration of microcomputers is likely to give as much concern to spatial planners of the post-industrial future as unemployment patterns did in the industrial past.

The need to know much more about the spatial impacts of new technology does not require to be spelt out, but the impact on attitudes is more complicated than on work and production. Bell's (1973) vision of post-industrial society being dominated by a scientific-technical élite suggests that computer literacy is an all-important index in measuring adaptability, resistance to change and related attitudinal values which condition the development of vibrant high-tech industries. We will examine the spatial impact of both attitudes to and ownership of personal computers as well as the impact of microcomputer software and its marketing. These questions will be developed in several ways. First we will explore the meaning of computer culture in general terms, briefly describing the development of computing and its impact on society at large during the last decade. Our reasons for dwelling so much on computer culture relate to the impact computers are having on attitudes to a variety of leisure, work and domestic pursuits; although these influences are subtle and bound up in notions of literacy and awareness, they are important to broader questions concerning information-rich and information-poor sectors, classes and regions.

We will relate emergent hard and soft computer cultures to the degree and type of computer awareness and to the spatial distribution of information-rich and information-poor areas. As we have already implied, conventional wisdom tends to equate the information rich-poor distinction with that between work and non-work, and early speculation has followed this assumption. However, our observations on Wigan already cast doubt on this. As a consequence of this debate, we will develop the notion that post-industrial and industrial societies might exist in the same nation, region or even city, side by side, certainly sectorally and in class terms but also spatially. And in Britain, as we shall see, there is a further twist to this argument for, as Bellini (1982) argues, the post-industrial future may contain areas which regress to the pre-industrial past.

The more empirically-grounded sections of this chapter deal with the mapping of the spatial impacts of personal-computer ownership and software in mainland Britain (England, Scotland, Wales). There is no published data on the spatial distribution of personal computers and thus we have been forced to construct our information-rich/information-poor map of Britain from less conventional sources. First, we have constructed a mental map of what we (two experts) consider to be the distribution of personal computers based on general awareness and perceptions of the spread of micros. Then we have examined three spatial patterns culled from an intensive study of locational information published in popular computer magazines in 1983-1985. We have been able to map the spatial distribution of readers' letters, games and business software for microcomputers using these sources. The patterns we find are both surprising and expected. The traditional north-south division stands out markedly but the area of intensive urban-industrial activity between south-east and north-west Britain, largely the product of the Industrial Revolution, appears to be weakening. Southern Britain more generally seems to stand out as the information-rich region with a small pocket of activity in the north-west. We will draw conclusions from these findings but the argument remains speculative and this essay simply points the direction to ways in which we might begin more considered research.

THE PERSONAL COMPUTER REVOLUTION

The development of computers appears to have followed the typical growth and diffusion path displayed by many technologies. Initially, invention occurs simultaneously and independently in several places, and the early technology is commanded by an elite group restricted to scientific, usually non-commercial contexts. Technological innovation then proceeds apace, and the technology becomes more reliable and less costly to produce. A mass market appears which is then penetrated, ultimately to the point of saturation, at which point the market comes to depend upon replacement and on further innovations to start the cycle again, at perhaps a different level. But there is one major difference between computers and most other technologies relating to both market and product. The market until the late 1970s (that is for twenty-five years or more) was commercially orientated to the industrial sector, but due to a major change in technology—the microprocessor—the market expanded to embrace the domestic sector. In fact, this change is not solely due to the micro revolution, for it was only made possible by the fact that the computer is a truly universal machine which can find a use everywhere. In this sense, then, the impact of computers is potentially a very complicated affair for it has the potential to penetrate and influence every product market.

The dramatic shift we will be focusing on here is the development of the personal computer from 1974 which is directly attributable to the invention of the microprocessor at Intel and Texas Instruments in 1971. Social impacts of computers prior to the mid 1970s were always conceived in terms of automation, in terms of work and production, and insofar as the population at large were influenced, this was through work practices, not through education or leisure. The micro revolution has changed all this. Naisbitt (1982) predicted that in 1985, 75 per cent of all jobs in the US would be influenced by computer use. Since the mid 1970s and certainly since 1980, the microcomputer has penetrated the much larger market of consumers in leisure and education as well as the service sector generally. There is obvious impact in terms of automation on these sectors but the more difficult impact is on a host of practices and attitudes towards learning, difficult to identify but critically important to the post-industrial future. This broad impact of computers on society we will refer to as 'computer culture', which we will discuss in the next section.

Computers have been developed in waves of technological innovation, sometimes referred to as generation. Conceptually, the idea of a computer goes back to prehistory, but significant advances occurred in the 1930s with the development of binary code and electrical transmission of information which were first realised in the vacuum tube computers of the immediate post-war years. Commercial exploitation came slowly in the 1950s but it was spurred, indeed probably only made possible, by the invention of the transistor in 1947. The huge dinosaurs of the 1950s were still restricted to rather specialist contexts; it was not until the development of the integrated circuit in the late 1950s that the revolution really got under way.

The 1960s saw miniaturisation proceeding apace. In 1971, the microprocessor or 'computer-on-a-chip' was invented: the era of very large-scale integration had begun (Braun and Macdonald, 1982). No one knew that an enormous market was ready and waiting for this product, and more particularly, no one realised that the biggest and arguably the most profitable market of all concerned the individual rather than the corporation or industry. The first home computer was marketed in 1974. It was called the Altair (Freiberger and Swaine, 1984). This machine and its immediate competitors attracted an enormous market of hobbyists who were essentially electronics buffs. It is doubtful whether these machines really worked at all but they were quickly followed by software houses writing higher-level programs for the machines and by other companies manufacturing peripherals making communications with these primitive microcomputers much easier. Indeed, one of the first software houses was Microsoft, which began by producing a BASIC for the Altair.

The growth in this market surprised everybody concerned. In fact, it was not until the late 1970s that microcomputers became reliable enough for very serious applied usage; it was then that the names of Apple, Commodore and Tandy became known almost overnight. The growth of ownership in personal computers began in the US in about 1977 and has continued apace since then, diversifying into specific markets, particularly education and services. In the UK, the microcomputer boom began in 1980 and really took off in 1982. It is instructive to look in a little more detail at this. From an industrial perspective, the main feature is the boom-bust phenomenon of small computer firms and the speed of their growth (and decline). In a consumer sense, the early market is clearly hobbyist dominated, where the line between consumer and producer is blurred. As the market grows, hard interests turn to soft and a much larger section of the population is affected. As yet there is little sense of how the market will respond to replacement of computers and further innovations relating to size and type of machine.

Table 13.1 presents the growth of the home market in Britain since 1981. Quite clearly the volume of hardware sales has stabilised while

Table 13.1: Sales of Home Computers, Software and Peripherals in Britain, 1981-1985

Year	Value in £M.							
	Hardware Sales (000s units)	H'ware	%	S'ware	%	Prphrls.	%	Total Value in £M.
1981	180	28	75	6	16	3	9	37
1982	650	95	68	28	20	16	12	139
1983	1600	220	67	60	18	48	15	328
1984	1700	235	57	110	26	70	17	415
1985	1750	245	43	190	33	140	24	575

Source: Retail Business (1985).

software is taking an increasing proportion. The size of the market, combined with its rapid growth, and the development of multipurpose applications in virtually every area of domestic and business life, makes prediction hazardous. In the US, there appear to have been several waves of computer purchase but these may simply relate to different sectors of the market. In the UK, the first boom now seems to be over. It is possible that the pattern of growth will follow that displayed in the video-cassette recorder (VCR) and other such markets, but computers are very different and the future is still largely unpredictable. Nevertheless, the fact that the first home computer boom seems to have run its course makes this a good point at which to examine spatial impacts, notwithstanding the fact that the sources we will use are from the period 1982-1985 when this most rapid growth occurred.

COMPUTER CULTURES

As computers are all-purpose, universal machines, it is not really surprising that they will penetrate both work and home. What is perhaps more surprising is that this point was not widely anticipated. In fact, the history of personal computers is full of stories about those closest to the computer industry in the 1970s rejecting the idea of the personal computer as even having a market among hobbyists (Freiberger and Swaine, 1984). Part of this problem relates to the fact that although the idea of a computer is universal, the reality of any particular computer is not. Machines are designed for particular markets. In the 1950s and 1960s, IBM designed for markets with an emphasis on both accounting and scientific use, while DEC designed for laboratory use, and so on. Computers have to be designed with markets in mind and this is what restricts established firms from developing new products despite their expertise. In the 1950s, computers were thought of as akin to calculators, machine tools, microscopes, etc., while the micro revolution of the 1970s and beyond has forced people to think of them as more like television, radio, cameras, etc. This in itself suggests that the computer represents a powerful social metaphor which will influence attitudes and hence change our culture.

Raymond Williams (1981) defines culture as 'the active cultivation of the mind' and more generally, suggests that such cultivation applied in a widespread sense to society at large, can develop characteristics in the population which are also referred to as culture. Computers are especially fertile soil on which to explore the influences on culture, and from the early days of computer development, there have existed subcults who have devoted themselves to various practices of computing made possible by digital computer systems. It was Weizenbaum (1976) who first drew attention to compulsive programmers—'bright, young men of dishevelled appearance, often with sunken, glowing eyes . . . sitting at computer consoles, their arms tensed and waiting to fire their fingers . . .' Some have referred to such programmers as hackers (Levy, 1984) although such hacking has become respectable as areas such as artificial intelligence have become institutionalised (Rose, 1985).

What is clear is that attitudes towards many areas of formal knowledge are being changed by computers, while it is less clear how the much greater areas of personal and informal knowledge are being influenced. In fact, Papert (1980) suggests that the role of the computer 'is that of a carrier of cultural "germs" or "seeds" whose intellectual products will not need technological support once they take root in an actively growing mind.' In one sense, the universal nature of computers offers immense - scope for the development of subcultures; computers can be used for hard or soft purposes, for control or for design, and so on. These are beginning to be reflected in the way people articulate and certainly verbalise common problems and tasks.

Studies of computer culture so far have been non-spatial, and have largely been restricted to hackers and to children. Levy's (1984) study of hackers shows how important these individuals are to the computer revolution in terms of innovation. Papert's (1980) study of learning via the computer is largely a prologemena for the use of computers in child learning, but his important message that what is emerging is a world of computer haves and have-nots, is relevant here. Turkel's (1984) study of a variety of computer cults is the most comprehensive and shows widely-differing attitudes to computers but also shows that computers enable a new softness and flexibility to be introduced into a variety of social and learning situations. These tend to be studies of cults rather than cultures but they do provide anecdotal support for the need for further study of spatial and sectoral impacts.

Computer literacy is not the same as numerical or mathematical literacy, although it does depend on this as well as on verbal literacy. Just as the parallel between work and non-work, information-rich and information-poor cannot be drawn, nor can the analogy of Snow's (1965) 'two cultures' be applied to the distinction between computer literacy and illiteracy. Computers are increasingly useful in non-numerical areas. Indeed, a decade or more ago, 90 per cent of computers were used to process numbers; today 90 per cent of computers are used to process text and graphics. This is already reflected in studies of the influence of computers on learning. Turkel (1984) refers to children's attitudes to the computer as displaying 'hard mastery' or 'soft mastery' over the machine—hard representing controlling the machine, soft reflecting its exploitation for design. Both involve creative pursuits.

The distinction between hard and soft in a sense reflects the distinction between hardware and software, between hackers and the cults growing up around softer uses of machines. The oldest cult of the hobbyist involves both hardware and software, whereas newer, less specialist cults reflect programming using high-level languages. But perhaps the most pervasive influences on culture are reflected in the use of machines for games, and business use—the use of software such as word processing, spreadsheets and the like.

The spatial study of computer culture must attempt to identify the various cults which make up the specific structure of the culture while at the same time attempting to provide more general indices of the pervasiveness of computers. Traditional indices of class and occupation as well as education are likely to be relevant here as well as spatial and sectoral measures of work and non-work. Computer culture thus represents the context in which areas and sectors of information-rich and information-poor populations exist and, in a sense, represents the backcloth to a study of the penetration of personal computers for leisure and educational use. These diverse issues are those against which our exploration of the spatial spread of personal computers in Britain is set.

CONCEPTUAL MAPS AND METHODOLOGIES OF SPATIAL IMPACT

The transition to an information society which is coincident with post-industrial society has been well charted by Bell (1973), Toffler (1980) and

Jones (1982). Just as the collapse of employment in agriculture was complemented by growth in manufacturing employment, both due to the technological innovations of the industrial era, the post-industrial era is characterised by a collapse in manufacturing employment and consequent growth in services: tertiary (banking, retailing, etc.), quaternary (knowledge-based services), quinary (home-based services) and so on. The optimistic scenario suggests that expansion in these sectors depends upon wealth created in the automated manufacturing sectors as well as an ability to sell services. In this way, it is assumed that post-industrial society will be richer than industrial society, notwithstanding the painful transition required in industrial regions and sectors of the population who can only adjust through reskilling.

This optimistic vision depends on the ability to control the wealth-creating sectors but what the vision (which is the usual one expounded) fails to deal with is the question of spatial inequality which invariably depends upon control over the means of production. The shift to tertiary, quaternary and quinary is only possible if automation in the manufacturing sector yields a level of surplus able to support a growing service sector and/or if the burgeoning service sector produces services for export. Both these conditions seem true of the US at present but not so for the UK where manufacturing industry has simply collapsed and where manufactured goods are now largely imported. In the UK, it appears that the investment in knowledge required to support the scientific elite which Bell (1973) argues must underpin post-industrial society, is simply not taking place; thus it is doubtful whether a large, vibrant, service sector can emerge to complement the decline in manufacturing employment. It is much more likely that the UK post-industrial future will contain pockets of post-industrialism, information-rich communities in a sea of industrial decline characterised by a poverty in information. This in fact is close to Bellini's (1982) vision of a society moving 'backwards into the future'.

This scenario, which is likely to hold for the UK, and perhaps other parts of Europe and the US, can be elaborated. Areas with no industrial heritage are likely to adapt well and the transition to post-industrialism will prove painless. Areas of heavy industrialism are likely to remain industrial or simply collapse into areas of high unemployment and underemployment which simply persist and are supported centrally at minimal levels. Work and non-work overlaying the pattern of information-rich and information-poor regions will be the order of the day with little correspondence between these two patterns. Complicating the picture are the effects of global restructuring, the loosening of locational ties to the cities, the movement from north to south for climatic and environmental reasons, the drift from central city to suburbs and beyond. In the UK, it is the drift from north to south and from city to exurbia, to small towns, which suggests that the most favoured areas are south and west of London. In the US the scale is different but the drift from frostbelt to sunbelt is equivalent to the north-south divide while the drift from city to small town is also observable (Clark, 1985).

This general picture can now be used as a backdrop on which to explore more detailed indices of the spatial impact of computer technology. Singling out spatial impact in so all-pervasive a phenomenon as computer culture may not reveal the most deep-seated differences which distinguish information-rich from information-poor groupings. Recent studies of spatial unevenness in the UK for example have moved away from simplistic regional distinctions to sectoral and class distinctions which compose these regional differences.

The picture we shall assemble is set against the emergent regional structure of the UK in which the traditional north-south divide is being reinforced as the industrial heritage disappears, and in which cities are increasingly less attractive locations in comparison with suburbia, exurbia and small towns. We have not considered the location of high-tech production in assembling this picture and thus the spatial impacts we deal with are extremely partial and limited. We have classified impact at rather a coarse spatial level: we have used the regional divisions and subdivisions adopted by Spence and Frost (1983) in their study of regional employment patterns. These are based on the sixty-one planning subregions of mainland Britain which in turn are based on pre-1974 local government units. What this level of spatial aggregation omits is the possibility of much more local distinctions particularly between city and suburb, small town, rural area, industrial region and so on.

As we have already cautioned, computer culture is an all-embracing concept and insofar as it can be defined in any measurable sense, this must involve synthesising a variety of indicators. All we have done here is to choose some measures which, in the spirit of this discussion, would seem to indicate aspects of computer culture. We have explored two approaches. First, we have attempted to construct a mental map of the ownership of personal computers. In essence, the construction of this map has been achieved by a systematic pair-wise comparison technique enabling the importance of regional subdivisions to be established in relative terms. Two experts, with the author acting as prompter and arbiter provided these paired comparisons in argumentative fashion, using something akin to a Delphi technique to resolve conflict and reach consensus.

The second and more direct set of measures we have compiled are based on data taken from popular computer magazines using complete runs of magazines selected between mid 1983 and mid 1985. We have examined the spatial spread of readers' letters which implies something about the use of personal computers, if not ownership. We have also looked at new releases of games and business software by centres of production and marketing; these provide another picture of spatial distribution. There are many problems associated with these data, not least of which is the difficulty of generalising from such partial sources. Nevertheless, they do provide a means of getting started, and at least provide a basis for disagreement, discussion and further research.

A MENTAL MAP OF PERSONAL-COMPUTER OWNERSHIP

In a situation where there is no primary data but where some measure of a phenomenon is required, it is often useful to systematically elicit perceptions of the distribution of the phenomenon in question. In spatial terms, such perceptual information is often referred to as a mental map, which is regarded as a unique and non-generalisable data source (Gould and White, 1974). Nevertheless, as Saaty (1980) has shown, there are many instances where such perceptions can represent highly accurate images of reality. For example, when the perceptions of an expert group—a set of professionals whose knowledge of the subject area is extensive—are systematically elicited, these perceptions may be as good, if not better than, any primary data subject to other biases.

The method we have used to construct a mental map of the spatial distribution of personal computers involves systematic comparison of every pair of subregions or zones in question, the establishment of the dominating zone, and a measure of this dominance. For example, for any two of our sixty-one subregions, we pose the question: 'Which of two areas, A and B, do you consider to contain the greatest number of personal computers?' This is then followed by the question: 'By what degree do you consider the chosen area to dominate the other?' The degree of dominance is measured on a 9-point ratio scale in which the value of 1 reflects equality between the areas in question, the value 5 involves one area being significantly different in scale or importance from the other, and the value 9 reflects dramatic dominance of one area over the other. An elaboration and justification of the scale is provided by Saaty (1980).

For meaningful comparisons, the number of objects in the set being compared must be between seven and two with five being probably an ideal number. Thus an obvious way to effect sixty-one paired comparisons which would be impossible in any direct manner, would be to partition the set of sixty-one into subsets of areas which met the conditions posed. A threefold hierearchy might involve dividing the UK regional space into four, then four again and so on, in this way enabling no more than four subdivisions to be compared at any one time at the lowest level. However, we have used the regional division and subdivisions shown in Figure 13.1 based on those used by Spence and Frost (1983). This is a two-level hierarchy in which there are ten regions at the highest level and up to eight

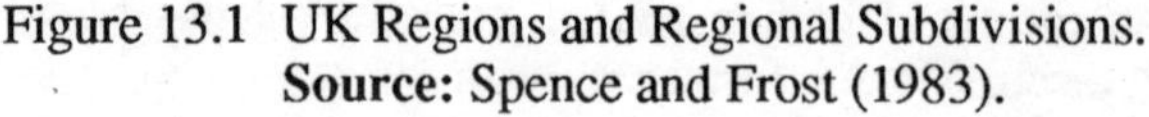

Figure 13.1 UK Regions and Regional Subdivisions.
Source: Spence and Frost (1983).

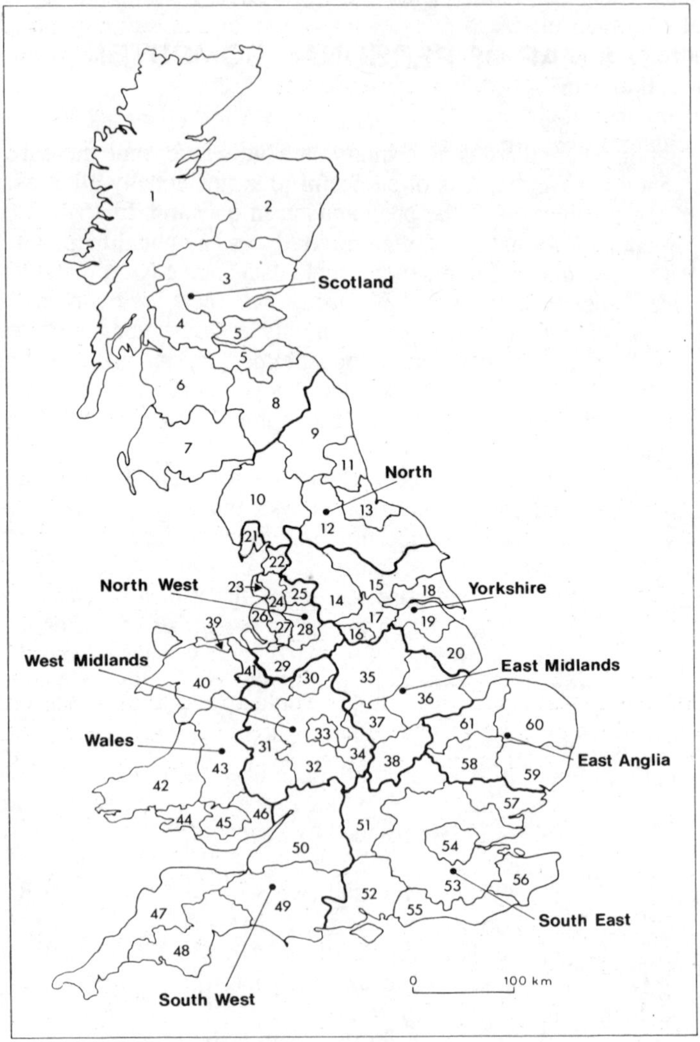

subdivisions in the ten regions at the lowest level. To use the method of paired comparisons according to this hierarchical structure involves making paired comparisons between the regions, extracting the weights or importance of each region, doing the same for each set of regional subdivisions, and then producing the global picture by concatenating the regional with the (relative) subregional weights. In this way, the global importance of each subdivision is established. Full details of the method are given in Batty and Spooner (1982) and Saaty (1980).

The number of comparisons involved at each level does exceed slightly, in a couple of cases, the guidelines noted above. However, for the experts who made the comparisons (work colleagues of the author), it was felt that their expertise in the regional and industrial geography of the UK, together with the strong sense of regional identity posed by the regional divisions in question, would enable systematic comparisons to be made. The two experts quickly reached agreement about the comparisons and, at the regional level where the ten regions were compared, the consistency of the comparisons was nowhere lower than 80 per cent. Saaty (1980) suggests that if consistency is lower than this value, the comparisons should be reworked in an effort to raise consistency. Using the method of concatenation referred to above, the overall importance scores for each region and the regional subdivisions were computed, the regional distribution being shown in Figure 13.2.

These scores are scaled to sum to 1000 and the regional percentages shown in Figure 13.2 indicate absolute, not per capita, levels of ownership of personal computers. The questions and comparisons were so phrased to reflect scale, and the strong sense of geographical identity of regions and the familiarity of regional thinking on the part of the two experts involved helped ensure the accuracy of these perceptions. In fact, a per capita ownership map was considered but this would appear to be a considerably more hazardous exercise to undertake. Figure 13.2 does covary with the distribution of population as one would expect although there is a higher level of ownership in the north-west and south-east, and a lower one in peripheral regions such as Wales and Scotland, than the map of population would imply. It is important however to continually emphasise that the picture presented, although a 'consistent' representation of 'expert' perceptions, cannot be thought of as a 'correct' distribution of actual ownership. It may be, and evidence of the most anecdotal kind suggests it is, but this remains speculative.

Our real interest here is in the spatial distribution at the subregional level which is mapped in Figure 13.3. Several points stand out. The dominance of the south-east, western corridor and south coast are clear, while there is some evidence of the Bristol-Cambridge corridor referred to in studies of high-tech production locations. The north-west is part of the higher ownership regions but the Midlands, Yorkshire and the north are

Figure 13.2 Percentage Ownership of Personal Computers in the UK by Region. **Source:** Batty and Spooner (1982), Saaty (1980).

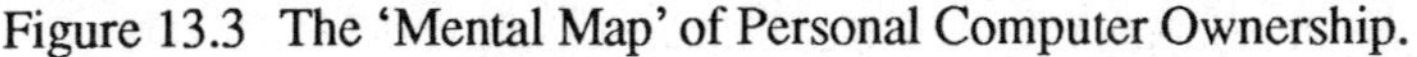

Figure 13.3 The 'Mental Map' of Personal Computer Ownership.

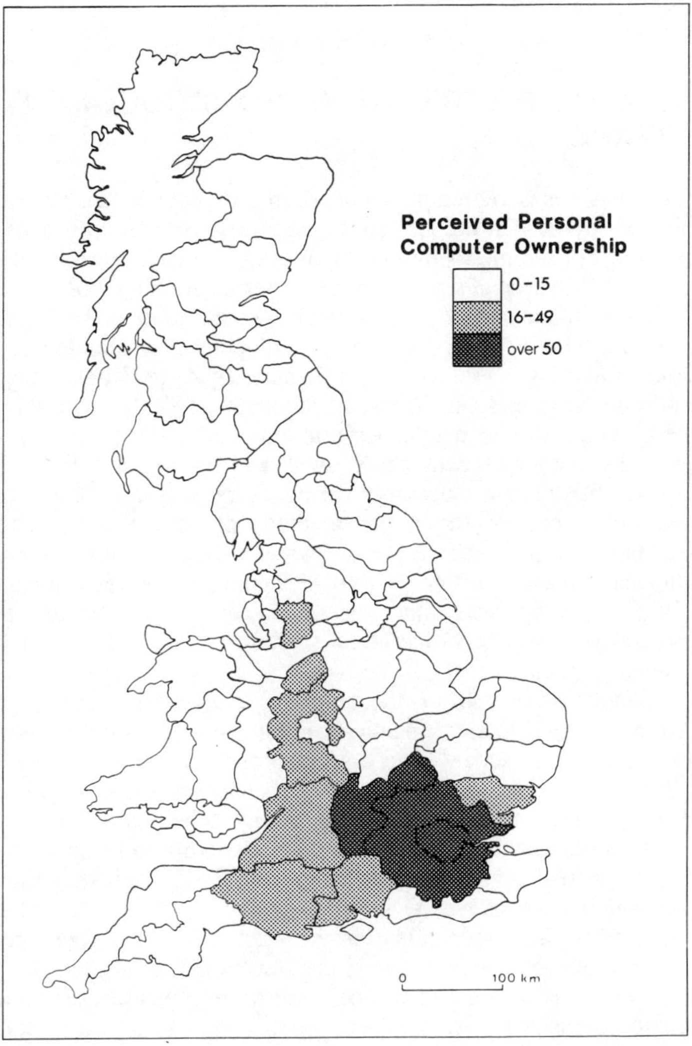

lower than one might expect. The peripheral regions of Wales and Scotland are exceptionally low in ownership. This in a sense is our first map of post-industrial Britain, of information-rich and information-poor regions, but we must be exceptionally cautious. If anything, the distributions may be too even. The technique tends to average out perceptions and reduce extremes.

THE SPATIAL DISTRIBUTION OF MICROCOMPUTER SOFTWARE

Computer magazines represent a rich source of data in the absence of more formal surveys. In the US, such magazines were simultaneous with the beginnings of the microcomputer industry, the magazine *Byte* first appearing in 1975. The real boom, however, began in about 1980 when, as Wayne Green, founder of *Byte* observed, 'A new kind of hobbyist had emerged, one who still liked to use the equipment, but tended to shun mechanical tinkering.' (Freiberger and Swaine, 1984).In Britain, computer magazines began to make a significant impact in 1982; between then and late 1984, the number of magazines (monthly and weekly) on display in the national-chain newsagent shops such as Menzies and W. H. Smith grew to over fifty. These magazines act in several ways. They are forums for mail order, one of the major methods of selling microcomputer products, but they also provide free software, advice and lighter comment about the state of the industry. In so dedicated a market, these magazines reflect a purer enthusiasm than other media and thus constitute useful sources of information in which editorial bias is likely to be rather lower than in other popular journals.

Microcomputer ownership is difficult to elicit from these sources but it is possible to gauge interest in microcomputers which implies usage of machines and hence, access. The first locational source we have used is based on readers' letters in which the writer's address is always published. This is an obvious area of editorial bias through selection of topic discussed, but for a reasonably large sample this is unlikely to be spatially systematic. The second source we have taken from these magazines relates to software which is classified by place of marketing rather than place of production. More bias is contained in this than in letters but we have examined two sources which show strong locational similarity. We have looked at games software and business software which are orientated to microcomputers of a similar kind to those referred to in the letters. The most modest machine covered in our surveys is the Spectrum Plus (Sinclair); the highest in the range being the IBM-PC. Many cautions are required to interpret this data, and the data do not necessarily imply ownership; rather they imply access. Nevertheless what follows does provide indicators of information-rich and information-poor areas and sectors.

To monitor readers' letters expressing interest in microcomputers, we have selected two magazines, *Acorn User* and *The Micro-User*, which deal with the BBC micro. This machine first appeared in 1981 but did not become widely available until late 1982. The growth of interest in the machine can be measured by membership of the BEEBUG national user-group which began publication of its newsletter in April 1982. Figure 13.4 shows the growth in this membership which appears to be levelling off in

Figure 13.4 Growth in Membership of the BEEBUG National User Group.

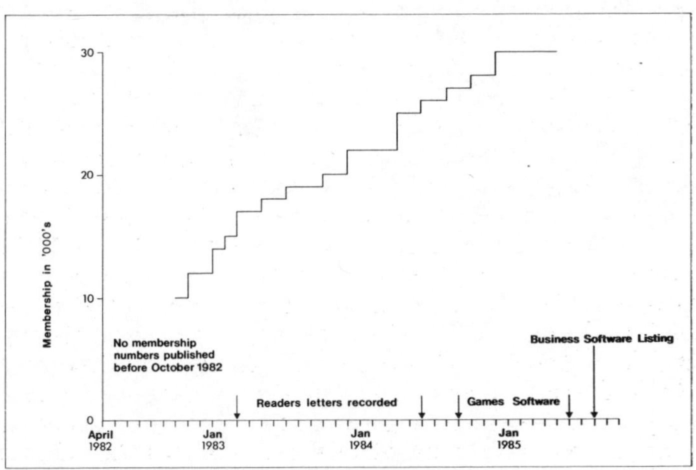

the same way the home market is behaving, as shown in Table 13.1. However, our samples of letters and software are taken from magazines whose circulation was growing at a rapid rate during this period. The growth itself is likely to have spatial implications—possibly diffusion from major centres of population—and thus there is an inbuilt bias in the surveys which reflect the fact that the data samples are drawn from a population which is continually changing.

This effect may be countered slightly by the fact that we have drawn our samples from issues of these magazines between March 1983 and June 1984 after the boom had stabilised. Second, one of these magazines is published in Manchester (the north-west), the other in London (the south-east). About 400 letters have been examined from sixteen monthly issues of both these magazines, and these are mapped in Figure 13.5 where it is clear the south-east, London and the western corridor dominate. Manchester and Liverpool are the only other significant centres but these have not reached the level of interest shown by readers in the south. The peripheral

Figure 13.5 The Spatial Distribution of Readers' Letters to *Acorn User* and *The Micro User*, March 1983-June 1985.

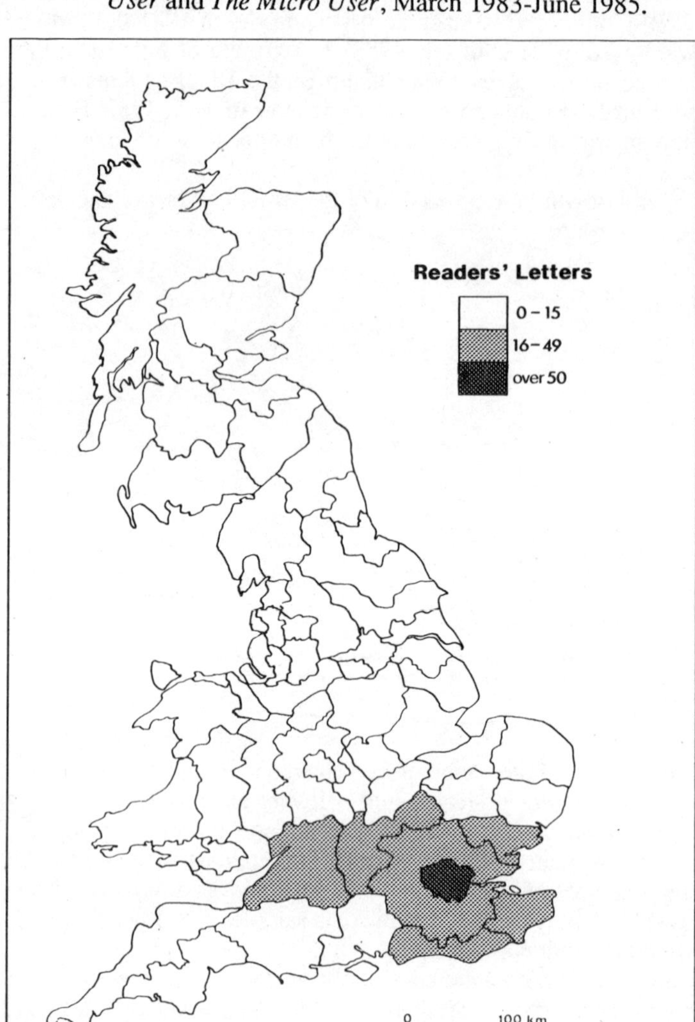

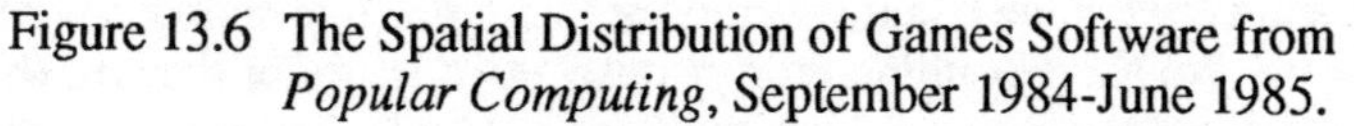

Figure 13.6 The Spatial Distribution of Games Software from *Popular Computing*, September 1984-June 1985.

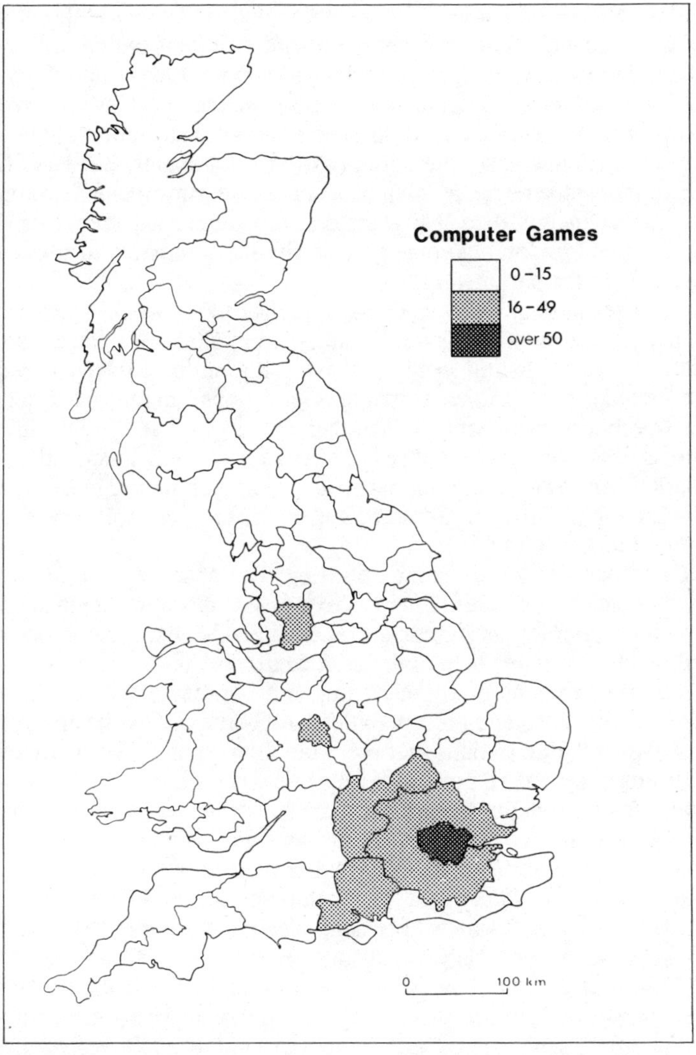

Computer Games

0 – 15

16 – 49

over 50

0 100 km

regions of Scotland and Wales show very low levels of interest.

The next source we have examined relates to the place of marketing of games software. We have examined forty consecutive issues of the weekly magazine *Popular Computing* which contains a regular section dealing with new releases of games. The games included involve both entertainment and education. The obvious qualification is that the games may be written in locations remote from their marketing. Three hundred and fifty games were sampled, about ten were new releases each week, over the period September 1984-June 1985. Their spatial distribution is shown in Figure 13.6. In this case, the south-east still dominates, but Manchester and Birmingham emerge as distinct centres, possibly demonstrating the market effect in this data. Nevertheless, the previous patterns of dominance west of London also emerge and similar comments to those made above apply to the peripheral regions.

The last set of data we have extracted relate to the market centres used to distribute business software for microcomputers. We have used the specialist software listing in the August 1985 edition of *Micro Decision* which includes professional software applications ranging from accounting to warehouse management. The data are mapped in Figure 13.7. The picture is less concentrated than games software and letters, the major population centres standing out but with strong domination in London, the south-east and south-west. Some 660 items of business software are used to compile these data sources.

In these maps there are some surprises. The absence of a strong software and microcomputer interest expressed through magazines and centred on Cambridge is significant. The Midlands, too, seem under-represented, while the dominance of the comparatively non-urban, environmentally-attractive regions of southern England is noteworthy. In one sense, our prior perceptions and preconceptions are borne out by the analysis. We have computed correlations between the four spatial distributions and these bear out the visual correspondence between Figures 13.3, 13.5-13.7. All the intercorrelations are greater than 80 per cent and the variance explained ranges from 70 per cent to 80 per cent. The amount of variance explained (r^2) can be listed as follows: between the mental map of ownership and letters it is 78 per cent, ownership and games 73 per cent, ownership and business software 80 per cent. The r^2 between letters and games is 76 per cent, between letters and business software 82 per cent. These levels of correlation do, however, raise our confidence in the fact that the data sources used here reflect the dominant trends in the spread of interest, access and ownership in personal computers; and thus indirectly reflect the spread of computer culture in the UK.

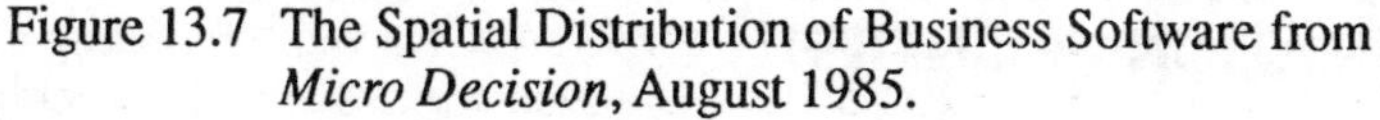

Figure 13.7 The Spatial Distribution of Business Software from *Micro Decision*, August 1985.

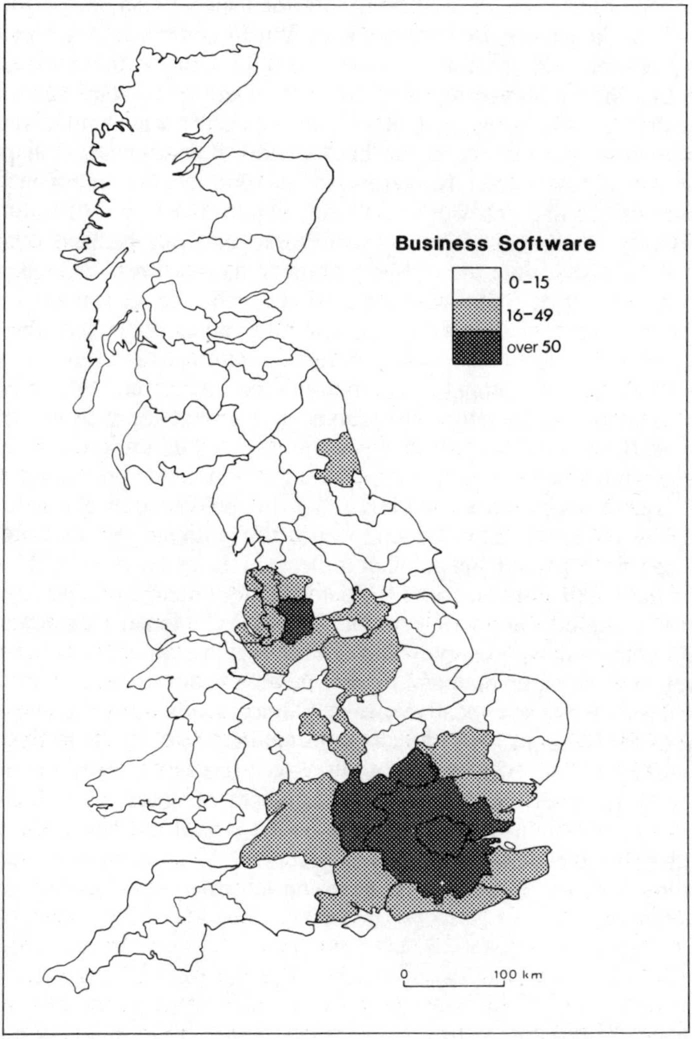

CONCLUSIONS: DEFINING THE SPATIAL INFORMATION FRONTIER

This chapter has been an exercise in informed speculation; the various pitfalls of the indicators used are obvious. But in constructing a map of information-rich and information-poor areas of Britain, these indices do show that the traditional regional distinctions are being reinforced in one sense and are changing in another. The age-old north-south distinction once again is reinforcing as the brief period of industrialisation passes, while climatic and amenity factors are increasingly important in where people wish to live and work in a world where industry no longer relies exclusively on the extraction of basic resources from the land. The next steps in this study are clear. More primary data are required. Ideally, a sample survey from the whole population is required so that the level of computer awareness, ownership and such like can be established, by social class, spatially and by other categorisations. Alternatives to this might involve surveys of computer companies, and more controlled content analysis of microcomputer publications. The production side needs to be consistently related to consumer patterns as does education, and various indices synthesised to produce more comprehensive interpretations.

The quote which introduced this essay was published as part of a series of articles on Britain as it headed towards the ominous year of 1984. That year has now passed but another milestone is reached in 1986 which marks the 900th anniversary of William the Conqueror's *Domesday Book*, the first complete survey of Britain. The BBC Computer Literacy project has mounted a national survey through schools in which a new Domesday Survey is to be completed and made available to home computer owners. As Robson (1984) so evocatively says, 'Almost twenty years after the Harrying of the North in 1069, the *Domesday Book* noted tersely of the region *hic wasta est.*' The BBC project will produce one of the most comprehensive surveys of Britain ever and if the indices revealed in this essay are borne out, post-industrial Britain will be the Britain of the south. And in the quest to develop an explicable precursor to contemporary regional planning, indices of computer culture and information technology are all-important.

REFERENCES

Batty, M. and Spooner, R. (1982) 'Models of Attitudes towards Urban Modelling', *Environment and Planning B*, 9, 33-61.

Bell, D. (1973) *The Coming of Post-Industrial Society: A Venture in Social Forecasting*, Basic Books, New York.

Bellini, J. (1982) *Rule Britannia: A Progress Report for Domesday 1986*, Abacus Books, London.
Braun, E. and Macdonald, S. (1982) *Revolution in Miniature. The History and Impact of Semiconductor Electronics*, Cambridge University Press, Cambridge, England.
Clark, D. (1985) *Post-Industrial America: A Geographical Perspective*, Methuen, New York.
Freiberger, P. and Swaine, M. (1984) *Fire in the Valley: The Making of the Personal Computer*, Osborne/McGraw-Hill, Berkeley.
Gould, P. and White,R. (1974) *Mental Maps*, Penguin, Harmondsworth.
Jones, B. O. (1982) *Sleepers, Wake! Technology and the Future of Work*, Wheatsheaf Books, Brighton, Sussex.
Levy, S. (1984) *Hackers: Heroes of the Computer Revolution*, Doubleday, New York.
Michie, D. (1974) *On Machine Intelligence*, Edinburgh University Press, Edinburgh.
Massey, D. B. (1984) *Spatial Divisions of Labour: Social Structures and the Geography of Production*, Macmillan, London.
Naisbitt, J. (1982) *Megatrends: Ten New Directions Transforming our Lives*, Warner Books, New York.
Osborne, A. (1979) *Running Wild: The Next Industrial Revolution*, McGraw-Hill/Osborne, Berkeley.
Papert, S. (1980) *Mindstorms: Children, Computers and Powerful Ideas*, Basic Books, New York.
Retail Business (1985) 'Special Report No. 1: Computers', No. 329, 17-23.
Robson, B. (1984) *Where is the North? An Essay on the North/South Divide*, North of England Regional Consortium, Manchester.
Rose, F. (1985) *Into the Heart of the Mind*, Century Publishing, London.
Saaty, T. L. (1980) *The Analytic Hierarchy Process Planning, Priority Setting, Resource Allocation*, McGraw-Hill, New York.
Snow, C. P. (1965) *The Two Cultures and the Scientific Revolution*, Cambridge University Press, Cambridge, England.
Spence, N. A. and Frost, M. E. (1983) 'Urban Employment Change', in J. B. Goddard and A. G. Champion (eds) *The Urban and Regional Transformation of Britain*, Methuen, London.
Toffler, A. (1980) *The Third Wave*, Bantam Books, New York.
Turkel, S. (1984) *The Second Self: Computers and the Human Spirit*, Granada Publishing, London.
Weizenbaum, J. (1976) *Computer Power and Human Reason: From Judgement to Calculation*, W. H. Freeman and Company, San Francisco.
Williams, R. (1981) *Culture*, Fontana, London.

Chapter 14

TECHNOLOGICAL CHANGE AND THE BUSINESS ENTERPRISE

M. Taylor

The purpose of this chapter is to examine the enterprise context of technological change and technology transfer in an attempt to create a framework within which the spatial dimensions of these processes can be better understood. Existing research on technological change and the transfer of technology in both geography and economics has generated a mass of intricate, empirical detail which hangs together only poorly. It is argued here that this situation stems from the inadequate conceptualisation and specification of the firm in this research, especially the multi-site and multinational firm. Through the over-emphasis on establishment-level analysis and maximisation principles, the firm has been inadequately specified and effectively homogenised in these two disciplines. The introduction of well-established ideas on organisational structure from management science and organistion theory may allow this problem to be at least partly overcome. In this chapter, therefore, a specification of organisational structure is attempted based on ideas relating to organisation-environment interactions and inter-organisation and intra-organisational power relations drawn from the management science literature. The usefulness of this specification is then judged in the context of established regularities in the international and inter-regional transfer of technology.

In geography there have been three pre-occupations in the work done on technological change: establishment-level innovation (for example Thwaites, 1978, 1983; Gibbs and Edwards, 1983); high-tech small firms (Oakey, 1984); and R & D activity (Malecki, 1980a,b). In the establishment-level analyses there is an implicit assumption that the focus of decision making is always the establishment as part of a multi-plant, multi-site enterprise. By implication, therefore, studies of technological change in geography have tended to ignore the unequal relationships that exist within and between different types of enterprise (Benson, 1975; Clegg and Dunkerley, 1980), producing within them a form of homogenisation by omission. Organisational structure in these studies is reduced to no more than simple dichotomies such as establishments of single-plant or multi-plant enterprises. The implication is that while single-plant firms are somehow different to multi-plant firms, separate plants within multi-plant firms are all much the same. In contrast, pre-occupation with high-tech small firms and R & D activity implicitly promotes the primacy of certain types of enterprise or parts of an enterprise as engines of technological change, but without ever establishing the basis of that primacy. The outcome is a series of geographical studies on technological change and the transfer of technology built on a shifting sand of definitions of organisational structure which reduces them to a mass of interesting detail rather than combining to create some larger and more integrated edifice.

A similar problem of homogenisation exists in economics, caused not by omission but by over-rigid conceptualisation based on profit-maximising principles. The consequences of this maximising assumption show up most clearly in studies of multinationals and technology transfer which employ the currently-fashionable internalisation model of the firm developed by Coase (1937) and elaborated by Williamson (1975), Rugman (1982) and Caves (1982b). In this model, market failures occur under conditions of oligopoly when prices for the inter-organisational transfer of proprietary knowledge become impossible to establish. To deal in and maintain control over such information and knowledge, transactions are internalised. Allocation within the business organisation is thus assumed to occur according to which sub-unit can make most efficient use of the proprietary knowledge (and technology). In other words, all sub-units of an enterprise are endowed with common corporate qualities, and each is charged with the same goal of seeking maximised profits. As such, external markets in this model are replaced by internal markets, cost-maximising assumptions are retained, allocation within enterprises is based on efficiency, and every sub-unit within an enterprise is endowed with autonomy in decision-making.

This, in effect, is a form of corporate homogenisation by commission, since nothing needs to be known of the internal functioning of the multinational firm. No allowance is made for the fact that some parts of large organisations are used as 'cash cows' and some may be obsolete, that it is difficult to establish satisfactory internal performance standards in multinational corporations (Yonkers, 1982) and that many large organisations function not as internalised markets but as demand economies (McCulloch, 1985). The result of this approach in economics is a large number of statistical studies testing simple hypotheses. They yield such conclusions as: 'The larger the proportion of global sales made by subsidiaries, the more R & D is dispersed abroad', or 'The more R & D is decentralised' (Caves, 1982a). Again, what has been created is a mass of information that adds up to a massive confusion of detail. By endowing all sub-units of an organisation with common corporate qualities the internalisation model has also endowed the enterprise with internal homogeneity.

The firm, therefore, is a term that needs to be unpacked in studies of business organisations and technology transfer rather than being reduced to random handfuls of variables to be varied arbitrarily from study to study. Precisely this issue of firm structure has been the concern of management science and organisation theory for a very long time, yet surprisingly little of this literature and research has found its way into either economics or geography (McDermott and Taylor, 1982). It is the purpose of this chapter, therefore, to attempt to develop a conceptualisation of the firm or business enterprise based on management-science ideas especially those contained in the strategic contingency or structural contingency theories.

CONTINGENCY THEORY AND THE ENTERPRISE

The structural-contingency model focuses on the fundamental role of environment in shaping the internal structure of complex organisations (Penning, 1975). For the present discussion there are three significant facets of this model: organisational structure, task environment and domain. Organisational structure is seen to be modified by pressures from the environment. These pressures stimulate differentiation or departmentalisation of the firm to meet homogeneous sections of a complex environment. The significance of the separate sub-units within the enterprise is then determined by the technical power they acquire together with positional power deriving from rank. This converts the organisation into an intra-organisational power network of the dominated and the dominant. In the model, however, environment is not a simple construct but has two interrelated facets; the task environment and the domain. The task environment is that set of organisations with which a focal organisation actually establishes exchange relationships (Thompson, 1967) and comprises customers, suppliers, competitors and regulators (including government and unions; Dill, 1958). Again there is asymmetry in the relationships that are developed, depending upon the resources the enterprise controls (Pfeffer and Salancik, 1978). The result is that an inter-organisational power network is created (Benson, 1975) again dividing the dominated from the dominant. The domain within which an enterprise functions is yet a larger environmental aggregate. It refers to the point within the societal environment with which a focal organisation could interact. Since it is set up to perform specific functions, an organisation stakes out a sphere of potential operations. To perform those functions it has to be recognised as legitimate and, therefore, requires domain consensus. Clearly this is a useful distinction because it goes beyond the cut-and-thrust environment of day-to-day business.

Organisational Structure and Intra-Organisational Power

Within organisation theory, organisations are seen to be doing two contradictory things at one and the same time. On the one hand, they are goal-seeking to serve the collective purpose of their members (to be, for example, a profitable engineering firm), while on the other hand, they are factionalised coalitions which respond to localised internal pressures and interests. This creates a constant state of tension within an enterprise as these two sets of forces play against each other.

However, not all sections of the firm or enterprise contribute equally to the organisational whole. The enterprise is seen as being functionally differentiated with separate departments and sub-units performing diverse and specialised functions (Figure 14.1). At the centre is the core technolo-

Figure 14.1 The Boundary Spanning Structures of an Organisation.
Source: Hower and Lorsch (1967).

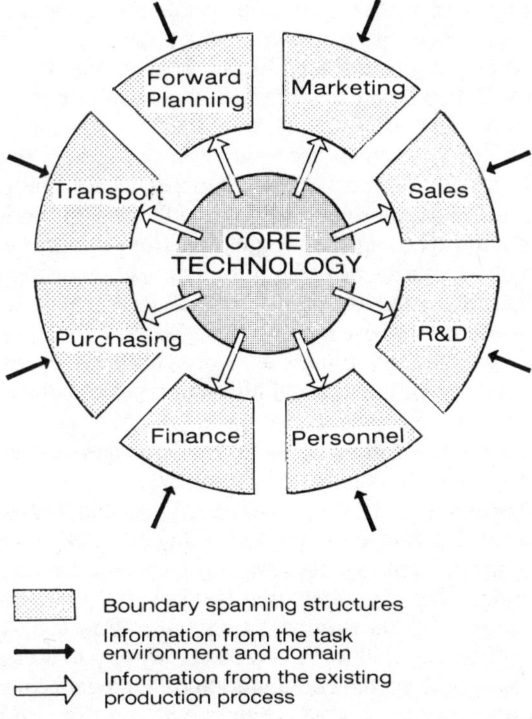

gy doing what the enterprise is known for, either making some product or performing some service. Surrounding this core are the parts of the enterprise that deal with its environment to gather the resources the organisation needs. These so-called boundary-spanning functions include finance departments, sales departments, personnel and transport departments, and so on. The knowledge and expertise these boundary spanners possess endows them with technical power within the enterprise as a whole. This technical power is also reinforced and modified by positional power—the power that comes simply from the office a person occupies and obedience to a superior which may only partly rest on greater technical knowledge. However, not all boundary-spanning functions are of equal significance to the firm (Hower and Lorsch, 1967). Sales may take precedence in one set of circumstances while finance or production may take precedence in another. What is defined, therefore, is an intra-organisational power continuum.

The spatial differentiation of firms into multi-plant and multi-site operations can be usefully added to this concept of intra-organisational power to define what amounts to the core and periphery of an enterprise. Following Mindlin and Aldrich (1975), centrality and peripherality in an intra-organisational power network depends on specific and measurable organisational traits. Thus, centrality increases with a plant's or sub-unit's interests being represented on increasingly higher boards of directors in an organisation. It will also be reflected in the numbers and types of boundary-spanning functions that are found at any particular site and whether functions that are not on-site are either brought in or supplied from elsewhere in the group. In addition, centrality will depend upon the ownership of equity in a plant. Greater equity holdings will tend to increase intra-organisational centrality along the progression from minority ownership to joint venturing to majority ownership to complete ownership. This is essentially a simple expression of differential technical power in an organisation but, unfortunately, it is perhaps the only measure used extensively in economics and geography. Intra-organisational centrality can also be calibrated in terms of the locus of different types of policy and decision making in a firm, mode of acquisition (with take-over increasing peripherality) and the length of time a sub-unit has been part of the organisation (with age and survival and increasing centrality).

There is clearly a great deal of congruence between this set of ideas on intra-organisational power relations and the geographical work on information flows undertaken by Tornqvist (1970) and others in the early 1970s. This work certainly recognised the internal subdivision of an organisation according to its members' communication tasks and their involvement in different contact networks. However, it was pre-occupied more with defining the spatial configuration of aggregate contact networks for countries and regions as a whole than with the exercise of positional and technical power within a single organisation to generate a corporate core and periphery.

However, not all organisations and enterprises are multi-plant, multi-site or even multinational enterprises. The majority of commercial enterprises are, indeed, single-plant or single-site small firms. At the intra-organisational scale they may show little or no internal differentiation and, in the terms used here, would have no internal periphery. However, they may, in fact, yield some of their intra-organisational centrality by using other firms to supply boundary-spanning functions. The most easily externalised functions are accounting, finance and wage payment by subcontracting; licensing and franchising may well lead to the externalisation of vital boundary-spanning function including purchasing and sales. Here, therefore, a firm's external relationships with its operations environment impinge upon intra-organisational structure and power.

Task Environment and Inter-organisational Power Network

An organisation's task environment comprises the constellation of organisations with which it actually interacts through competition, regulation or exchange. To survive in this environment it must acquire and maintain resources (Pfeffer and Salancik, 1978). These resources have been envisaged as including funds, personnel, information, products and services (Aldrich, 1972) which to Benson (1975) reduced to money and authority. Authority was the right and responsibility to carry out programs of action while money facilitated the execution of those programs.

Control of resources, therefore, endows interrelated organisations and enterprises with different amounts of power and binds them into networks of unequal relationships. Networks can be either highly centralised or decentralised with centrality having been identified as a measure of dominance. Owing to the significance of the resources they control, banks and financial institutions have been seen to dominate many power networks (Levine, 1972). However, the exercise of this power is conditioned by the rules governing the relationships between one organisation and another. The rules are manifest in the legal frameworks of societies, and sanctioned inequality is evident in company law, taxation and labour legislation for example.

Therefore, just as within organisation, there is a power continuum between organisations. Building on Taylor and Thrift's (1983) segmentation model, this continuum would run from global corporation at the core through multinationals, large national corporations, leader, loyal opposition and satellite smaller firms to craftsmen smaller firms and livelihood enterprises (from which the returns to the owner are no greater than could be gained from wage employment).

The networks, however, are never stable. Individual network participants will adopt strategies to maintain and secure supplies of resources. Such modification of network power can be achieved through codes of practice and joint venturing to reduce risks, through co-option to boards of outside personnel to ensure resources (especially finance), through lobbying and petitioning government and, in the final extreme, through take-over and merger. Taylor and Kissling (1983) have shown in detail the dynamics of these relationships in the inter-organisational power network linking the airlines of the South Pacific.

The position of a firm or enterprise in an inter-organisational network will also depend on its efficiency and effectiveness. Efficiency is usually measured as profitability and obviously affects an organisation's ability to secure resources. However, it should not be confused with effectiveness which is the continued ability of an organisation to secure those resources. It is clear from the histories of many small firms that efficiency does not necessarily bring effectiveness especially when it comes to raising capital. Equally, many large enterprises have remained effective long after their efficiency has evaporated. Thus, centrality or peripherality in an inter-organisational power network cannot be read-off from some simple econometric notion of efficiency.

Domain and Environmental Centrality

The environment within which an organisation functions is, however, wider than the tightly prescribed task environment. This has been shown clearly for sub-units of firms in India, for example, by Negandhi and Reimann (1973). A wider environment is also integral to the resource-dependence perspective of organisation which sees the powers of broader societal aggregates, such as governments, consumer and environmental groups, being called upon by business organisations to help them modify their positions within inter-organisational power networks. The domain, therefore, is a broad specification of the context within which a plant, establishment or sub-unit of a complex organisation functions rather than the narrow task environment of direct interaction and linkage. It is, in effect, a measure of inter-regional comparative advantage.

There is clearly a practical problem of calibrating the environmental domain of a plant. In a preliminary fashion, however, it can be equated with the national environment within which a plant functions which brings with it the advantage that there are readily available data. Clarke (1984) has attempted to specify environmental domains for separate plants of the paints division of ICI. National domains were described in this research in terms of six sets of variables: political attributes, economics attributes, international trade structure, international capital structure, industry factors, and sector factors (in this case the chemicals sector). Such an approach to calibration is derivative of the increasingly sophisticated systems that have been devised for assessing international financial risk and country risk (Taylor and Thrift, 1982; Robinson, 1981). In essence, it distinguishes between core and peripheral economics, in a sense similar to that used by Wallerstein where domains are arrayed along a continuum of greater or lesser centrality.

For the countries recognised by Clarke (1984) in his study of ICI Paints Division plants, the derived indices of domain centrality and peripherality are shown in Figure 14.2 together with the *Euromoney* financial risk-rating

Figure 14.2 Environmental Domain. Source: Clarke (1984).

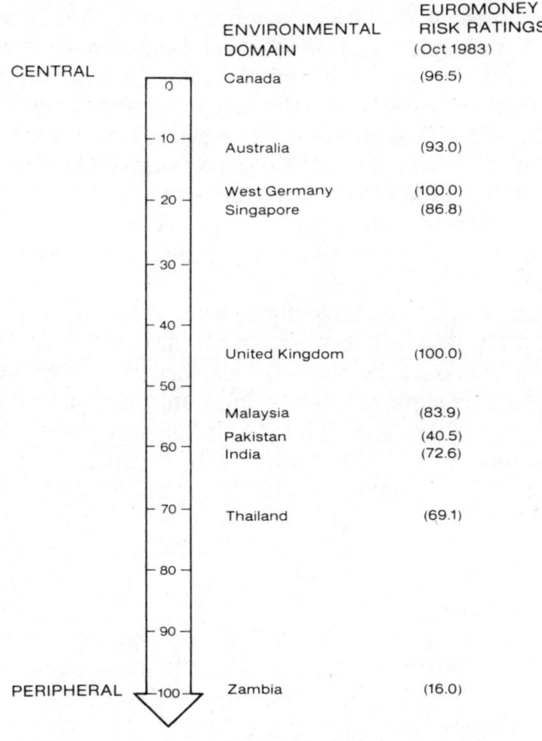

for the countries concerned in October 1983. It is possible in some instances that the most appropriate domain of an establishment may be regional rather than national. For the regions of a country, therefore, a national index of centrality might be looked on as a mean or median around which the regions are dispersed. Data limitations make sub-national domains extremely difficult to calibrate. Such calibration difficulties, however, do not undermine the basic concept of an environmental domain impinging and modifying inter-organisational and intra-organisational power networks.

THE TERRAIN OF ORGANISATION

These three facets of organisation theory add up to the definition and specification of three basic dimensions of enterprise and organisation embracing intra-organisational power and the milieu within which a whole enterprise or part of an enterprise functions. The foundation of this specification is the asymmetry and inequality involved in dealings and interactions within and between organisations and institutions. This is where the framework proposed here differs from frameworks already available in geography and economics. Asymmetry and inequality have rarely been recognised in economic geography; this deficiency is most striking in linkage and information-flow studies. The geographical literature, in its preoccupation with space, has tended to see organisational interactions in entirely neutral terms. Asymmetric relations have been reduced in economics to the implicit recognition of only technical power and the ability of such power to create efficiency. The exercise of positional power, probably better labelled political power within and between organisations, has been largely ignored.

When, for simplicity, these three dimensions of enterprise are combined as orthogonal axes, they define a terrain of unequal relationships across which any organisation can be mapped. Consider a US global manufacturing corporation for example (Figure 14.3a). It may well have its headquarters in New York, primary R & D in a US technology park, principal subsidiaries and secondary R & D in Europe, subsidiaries through Australasia and Latin America, joint ventures in the NICs and licensed production in other LDCs and micro-states. An Australian multinational such as ACI or TNT would map very differently onto this terrain: headquartered in a semi-peripheral environment; having begun with operations in New Zealand and the Pacific islands; more recently expanding through subsidiaries and joint ventures into South-East Asia and now beginning to invest in Europe and North America. A Third World multinational would again map entirely differently onto this terrain (Figure 14.3b).

National enterprises and single-plant small firms map equally easily onto this terrain and, in so doing, demonstrate the diversity of ways in which they are welded into the national and global economies through the unequal relationships they must necessarily enter into in order to function.

National economies obviously cut across the grain of this terrain in a totally different way since they contain mixes of corporate sub-units and single plant firms occupying various positions in inter-organisational and intra-organisational power networks which both reinforce and are reinfor-

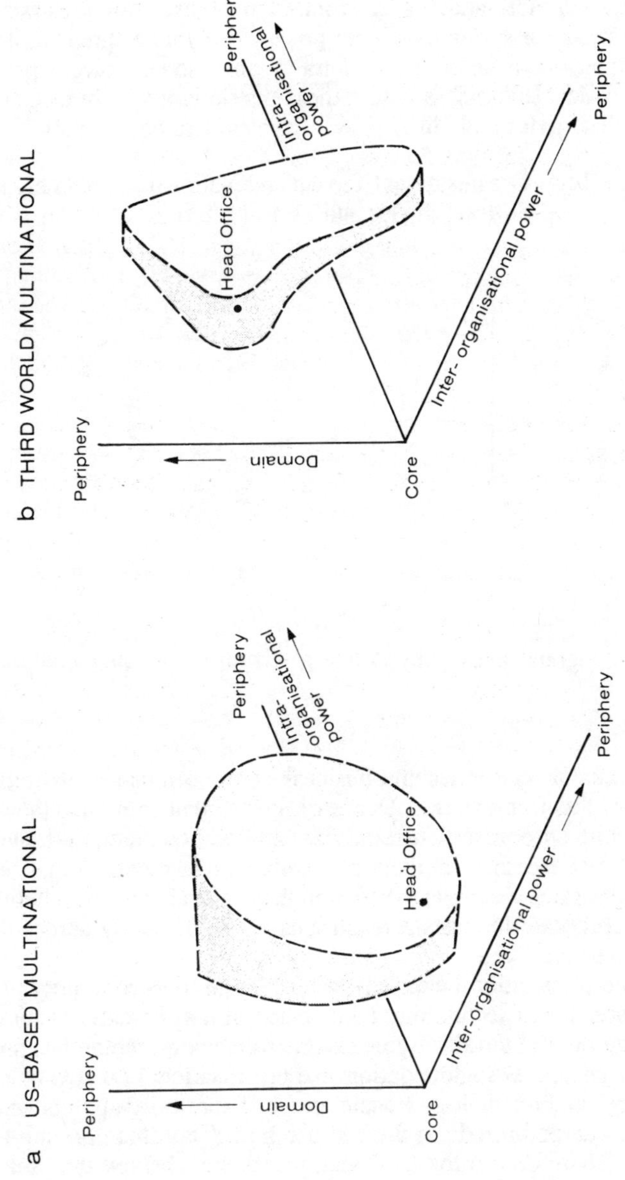

Figure 14.3 Organisational Terrain and Multinational Corporations.

ced by the domain characteristics of that economy. The hypothetical mapping of the UK and Fiji economies in Figure 14.4 illustrates this point. Enterprises or sub-units of enterprises need not be spread uniformly across the inter-organisational and intra-organisational power spectra of those economies. Lumpiness within these distributions is, in fact, the segmentation that Taylor and Thrift (1983) have elaborated in detail.

Figure 14.4 Organisational Terrain and National Economies.
Source: Taylor and Thrift (1983).

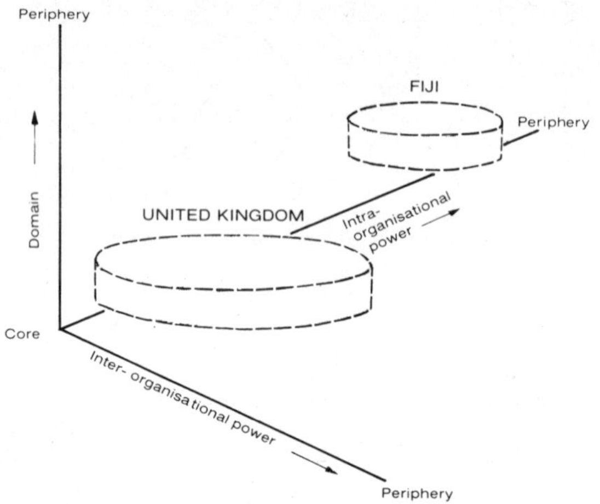

National economies and business organisations are obviously not static within this framework. LDCs seek investment to reduce their peripherality, while enterprises constantly expand and contract, take-over and close, and shift the mix of their productive investments. What this approach demonstrates clearly is the tension that must always exist between national and corporate interests as each cuts very differently across this organisational terrain.

The proposition being advanced here is that technological change and the transfer of technology takes place in a systematic fashion across this organisational terrain. By using this framework, which has unequal power relationships as its foundation, the organisational context of technological change and technology transfer can be more fully appreciated and greater order can be brought to the confusion of detail that currently exists in this area. Therefore, in the final sections of this chapter, the usefulness of the framework is examined in relation to multinationals and the transfer of technology and inter-regional technological change using the UK as an example.

MULTINATIONALS AND THE TRANSFER OF TECHNOLOGY

Before detailing the transfer of technology by multinationals it is important to establish some basic features of both technology and technology transfer. First, two types of technology can usefully be distinguished—product technology and process technology. Product technology entails the development of goods that have few substitutes, while process technology embraces the production and marketing of a product. For product innovations the R & D effort is large and visible and risks are great owing to market uncertainty. In these circumstances it is difficult to allocate R & D expenditure with precision to a successful new product. In contrast, process innovations are likely to be cheaper and less visible. They are also easier to price since there is always at least one substitute—the old technology. Thus, prices can be set with more certainty (Coughlin, 1983) with the result that process technology is likely to be more readily transferred than product technology.

Second, predicting the profitability of new technologies is extremely difficult and often wildly inaccurate. Beardsley and Mansfield (1978) have shown that for new products initial profitability estimates were more than double or less than half the actual discounted profits in 62 per cent of surveyed cases. The equivalent figure for new processes was 50 per cent.

Third, the transfer of technology is very costly. As Teece (1977) has shown, the average cost of transferring a production process is 19 per cent of the total costs of a receiving project, with a range in his survey of 2 per cent to 59 per cent. Moreover, first-time transfers, inexperienced recipients, new technologies and lack of competition all tended to increase these costs. Nevertheless, large US companies still expect to draw 29-34 per cent of the returns from any R & D project from overseas markets through subsidiaries, licensing or exports.

Quite clearly, therefore, there is an imperfect international market in technology which will promote internalisation in the sense used by Coase (1937) and Williamson (1975) as a goal which multinationals strive to achieve. This predisposition to internalise transactions can be matched with the organisational terrain framework to generate testable hypotheses on the international diffusion and transfer of technology through multinational corporations. Thus, it can be suggested that technology is developed at the core of the organisational terrain and released most preferably through the intra-organisational network into more peripheral environmental domains.

Such a proposition fits well with established facts at the international scale. Caves (1982a) reports a body of literature which shows that the R & D activity of US multinationals tends to stay close to home, especially that which deals with basic research rather than product modification. The R & D activities of foreign subsidiaries in countries away from the core, like Australia, for example, are more concerned with modifying technology from abroad (Parry and Watson, 1979). Moving to really peripheral environments, like Fiji, any form of R & D activity in foreign-owned companies is virtually absent (Taylor, 1983a).

In transferring technology abroad, multinationals obviously prefer to retain control through ownership, favouring foreign investment over joint ventures or licensing, not least because of the greater rent-extracting potential that it offers (Baranson, 1978; Telesio, 1979; Kroner, 1980; Caves, 1982a). In short, they strongly favour intra-organisational control. Licensing technology is preferred only in those situations where entry barriers deter foreign investment, markets are too small, the licensor lacks assets, rents to intangible assets are short lived, the technology being shifted is simple, the risk of technology leakage is low, and where reciprocal licensing is possible (Caves, 1982a). All of these factors bear directly on the configuration of organisational terrain.

Coughlin (1983) has shown that in similar domains (in this case countries at the same stage of development) government restriction of majority ownership affects the type of technology transferred. For Yugoslavia and Spain, where majority foreign-ownership was discouraged, 27.5 per cent of the transferred technology was the more vulnerable product technology. In Greece and Portugal, where there was no such restriction, 45.5 per cent of transferred technology was product technology. Clearly, ownership, as a surrogate for intra-organisational power, is an important determinant of technology transfer. Furthermore, Coughlin's figures can also be used to show that domain centrality is equally important to technology transfer. Comparing West Germany and France to Greece and Portugal, transfers of technology relative to the numbers of foreign subsidiaries in each country were greater in Germany and France than in more peripheral Greece and Portugal for both processes and product technologies (Table 14.1).

The speed at which technology is transferred also conforms with the shape of the organisational terrain that has been defined. In a study of US-based multinationals, Mansfield and Romeo (1980) have shown that the mean time-lag in transferring technology to subsidiaries in developed countries is six years and to subsidiaries in LDCs is ten years. Furthermore, transfers of technology from US parents to joint ventures or to arm's-length licences took an average of thirteen years. Coughlin's (1983) analysis tends to reinforce these results. Technology is transferred from the US more rapidly to central rather than to more peripheral environments (on average 8.9 years after first commercialisation for the shift of new-product technology to Germany, but 10.2 years to Greece and Portugal). It is also transferred more rapidly when stronger equity control is exercised (10.2 years to Greece and Portugal; 14.1 years to Yugoslavia and Spain).

The unequal relationships that underpin the organisational terrain defi-

Table 14.1: Transfers of US Technology to Different Environmental Domains

		(Transfers per 100 subsidiaries)	
	Recipient	Process Technology	Product Technology
1.	Germany/France	21.8	7.3
2.	Greece/Portugal	7.9	6.6

Source: Based on figures in Coughlin (1983)

ned here also ensure that new technology and innovation outside the core is garnered back to the core. Thus, Johns (1983) has exemplified the acquisition of Australian technology by US firms through the process of take-over, and Caves (1982a) has reported that Swedish firms have been taken over by foreign enterprises simply to gain technology.

THE INTER-REGIONAL TRANSFER OF TECHNOLOGY

The organisational context of technological change and technology transfer has been hinted at in research undertaken in the UK (Thwaites, 1983). The following discussion draws extensively on the interpretation by Taylor (1983b) of the work on technological change done at the Centre for Urban Regional Development Studies (CURDS) at the University of Newcastle-upon-Tyne. The model derived from this interpretation is presented in Figure 14.5.

In the body of survey research generated at CURDS two organisational terrains could be defined: a dependent development-area structure and a growth-oriented south-east structure. The development-area structure consisted of small regional groups, branch plants of large national and multinational enterprises and a largely non-innovative small-firms' sector. The branch plants had few on-site business functions such as marketing and purchasing, their R & D was restricted to amending existing technologies, and they were primarily recipients rather than generators of new technologies. By inference, therefore, development-area enterprises and sub-units of enterprises were both inter-organisationally and intra-organisationally peripheral.

Figure 14.5 Business Organisations and the Inter-regional Dimension of Technological Change in the UK.
Source: Research by Thwaites (1983) interpretation by Taylor (1983b).

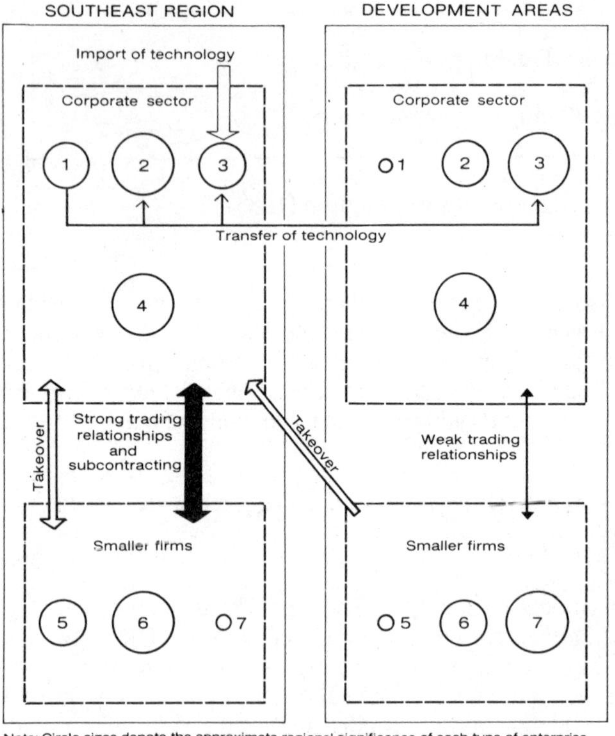

The south-east structure comprised highly innovative corporate establishments and a strongly innovative small-firms' sector, with some intrusion of foreign control. The innovative group plants had large on-site R & D facilities, employing up to fifty people; they also had greater numbers of on-site business functions. They were mainly concerned with product innovation but, in some cases, they received process technology from abroad. The strongly-innovative small firms were engaged in high-growth sectors of the economy and generated both process and product innovations. The plants of the south-east were, therefore, more central in both intra-organisational and inter-organisational power networks than those of the development areas. Nevertheless, their receipt of process technology from abroad would imply some element of intra-organisational peripherality even in the south-east when it is placed in a global context.

Very different patterns of technological change occurred across these different organisational terrains. The south-east was the main source of corporate sector R & D, although some technology came in from abroad through intra-group channels. There were, however, pronounced intra-group transfers of 'low-technology' process and product innovations (as judged by a panel of experts) from the south-east to the development areas. The least innovative group plants in the development areas were frequently foreign owned, which leads to the conclusion that these areas were being used as centres for the mass production of goods nearing obsolescence.

It was also suggested in Taylor's (1983b) interpretation of the CURDS research that the innovating corporate concerns of the south-east region developed strong trading and sub-contract relationships with small firms in the region, stimulating R & D in those single-plant enterprises. The subsequent take-over of these small-firm high fliers by corporate-sector organisations in the south-east served, in turn, to enhance the growth of the corporate acquirers. Set against this picture of growth was the picture of decline presented by the development areas. Corporate branch plants, depending on intra-group transfers of technology and with few commercial functions on-site, develop few and only minor contacts with other local enterprises. Therefore, in these regions there was no stimulus to small-firm development. The innovative small firms that did appear were prone to take-over as were their counterparts in the south-east. However, their acquirers were most likely to be from the south-east or from abroad. So, just as on the international scale, new technology was drawn back to the core.

Quite clearly, the relationship between the inter-regional transfer of technology and organisational terrain described in this section both parallels and complements the relationship inferred from fragmentary evidence on the international scale. Indeed, if the framework is to have any validity it must be able to integrate these two scales of analysis since the same organisational process operates on both scales.

CONCLUSION

What has been presented here is a tentative framework for understanding the organisational context of technology transfer. The essential propositions of this framework are that unequal power relationships, within and between organisations and within and between national economies, define a terrain of organisational inequality and asymmetry. When the market in information is imperfect, technological leaders necessarily attempt to internalise their technologies to maximise the returns they can extract. Under these conditions, the transfer of technology across this organisational terrain then appears to occur in an orderly and predictable fashion. The nature of this transfer has been described at both inter-regional and international scales.

From their peripheral positions in this organisational terrain, it is small wonder that countries like Malaysia complain of the obsolescence of the technology they receive and the high prices they pay for it (Abdul, 1984; Sheng, 1984). However, it has been too easy to ignore the organisational context of technological change and technology transfer. The framework developed here suggests simple and tenable hypotheses on the relationship between organisational structure and technological change, which are consistent with existing fragmentary evidence.

REFERENCES

Abdul, A. R. (1984) 'Joint Ventures Between Malaysian Public Corporations and Foreign Enterprises: An Evaluation', in I. L. Lim and P. L. Chee (eds) *The Malaysian Economy at the Crossroads*, Malaysian Economic Association and Organisational Resources, Kuala Lumpur.

Aldrich, H. E. (1972) 'Technology and Organisational Structure: A Re-examination of the Findings of the Aston Group', *Administrative Science Quarterly*, 17, 26-43.

Baranson, J. (1978) *Technology and the Multinationals: Corporate Strategies in a Changing World Economy*, D. C. Heath and Co., Lexington, Mass.

Beardsley, G. and Mansfield, E. (1978) 'A Note on the Accuracy of Industrial Forecasts of the Profitability of New Products and Processes', *Journal of Business*, 51, 127-35.

Benson, J. K. (1975) 'The Interorganisational Network as a Political Economy', *Administrative Science Quarterly*, 20, 229-48.

Caves, R. E. (1982a) 'Multinational Enterprise and Technology Transfer', in A. M. Rugman (ed.) *New Theories of the Multinational Enterprise*, Croom Helm, London.

———— (1982b) *Multinational Enterprise*, Cambridge University Press, Cambridge, England.

Clarke, I. M. (1984) 'The Spatial Organisation of Corporations: A Case

Study of the Multinational Chemicals Industry with Specific Reference to ICI', unpublished Ph.D. thesis, Australian National University, Canberra.

Clegg, S. and Dunkerley, D. (1980) *Organisation, Class and Control*, Routledge and Kegan Paul, London.

Coase, R. H. (1937) 'The Nature of the Firm', *Economica*, 386-405.

Coughlin, C. C. (1983) 'The Relationship Between Foreign Ownership and Technology Transfer', *Journal of Comparative Economics*, 7, 400-14.

Dill, W. R. (1958) 'Environment as an Influence on Managerial Autonomy', *Administrative Science Quarterly*, 2, 409-43.

Gibbs, D. C. and Edwards, A. (1983) 'Some Preliminary Evidence for the Interregional Diffusion of Selected Process Innovations', in A. Gillespie (ed.) *Technological Change and Regional Development*, Pion, London.

Hower, R. M. and Lorsch, J. W. (1967) 'Organisational Inputs', in J. A. Seiler (ed.) *Systems Analysis and Organisational Behaviour*, Irwin and Dorsey, Homewood, Illinois.

Johns, B. (1983) 'Australia', in D. J. Storey (ed.) *The Small Firm: An International Survey*, Croom Helm, London.

Kroner, M. (1980) 'US International Transactions in Royalties and Fees: 1967-78', *Survey of Current Business*, January, 29-35.

Levine, J. H. (1972) 'The Sphere of Influence', *American Sociological Review*, 56, 777-87.

Malecki, E. (1980a) 'Technological Change: British and American Research Themes', *Area*, 12, 253-9.

——— (1980b) 'Corporate Organistion of R & D and the Location of Technological Activities', *Regional Studies*, 14, 219-34.

Mansfield, E. and Romeo, A. (1980) 'Technology Transfer to Overseas Subsidiaries by US-Based Firms', *Quarterly Journal of Economics*, 92, 737-50.

McCulloch, R. (1985) 'US Direct Foreign Investment and Trade: Theories, Trends and Public-Policy Issues', in A. Erdilek (ed.) *Multinationals as Mutual Invaders: Intra Industry Direct Foreign Investment*, Croom Helm, Beckenham.

McDermott, P. J. and Taylor, M. J. (1982) *Industrial Organisation and Location*, Cambridge University Press, Cambridge, England.

Mindlin, S. and Aldrich, H. E. (1975) 'Interorganisational Dependence: A Review of the Concept and Re-examination of the Findings of the Aston Group', *Administrative Science Quarterly*, 20, 382-92.

Negandhi, A. R. and Reimann, B. C. (1973) 'Task Environment, Decentralisation and Organisational Effectiveness', *Human Relations*, 26, 203-14.

Oakey, R. (1984) 'Innovation and Regional Growth in Small High Technology Firms: Evidence from Britain and the USA', *Regional Studies*, 18, 237-51.

Parry, T. G. and Watson, J. F. (1979) 'Technological Flows and Foreign Investment in the Australian Manufacturing Sector', *Australian Economic Papers*, 18, 103-18.

Penning, J. M. (1975) 'The Relevance of the Structural Contingency Model of Organisational Effectiveness', *Administrative Science Quarterly*, 20, 393-410.

Pfeffer, J. and Salancik, G. R. (1978) *The External Control of Organisations: A Resource Dependence Perspective*, Harper and Row, London.

Robinson, J. N. (1981) 'Is It Possible to Assess Country Risk?', *The Banker*, 131 (659), 71-9.

Rugman, A. M. (1982) 'Internalisation and Non-Equity Forms of International Involvement', in A. M. Rugman (ed.) *New Theories of the Multinational Enterprise*, Croom Helm, London.

Sheng, A. (1984) 'Growth of the Malaysian Economy—Some Issues for the Eighties', in L. L. Lim and P. L. Chee (eds) *The Malaysian Economy at the Crossroads*, Malaysian Economic Association and Organisational Resources, Kuala Lumpur.

Taylor, M. J. (1983a) *Business Organisation Segmentation and the Functioning of the Fiji Economy*, Working Paper 32, Central Planning Office, Fiji Employment and Development Mission, Suva.

——— (1983b) 'Technological Change and the Segmented Economy', in A. Gillespie (ed.) *Technological Change and Regional Development*, Pion, London.

——— and Kissling, C. C. (1983) 'Resource Dependence, Power Networks and the Airline System of the South Pacific', *Regional Studies*, 17, 237-50.

——— and Thrift, N. J. (eds) (1982) *The Geography of Multinationals*, Croom Helm, London.

——— and ——— (1983) 'Business Organisation Segmentation and Location', *Regional Studies*, 17(6), 445-65.

Teece, D. J. (1977) 'Technology Transfer by Multinational Firms: The Resource Cost of Transferring Technological Know-How', *Economic Journal*, 87, 242-61.

Telesio, P. (1979) *Technology, Licensing and Multinational Enterprise*, Praeger, New York.

Thompson, J. D. (1967) *Organisations in Action*, McGraw-Hill, New York.

Thwaites, A. T. (1978) 'Technological Change, Mobile Plants and Regional Development', *Regional Studies*, 12, 445-61.

——— (1983) 'The Employment Implications of Technological Change in a Regional Context', in A. Gillespie (ed.) *Technological Change and*

Regional Development, Pion, London.

Tornqvist, G. (1970) *Contact Systems and Regional Development*, Lund Studies in Geography, Series B, Human Geography, 35, Gleerup, Lund.

Williamson, O. E. (1975) *Markets and Hierarchies: Analysis and Antitrust Implications: A Study of the Economics of Internal Organisation*, Free Press and Macmillan, New York.

Yonkers, P. J. (1982) *Transfer Pricing and Performance Evaluation in Multinational Corporations*, Praeger, New York.

Chapter 15

INNOVATION AND URBAN CHANGE

R. Funck and J. Kowalski

The main economic characteristics of current technological developments can be summarised as follows:

1. the share of research and development, management and other information content in the total input of industries increases as a result of changes in the relevant production processes,
2. consequently, and at the same time reflecting shifts in final demand, a higher share of services and other products with high information content in real national product occurs,
3. a substitution of heavy by light materials (e.g. of steel by plastics), and
4. a reduced input of material as a consequence of better construction or quality of materials takes place.

These tendencies will probably continue or even increase in the years ahead.

Based on wage differentials between the labour markets in highly-developed and in less-developed countries, combined with the results of international development policies and the effects of the related restructuring efforts in developing countries, heavyweight commodity production (iron, steel, metals, shipbuilding, etc.) have increasingly moved away from highly-industrialised to threshold countries.

In the highly-industrialised countries, the spatially relevant features of these developments are that the weight-per-unit-of-real-value ratio (real-value-per-unit-of-weight ratio) of many commodities has decreased (increased) drastically (and the same applies to the average unit of real national product).

As a result, classical location factors have been rapidly declining in importance, which means that:

1. production tends to become less dependent on material-oriented, transport cost-minimising locations;
2. the requirements on the commodity transport sector have clearly changed, and are increasingly focusing on small, fast, high-precision shipments of logistic quality (Funck, 1977; Eckstein, 1985);
3. labour markets which are able to supply a sufficient volume and broad variety of specialised labour are becoming highly important elements in location decisions; and
4. old industrial centres have become (more or less) obsolete.

At the same time, the high-technology, management, R & D-oriented sectors increasingly require a high-quality tie-in with the existing or newly developing passenger transport and communications systems. Additionally they require a high degree of urban quality as determined by the supply of consumer-oriented infrastructure services, sufficient to attract the necessary labour as well as supply of specialised industrial services providing the necessary economic environment.

Economic and thus spatial relevance requires that the technological developments are being implemented in innovative processes. Based on Schumpeter's classification, the following processes can be distinguished:

1. introduction of new commodities or services;
2. implementation of new technical procedures in production (technical progress);
3. disclosing of new resources or resource locations; and
4. finding new markets or demand strata.

Usually, the real-life processes of innovation will be a mixture of these characteristics. New production procedures, as a rule, will implicate the use of newly developed machinery, and as a result of reduced prices (or an increased benefit-cost ratio), new demand will open up.

SPATIAL ASPECTS OF INNOVATION

A distinct shift of viewpoints on the role of technological information in urban and regional development has occurred in recent years. While, during the seventies, the economic-technological content of information, the degree of access to it and the spatial patterns of information flow have been predominant as objects of research, in the early eighties the advent of new communication technologies became the centre of analytical interest (Funck and Kowalski, 1984).

Under the first approach, spatial access to innovation has been recognised as the most important factor for enabling or facilitating technology-based activities as a prerequisite for regional development, especially in backward, lagging regions. The peak of this type of approach to the role of information in regional development may be epitomised by the formulation of the 'innovation-oriented regional policy' and related proposals (Ewers and Wettmann, 1979, 1980; Kowalski 1980).

In contrast to this concept of indirect effects of information—through innovation in production techniques—on regional development, the second approach visualises a more direct impact of new communication technologies on locational requirements of production and on mobility patterns. At the same time, fears and anxieties concerning the negative impacts of these technologies come to the fore. The ease with which the friction of distance may be overcome with the help of these technologies is held to constitute danger for jobs and for the traditional role of cities, without leading by itself to a diminishing of the differences in the development levels between core and peripheral areas.

Concerning the first approach it must be noted that peripheral regions are indeed handicapped with respect to the presence of technologically-advanced, innovation-oriented activities. To a large extent this is due to the absence of or difficulty of access to innovation-relevant information. (For a comprehensive review concerning the Anglo-Saxon world see Malecki, 1983.)

In the German context three groups of studies may be considered to belong to the innovation-information point of view on regional policy:

1. Studies of the effects of large corporations on regional economic structure.
2. Propositions concerning innovation-oriented regional policies.
3. Studies on and proposals for decentralisation of public employment.

Let us consider these various aspects.

The first group of studies stems mainly from Bade (1979, 1980, 1981, 1984), and draws conceptually on the Swedish and English research concerning the spatial structure of organisations. Bade's main objective was to enquire whether the large, multiplant corporations which, on average, grow much faster than the rest of the firms in the Federal Republic of Germany and concentrate strongly through mergers, thus occupying a large share of production, sales and job opportunities in the economy, exert detrimental effects on the economic structure of the peripheral, lagging regions. If they do, it is mainly through spatial division of functional control and production, which in turn is supposed to result in a far greater sensibility of employment in the lagging regions to the economic cycle, and also in a deterioration of the innovation-adoption potential, of the innovative climate and generally in bad 'psychological industrial relations' in these areas. More recent studies (Bade, 1981, 1984) show the disparity between rural, lagging regions and the core regions in regional-functional specialisation. This means that the lagging regions are generally devoid of certain activities, and that, should some enterprise decide to locate workplaces there, they will most probably be in the routine, low-level category.

The proposals for innovation-oriented regional policies (Ewers and Wettmann, 1979, 1980; Giese, this volume) also draw conceptually on the writings of Törnqvist (1970, 1973) and other Swedish scholars. The main idea put forward is that the traditional regional policy measures such as tax reduction, low-interest loans, direct subsidies, etc., aimed at supporting the functioning of firms in lagging regions tend to create an artificial economic environment and artificial market conditions. If for these firms it is difficult to survive without support but possible to function with subsidy, an adjustment by the firms to the artificial conditions will follow. They then tend to remain less competitive than firms in other areas which do not obtain support. The authors argue for a policy aimed at increasing competitiveness of the medium and small firms which dominate the economy of the lagging regions. One way to achieve this goal is to increase the share of products which are new, which are a result of innovation. The major obstacle for innovative activities is the lack of highly-specialised technological and other information, of qualified personnel, and of a risk-oriented economic environment in peripheral areas.

The final group of studies deals with problems of location and relocation of public authorities of various kinds (Friedrich *et al.*,1984; Friedrich and Wonnemann, 1985; Funck and Blum, 1984; Ganser, 1979). The rationale behind the relocation of public-sector employment focused on the fact that workplaces in the public sector are mainly permanent, not dependent on the short-term or even long-term economic tendencies. They therefore provide a peripheral lagging region with a stable core of employment which will survive bad times, exerting primary and secondary income effects.

Apart from that, as already discussed above, it can be expected that the overall level of employment in manufacturing activities will decline in the future. This occurs because of closures of old plants and a continually diminishing number of new establishments of manufacturing plants in the Federal Republic of Germany during the last twenty years (see Ganser, 1979, on this point). This in turn leads to the conclusion that the spatial structure of manufacturing employment will, to a large extent, remain stable until the end of the century or beyond, with only small adjustments occurring in a limited number of locations. Thus, the major goal of regional policies—the provision of workplaces and an improvement of the quality of life in peripheral regions—can be realised almost exclusively by means of a relocation of employment in services, specifically in services provided by the public sector. Until now, in the Federal Republic of Germany, no comprehensive program of a relocation of public employees to peripheral regions has been undertaken, at least not comparable to the efforts which have been pursued—and already have sometimes failed—in Sweden, the Netherlands and Great Britain.

The possible spatial and employment impacts of the introduction of new information technologies came to the attention of spatial planners in the Federal Republic of Germany only very recently. In a forecast of the future spatial organisation of the country prepared by Prognos und Bundesforschungsanstalt für Landeskunde und Raumordnung (1983), several scenarios of future employment patterns include different assumptions concerning the growth of productivity of labour, but no spatial consequences of new communication technologies are discussed.

Referring to the long-term effects of these technologies on employment as a whole, no consensus has, so far, been achieved in the literature. Some studies maintain that the overall impact will be neutral or negligible, since the disappearance of workplaces will be compensated by the creation of others (Dostal, 1982). Other authors are more pessimistic in predicting considerable drop in the employment level (e.g. Henckel *et al.*, 1984), while still other reports recognise that the results of forecasts are extremely sensitive to the underlying assumptions, so that very little in the way of hard facts can be achieved (OECD, 1981). Generally, the overall effects of new technologies on society and its functioning can be estimated only in a value-loaded way (OECD, 1983).

It seems obvious that the advent of new communication technologies will lead at least to serious structural changes in the labour market. Many routine jobs will vanish and certainly others will be created. But with respect to the possible overall decrease in the number of workplaces some fears seem exaggerated, at least in a long-term perspective, considering the present demographic tendencies in industrialised societies.

As mentioned above, new technologies also influence locational behaviour. Generally they increase the degree of freedom of locational choice of firms and individual households. At the same time they may widen the attractiveness gap between core and periphery should the latter for some reason remain outside the umbrella of new information-networks. Also, regarding some types of service activities (banking, insurance, etc.), new information technologies—by permitting significantly increased scales of firms, and by an increase in the degree of complexity of operation—may lead to a stronger reliance on locations within the large agglomerations.

In a recent study of the spatial effects of new communication technologies on urban and regional development a team of German authors (Henckel *et al.*, 1984) investigated these issues in a rather speculative manner. They conducted questionnaire analyses and interviews with experts in information technology-dependent enterprises. As the authors state, their conclusions must be treated with caution due to the incomplete data and the particular methodology. They may be summarised as follows (*op. cit.*, pp. 161-62):

1. Economic factors and increased acceptance levels will lead to a relatively rapid diffusion of new information technologies.
2. Advanced information technology does not in itself cause spatial change. However, it can reinforce existing migratory trends.
3. Availability of advanced information technology is a necessary but not a sufficient condition for the acceleration of spatial deconcentration.
4. Spatial effects of information technology are not uni-directional. Both centripetal and centrifugal effects are discernible.
5. No equalisation of development levels between agglomerated areas and the countryside should be expected. In the foreseeable future the spatial effects of new communication technologies will be the strengthening of the advantages of metropolitan areas and local centres as compared to the periphery.

The conclusion can be drawn that the recognition of possible spatial effects of the adoption of new information technologies is a necessary complement to the proposals concerning information-based, innovation-oriented, regional policies. It must be considered that the explosive spread of these technologies, while in principle making it easier to provide information facilities and to enable the processing and transfer of information in practically every location in peripheral areas, will also lead to tensions concerning the viability of the centres of large cities and increase the relative backwardness of the periphery.

At present, a real danger exists that the spread of new information technologies will be directed predominantly to already developed areas whereas the peripheral regions will stay largely devoid of information transfer networks (Hoberg, 1983; Schulz-Trieglaff, 1982). Thus, the introduction of new communication media does not diminish the relevance of regional policy measures including the provision of innovation-relevant information and conditions. On the contrary, the growing importance of information technologies in the reshaping of society will enhance the potential role of regional policy in the future.

SPATIAL INNOVATION POTENTIAL

Determinants of regional innovation potential are manifold, including such factors as the structure of the regional economy as described by the sector, size and control structure of enterprises; the locational specifics of the region as to the quality of its transportation and communication links; the quality and diversity of regional labour markets; the know-how and risk-acceptance of managerial personnel; the availability of risk-capital; the availability of technological information; and the presence and the level of activity of innovation promotion agencies in the public or private sectors and other innovation supporting public programs.

An evaluation of the regional effects of some of these public programmes was part of a research project carried out by the Institut für Systemtechnik und Innovationsforschung (ISI), Karls-ruhe, which tried to identify regional innovation deficiencies (Meyer-Krahmer *et al.*, 1984) based, in particular, on data about the innovation performance and structure of small and medium-sized enterprises which received public grants for the employment of R & D personnel. The most important indicator in this study was the so-called regional innovation intensity, measured by the share of innovating firms (defined as those which received the grant) in the region. Results of the study indicated the following:

1. There are regional differences in innovation intensity. It found a relatively higher share of innovating firms in central, agglomerated areas than in peripheral, rural areas.
2. Innovating small and medium-sized enterprises seem to produce a higher positive contribution to regional employment than firms which do not innovate.
3. Differences in the innovation activities between similar innovating firms in central and peripheral regions are rather small and statistically significant only with respect to a small proportion of the indicators.
4. In general, bottlenecks in the innovation process do not depend on location factors (except in respect of the labour market and research facilities).

Concerning the different factors which determine the innovation potential of a region, studies on the role of relevant information should be regarded. Based on an investigation of indicators of information activities for ninety-one major cities (kreisfreie Städte) in West Germany (Funck and Kowalski, 1984) certain spatial impacts may be expected from these developments with regard to the city hierarchy, the urban and regional structure, less developed regions, and old industrial centres.

In particular, it may be expected that large cities will gain in position in the information hierarchy determined by the intensity of information activities per head while regional centres in lagging regions, and old industrial centres, as a rule, will stay very low in the information hierarchy. Medium-sized and smaller centres, while not showing a clear pattern, will probably perform favourably if the determinants listed above can be supplied. Further, the overall superior position of Southern Germany as against Northern Germany will be maintained if not further ameliorated.

Consequently, the following policy measures may be suggested for peripheral areas. Firstly, an elaboration and extension of services of innovation-advisory units, provision of access to technical data banks and other sources of innovation-relevant information. Secondly, an expansion of institutions of higher learning and research (universities, institutes of technology) in non-centre areas. This should take place in medium-sized cities performing central place roles in the periphery. Of course the establishment of a university or an institute will not by itself necessarily lead to development through an expansion of information dependent, innovative activities. But in the long run the presence of such institutions of higher learning and research changes the intellectual milieu and level of technological know-how in their environment, gradually creating important preconditions for development (Fürst, 1982). Thirdly, the relocation from core to periphery of selected units of public administration. Finally, provision of technology centres (technology parks).

Similar recommendations can be made with respect to old industrial areas. They can usually offer good traffic and communications networks, ample supply of land and building structures for redevelopment towards use as technology centres, and good provision of institutes of higher learning and research. At the same time, however, they often suffer from unattractive residential environments and low-quality, consumer-oriented infrastructure and a surplus supply of labour inadequately skilled for high-technology employment. Thus, under the requirements of technological change, the need is for adequate re-structuring in industrial (sectoral) structure, size structure of enterprises (to medium-sized and small-sized), better infrastructure supply in old centres and medium-small-sized centres, and tie-in with high-quality transport and communications networks.

Finally we suggest that although the advent of the era of high technology and universal communication heralds to some extent the decline of the traditional regional economic policies stressing accessibility variables, regional planning will retain its importance. True, it will have to re-orient its emphasis. The diminishing importance of physical accessibility as a location variable means that the specific features of a given place as a location for a concrete activity will come to the fore. Regional planning may help to create this specific quality, for example by improving cultural-recreational features of a location, or by provision of specialised services, high quality personnel, etc. (Bökemann, 1982). Also, specific measures aimed directly at promoting technological innovation will constitute a very important part of planning activities.

REFERENCES

Bade, F.-J. (1979) 'Funktionale Aspekte der Regionalen Wirtschaftsstruktur' (Functional Aspects of Regional Economic Structure), *Raumforschung und Raumordnung*, 6, 259-68.

——— (1980) 'On the Influence of Large Corporations on Regional Economic Structure', Discussion Paper, International Institute of Management, Berlin, (mimeo).

——— (1981) 'Zweiter Zwischenbericht zum Forschungsprojekt MFPRS 78/19 Konzerninterne Standortstrategien' (Second Interim Report on the Research Project MFPRS 78/19 Intra-Corporation Location Strategies), International Institute of Management, Berlin, (mimeo).

——— (1984) 'Die Funktionale Struktur der Wirtschaft und ihre räumliche Arbeitsteilung' (The Functional Structure of the Economy and its Spatial Division of Labour), International Institute of Management, Berlin.

Bökemann, D. (1982) *Theorie der Raumordnung, Regionalwissenschaftliche Grundlage für die Stadt-, Regional- und Landesplanung* (Theory of Spatial Planning), Oldenbourg Verlag, München, Wien.

Dostal, W. (1982) 'Fünf Jahre Mikroelektronik Diskussion' (Five Years of Discussion on Microelectronics)', in *Mitteilungen aus Arbeitsmarkt- und Berufsforschung*, 2.

Eckstein, W. E. (1985) 'Kooperative Systembildung in der Transportwirtschaft' (Co-operative Systems in Transportation), in R. Funck (ed.) *Karlsruher Beiträge zur Wirtschaftspolitik und Wirtschaftsforschung*, von Loeper Verlag, Karlsruhe.

Ewers, H.-J, and Wettmann, R. W. (1979) *Innovationsorientierte Regionalpolitik* (Innovation-Oriented Regional Policy), Schriftenreihe des Bundesministers für Raumordnung, Bauwesen und Stätebau, Bonn-Bad, Godesberg.

——— and ——— (1980) 'Innovation Oriented Regional Policies', *Regional Studies*, 14, 161-80.

Friedrich, P., Buckel, E. and Wonnemann, H. G. (1984) *Erfahrungsbericht über Behördenver-lagerung vorwiegend in Europäischen Ländern unter Berücksichtigung der für den Freistaat Bayern verwertbaren Erkenntnisse* (Report on the Experiences with Administration Relocation, Mainly in European Countries, with Consideration of Knowledge Applicable for the Free State of Bavaria), Bayerisches Staatsministerium für Landesentwicklung und Umweltfragen, München.

——— and Wonnemann, H. G. (1985) *Financial Effects of Establishing a Public Enterprise*, Discussion Paper 10/1985, Lehrstuhl für Finanzwissen-schaft, Universität Bamberg, Bamberg.

Funck, R. H. (1977) 'Zukunft des Güterverkehrs—Angebot und Nachfrage, Investitionen, Wettbewerb' (The Future of Commodity Transport: Supply and Demand, Investment, Competition), *Schriftenreihe B der DVWG4*, 41.

——— and Blum, U. (1984) 'Zentralörtlicher Status, Verkehrsinfrastruktur und regionales Einkommen' (Central-Place Status, Transport Infrastructure and Regional Income), in H.-J. Ewers and H. Schuster (eds) *Probleme der Ordnungs- und Strukturpolitik*, Vandenhoeck und Ruprecht, Göttingen.

——— and Kowalski, J. S. (1984) 'The Role of Information in Regional and Urban Development', in *IARUS4*, 11th Meeting, Copenhagen.

Fürst, D. (1982) *Informationswirkungen einer Hochschule auf ihre Region* (Information Impacts of a University on its Region), Diskussionsbeitrag 2/1982, Universität Konstanz, Sozialwissenschaftliche Fakultät, Konstanz.

Ganser, K. (1979) 'Zur Dezentralisierung öffentlicher Arbeitsplätze', *Informationen zur Raumentwicklung*, 5, 257-72.

Henckel, D., Nopper, E. and Rauch, N. (1984) *Informationstechnologie und Stadtentwicklung*, Kohlhammer Verlag, Stuttgart.

Hoberg, R. (1983) 'Raumwirksamkeit neuer Kommunikationstechniken—innovations-und diffusionsorientierte Untersuchungen am Beispiel des Landes Baden-Württemberg' (Spatial Impacts of New Communication Techniques—Innovation and Diffusion Oriented Analyses for the State of Baden-Württemberg), *Raumforschung und Raumordnung*, 5-6, 211-22.

Kowalski, J. S. (1980) 'Development Pole Theory in the Perspective of Studies of the Contacts of Central, Sectoral and Regional Systems', in A. Kuklinski (ed.) *Regional Dynamics of Socio-Economic Change*, Finnpublishers, Tampere.

Malecki, E. (1983) 'Technology and Regional Development. A Survey', *International Regional Science Review*, 8(2), 89-126.

Meyer-Krahmer, F., Dittschar-Bischoff, R., Gudrum, U. and Kuntze, U. (1984) *Erfassung Regionaler Innovationsdefizite* (Understanding and Measuring Regional Innovation Deficits), Schriftenreihe 06 "Raumordnung" des Bundesministers für Raumordnung, Bauwesen und Städtebau, 06.054, Bonn.

OECD (1981) *Micro-electronics, Productivity and Employment*, OECD, Paris.

——— (1983) *Assessing the Impacts of Technology on Society*, OECD, Paris.

Prognos und Bundesforschungsanstalt für Landeskunde und Raumordnung (1983) *Kurzfassung der Raumordnungsprognose 1995* (A Concise Version of the Forecast of Spatial Organization in 1995), Prognos BfLR, Basel, Bonn.

Schulz-Trieglaff, M. (1982) 'Können neue Medien einen Beitrag zum Ausgleich von Entwicklungsunterschieden einzelner Regionen leisten?' (New Media, Can They Contribute to an Equalization of Development Levels of Regions?), *Informationen zur Raumentwicklung*, 3, 249-50.

Törnqvist, G. (1970) *Contact Systems and Regional Development*, Lund Studies in Geography, Series B, Human Geography, 35, Gleerup, Lund.

——— (1973) *Contact Requirements and Travel Facilities—Contact Models of Sweden and Regional Development Alternatives in the Future*, Lund Studies in Geography, Series B, Human Geography, 38, Gleerup, Lund.

Chapter 16

THE DEMAND FOR INNOVATION-ORIENTED REGIONAL POLICY IN THE FEDERAL REPUBLIC OF GERMANY: ORIGINS, AIMS, POLICY TOOLS AND PROSPECTS OF REALISATION

E. Giese

After the economic miracle of the fifties and sixties, the economy of the Federal Republic of Germany for the last fifteen years has been characterised by stunted growth and even temporary stagnation. The reasons for this are firstly, that market saturation has occurred in the area of standardised goods destined for mass consumption. The marketing potential of standardised, technical consumer goods (refrigerators, washing machines, etc.) is limited. At the same time a shift in demand towards high-quality products arose. Supply structures have not fully adjusted to this change in demand. Secondly, changes in the international distribution of labour have made themselves felt through increased competition. The increasing costs of wages, raw materials and environment policies in the industrial countries have enabled the developing countries to compete more favourably in the markets for standardised products. This competitive pressure bears very heavily on those firms in industrial countries who employ a high percentage of low-skilled labour. In order to meet this competitive challenge, more and more investments were directed towards plant rationalisation rather than expansion (Rothwell, 1982, p. 363).

From a regional point of view, this change in economic growth threatens existing employment in lagging regions, as an excessively high proportion of lower-skilled jobs are found there (Bundesminister für Raumordnung, Bauwesen und Städtebau, 1980, p.14; Giese and Nipper, 1984, p. 203; Ewers and Wettmann, 1980, p. 162). However, the 1965

West German law pertaining to regional organisation commits the Federal Government to a regional policy of developing similar economic, social and cultural conditions throughout the country and to the provision of equal living and employment opportunities in lagging regions. This commitment not only includes the improvement of traffic and public utility facilities, but also the provision of an adequate number of jobs at a reasonable distance from the homes of prospective employees.

In conformance with this aim, regional policy in the fifties and sixties was concerned with the quantitative growth of employment opportunities in lagging regions. The success of this policy was attributed to inter-regional capital mobility which was particularly widespread at that time owing to the shortage of space and manpower in the metropolitan areas. It was only necessary to take this existing trend towards capital mobility, intensify it and then direct it by way of incentives towards certain regions (Bundesminister für Raumordnung, Bauwesen und Städtebau, 1980, p. 14; Giese and Nipper, 1984, p. 203). The critical factor in regional competitiveness was attributed to costs created by peripheral locations. Thus, aid was primarily directed at lowering production-related costs.

The vehicle for doing so was, and still remains, the Gemeinschaftsaufgabe zur Verbesserung der regionalen Wirtschaftsstruktur (GRW), a Federal-State Agreement for the Improvement of Regional Economic Structures. This program aids the establishment or relocation of companies in the lagging regions of West Germany (see Figure 16.1). The assistance is primarily in the form of investment allowance and the development of industry-related infrastructure. This strategy can only be successful if the type of production involved is cost-sensitive, has few special requirements as far as location is concerned, and cannot be set up in metropolitan areas on account of the higher production costs (location and labour) (Bundesminister für Raumordnung, Bauwesen und Städtebau, 1980, p. 14; Ewers and Wettmann, 1978, p. 468).

The result of this regional policy was the establishment and relocation of subsidiary firms, in particular, shrinking or stagnating branches of industry requiring a large amount of low-skilled labour and standardised production. Such establishment may be regarded as economic buffer zones which can be set up as extended workbenches in times of prosperity and closed down when demand is on the wane (Flore, 1978, p. 506; Giese and Nipper, 1984, p. 203; Glaubitz and Priewe, 1976, p. 735). For this reason, subsidiaries seldom have their own marketing or management departments, so decisions are taken outside the region and connections to the local market and other companies in the area are limited (Thwaites, 1978, p. 458). Due to changes in the economic situation, the mobility of capital and, tied with it, the inter-regional movement of companies, decreased, factors which had been the instruments for carrying out traditional regional policies.

Figure 16.1 Metropolitan Areas and Assisted Regions in the Federal Republic of Germany, 1984. **Source:** Gatzweiler and Runge (1984).

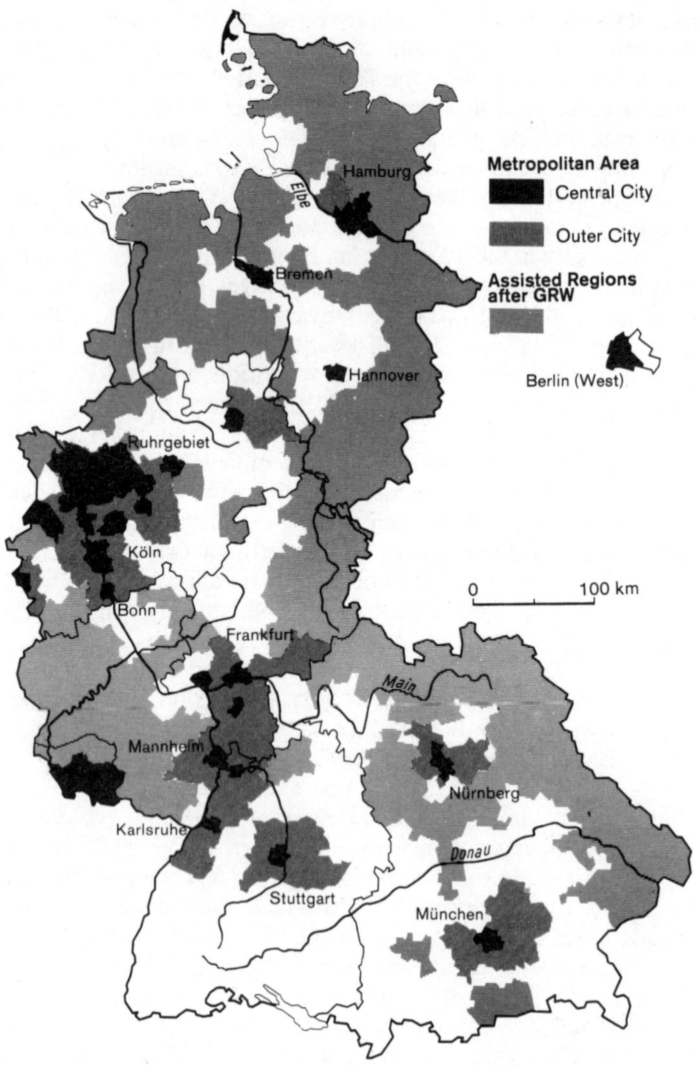

As the more competitive sectors of industrial countries enter into the production of skilled, labour-intensive products, new jobs must be created in this sector by lagging regions in order to compensate for the loss of former markets by winning over new ones. This is only feasible for a region if sufficient manpower and scientific and technological know-how are available (i.e. regional innovation potential). The inadequacy of such factors creates a bottleneck (Bundesminister für Raumordnung, Bauwesen und Städtebau, 1980, p. 17; Ewers and Wettmann, 1978, p. 470).

THE CONCEPT OF INNOVATION-ORIENTATED REGIONAL POLICY

New concepts in regional politics have come to the fore in order to overcome this bottleneck in the restructuring of the marketplace. Of these, innovation policy is the most important. This is because the opening up of new markets and the rapid creation and assimilation of technological innovations are crucial factors for the growth and survival of many industries (Ewers and Wettmann, 1980, p. 162). Research results can only be of economic value when they are put to a practical purpose (i.e. the basis of an innovation which is subsequently diffused). If one assumes that these innovations are aimed at industry and not at the household sector, then the companies themselves must be tools of innovation and thus targets for a newly-conceived regional policy (Thwaites, 1978, p. 449). In this regard there have been calls for a re-orientation of traditional regional politics for some years now. This new concept is based on the idea that regional policy must concentrate on companies already located in the region, that is, on the endogenous potential. The emphasis on endogenous potential arises from the evidence that a shift in production towards labour-intensive products is taking place in metropolitan areas. Due to the fact that such firms are not expected to relocate from the metropolitan areas (Bundesminister für Raumordnung, Bauwesen und Städtebau, 1980, p. 16), the lagging regions become even more disadvantaged.

The strategy of encouraging innovative investments in new technology and markets in lagging regions is known as innovation-orientated regional policy (Bundesminister für Raumordnung, Bauwesen and Städtebau, 1980, p. 57; Ewers and Wettmann, 1980, p. 165). It is geared to initiating the modernisation of economic structures in the problem regions by stimulating research, development and innovation, particularly in those small and medium-sized companies which are predominant in such areas. In this way, it should then be possible to bring about positive results in certain areas (Bruder, 1980, p. 235). Innovation-orientated regional policy is aimed at the individual companies, because they themselves are responsible both for qualitative changes in the production program and for the quantity and quality of the jobs they provide.

In any discussion of the promotion of endogenous potential in a region, consideration must be given wherever possible to companies producing high-technology goods. For one thing, they satisfy the demand for labour-intensive production. The rate of growth of turnover, profits and jobs also tends to be high.

As before, the task of regional policy in West Germany is still to reduce disparities between the regions. The crucial difference between traditional and innovation-orientated aid to the regions is that the latter attempts to create long-term, secure employment and greater regional growth rates through the provision of more desirable job opportunities in an effort to increase the total number of jobs.

POTENTIAL TOOLS FOR INNOVATION-ORIENTED REGIONAL POLICY

At this point, the question arises as to which tools are available for regional policy of this type. This section will attempt to describe those which are available and to outline their present use in West Germany. They will be judged by the following parameters:

- fiscal measures (tax measures, appropriation of capital, government contracting),
- infrastructure (technology-transfer organisations, venture centres), and
- human capital (education and university policy).

Tax Measures

The degree of solvency of a company can be increased by concessionary tax measures (capital), which in turn should enable the firm to invest more (Bräunling and Harmsen, 1975, p. 139). As tax measures may be applied to various kinds of investment (e.g. R & D investments) and also with a degree of regional differentiation, in principle, the possibility of their being used within the framework of innovation-orientated regional policy is feasible.

In West Germany one can categorise the various forms of tax concessions as being (a) branch-specific, (b) specific to the various regions, and (c) innovation-orientated. Branch-specific measures are employed to back up endangered industries (e.g. special allowances in the coal and ore-mining industries). This course of action is, in fact, exactly the opposite of an innovation-orientated policy. Moreover, although the various types of regionally-specific tax concessions are aimed at the problem areas, they actually promote any kind of investment. This practice has its origins in the traditional regional policy (e.g. the investment allowance law as specified in the GRW). The only type of tax concession which may be classed as innovation-orientated is an allowance for R & D investment on personnel, regardless of regional disparities (Bundesminister für Forschung und Technologie, 1981, p. 97).

Appropriation of Capital

The government has a wide range of choices as far as appropriation of capital for innovative purposes is concerned. These range from granting credit at low rates of interest and providing surety and guarantees, to the shouldering of full responsibility for a firm. The end result of any measure is a greater degree of solvency and a reduction of the R & D risk for companies, some of which may be achieved through a regional component. Capital appropriation in West Germany, as is the case with tax measures, generally tend not to be regionally orientated.

Innovation-orientated promotion is aimed at economic targets whether or not they are related to regional goals. Thus, the question of how worthwhile it is to promote an innovation as the first step of an innovation promotional program is decided by the product's prospects of economic success. The government-backed Deutsche Wagnisfinanzierungsgesellschaft mbH (Venture Financing Co.) specifies that a prototype be submitted and that the product be brought onto the market within two years before they agree to participation in the project. It further guarantees its own security by concentrating on already successful companies.

Government Contracting

Research projects can be directly initiated and promoted by government contracting. The main difference between this type of assistance and appropriation of capital is that besides financial aid, product demand assured by the government reduces the market risk for a company.

Government contracting not only helps large, established companies, but also encourages the formation of new, innovative enterprises, either directly or by way of subcontracting (Eiff and Wagner, 1984, p. 5; Mansfield, 1968, p. 225; Roberts and Wainer, 1968, p. 81). It is becoming obvious that government contracting is an effective vehicle for innovation promotion (Bollinger *et al.*, 1983, p. 9). However, in West Germany it is not used in accordance with regional policy aims. The problem is that government R & D contracts are allocated by the West German Chamber of Commerce, the Department for Research and Development and the Ministry of Defence and that no adequate channels exist between them and the Department for Building and Construction which could be used for project co-ordination.

Technology-Transfer Organisations

In order to encourage the spread and use of new technology, more and more government or government-backed technology-transfer organisations are being set up. Among the tasks which these infrastructural elements of innovation promotion have been set up to perform are arranging contacts between potential buyers and suppliers of technology and facilitating the adaptation of new technological developments.

West Germany has a wide range of these transfer organisations. One type is solely concerned with the provision of information. They prepare material on topics such as nuclear technology (Zentralstelle für Atomenergie-Dokumentation, the Nuclear Power Documentation Centre) or patents for government-backed institutions (Arbeitsgruppe Patentverwertung, a study group for patent exploitation). This documentation is then made available to interested parties. Moreover, advisory schemes are run by several institutions in the larger German cities (the Chambers of Industry and Commerce, the engineering union). The most comprehensive service is provided by the Rationalisierungskuratorium der Deutschen Wirtschaft e.V. (a controlling board concerned with rationalisation) which, besides giving advice and guidance, organises management training, negotiates sources of capital and even undertakes the running of whole departments (e.g. accountancy) of a company (Eickhof, 1982, p. 252; Rupp, 1976, p. 81; Steigerwald, 1980, p. 85). As is the case with the above-mentioned policy tools, the use of technology-transfer organisations is not directed by regional policy.

Venture Centres

Venture centres constitute a further vehicle for the promotion of innovations. They are similar to centres such as the Advanced Technology Center in Atlanta, which was founded on the campus of the Georgia Institute of Technology and which represent a form of incubator for high-tech enterprises (Hort, 1985, p. 10). These centres are usually housed in empty factory buildings and the individual rooms let to newly-formed, technology-based companies for a small fee. Close relations are maintained to nearby universities which offer the use of their equipment, laboratories, computers and may also be called upon for advice in matters of industrial management. Furthermore, the centres themselves often provide office services for communal use so as to further reduce overheads for the individual venture companies. Once a company has survived its innovative phase with the help of these measures, it must abandon the centre and may set up in business in a neighbouring industrial park, an area especially designed for just such a purpose (Knauer and Meffert, 1984, p. 113; Lindlar, 1984, p. 35; UNI, 1984, p. 22).

The three main aims of the venture centres are to promote closer co-operation between research institutes and industrial firms, to facilitate the setting-up of new companies, and to create jobs. Because the setting up of venture centres is the responsibility of local authorities, this is a trend which is sweeping the whole of West Germany. These endeavours are further stimulated by assistance programmes in the States of Lower Saxony, North Rhine-Westphalia and Baden-Württemberg. In the face of this development, one must ask if this network of venture centres over the whole Federal area is such a good thing. So far, no estimates have been made as to the potential of innovative venture companies and subsequently it is likely that, if all existing projects are realised, more centres than are actually necessary will be set up whose capacities will then not be fully exploited. This situation could develop to a stage where there is keen competition between the various local authorities to attract venture companies. One must also ask if it is such good policy for all the communities which have expressed an interest in such centres to actually set them up. It is most likely the case that the potential of venture companies and technology available at any one location will not always be sufficient to run such an installation.

Furthermore, it seems that venture centres have concentrated their efforts too exclusively on micro-electronics and are not exploring other areas of technology with good development prospects. Nor is much known about the conditions under which such venture companies best flourish.

Educational Measures

Educational measures can indirectly influence future labour potential, therefore they can also be implemented in order to enhance a region by providing it with a pool of skilled employees. In West Germany the situation varies greatly from State to State, as each State is responsible for its own education policies. In the present section we consider only the universities, the last step on the ladder of the process of creating human capital.

The role of universities as tools for innovation-orientated economic policy has been increasingly a matter for discussion since the end of the seventies. They have been allocated the task of stocking up both the pool of labour and of new technology available in a region. They are expected to carry out more applied research in order to facilitate contacts with industry and consequently the implementation of research results for a practical purpose. Besides ensuring the transfer of information relevant to innovation, it is their job to train students for highly-qualified jobs. Such people are in great demand for research, development and innovation. Thus, those regions which have differentiated labour markets at their disposal are very popular site locations for high-tech industries.

The importance of universities for innovation-orientated regional policy is greatly over-estimated in West Germany. A large number of them can hardly be described as being in a position to increase the flow of technology into the regions, on account of their traditional range of courses and primary emphasis on basic research. Apart from the role which every university plays in improving the supply of future employees, one can say that only technical colleges and technically-orientated polytechnics can really be expected to improve the flow of technology at this level. Aachen possesses the largest and most prestigious technical college in the Federal Republic of Germany. One would expect the Aachen region to be the home of a large number of high-tech enterprises. Unfortunately, this is not the case.

The government would like to see more co-operation between science and industry. This is desirable for the purpose of feeding more new technology into industry faster than ever before. The field of research-subcontracting, that is, research carried out under contract to certain companies, should also be developed and intensified. As a result, it is hoped that the innovative process will be stimulated and that more third-party grants will be channelled into the universities so that they will be able to extend their research activities beyond the present level. An attempt is also being made to induce university professors to become more practical as to the type of work they do in order to provide an incentive for collaboration with industry and to encourage competition between the various universities and individual lecturers.

Recently, an array of so-called transfer agencies have been set up at various universities to develop contacts with industry. Their purpose is to intensify the transfer of know-how from universities to industry. It seems that it is becoming a question of prestige to have such an agency. In October 1984, forty-four universities and polytechnics already had one at their disposal (Blohm, 1984). They function as intermediaries: a company looking for a partner with whom to collaborate in a certain project contacts the agency, who in turn puts it in touch with the type of researcher it requires. The target group for this kind of transfer negotiation consists mainly of medium-sized businesses.

FUTURE PROSPECTS FOR INNOVATION ORIENTATED REGIONAL POLICIES

What are the prospects for innovation-orientated regional policies in West Germany? Are steps being taken to promote such a development?

It is the small and medium-sized businesses which determine economic growth and employment opportunities in lagging regions of West Germany (see Figure 16.1) and so it is here (<500 employees; less than DM 200 million turnover) that the government is aiming at improving the level of innovation. Small-sized and medium-sized firms do not usually have any self-contained R & D departments of their own which are independent of the production process. It is also quite difficult for them to gain access to R & D data and to recruit highly-skilled employees. Many R & D projects are ultimately abandoned on account of the high financial risk they incur. As far as the costs involved are concerned, investment in an R & D project constitutes a much greater risk for a small or medium-sized firm.

Thus, an attempt is being made to improve the R & D performance of this type of company by means of a whole range of measures. A summary of them can be found in the 1979 report entitled 'An Overall Concept of Political Measures taken by the Government of the Federal Republic of Germany for the Promotion of Research and Technology in Small- and Medium-Sized Companies'. The main points to note are as follows: a program for the subsidisation of personnel costs; encouragement of external research sub-contracting; and encouragement of the setting-up of technology and innovation transfer agencies.

Both government and local authorities attach great importance to the latter. The idea is that they should activate existing potential for new technology and modernization in small-sized and medium-sized companies and help to realise innovation projects. It is, however, doubtful if this can successfully combat the disparity in regional development in West Germany, as one must bear in mind the following facts:

1. Financial assistance afforded to small- and medium-sized firms amounted to DM 0.8 billion in 1981. This is only a fraction of the sum spent by the government on R & D each year. In 1981 a total of DM 41.5 billion was spent on R & D projects. Fifty-four per cent of this sum was provided by private enterprise and 43 per cent by the government (see Figure 16.2), the latter figure revealing the extent of influence that government exercises upon innovation projects in West Germany.
2. Eighty per cent of the research institutes financed by government and local authorities (universities, polytechnics, etc.) are situated, as can be seen in Figure 16.3, in the metropolitan areas. This is also almost exclusively the case with the large-scale enterprises which run their own research departments (90 per cent; Giese and Nipper, 1984, p. 210). Therefore it is not surprising that by far the largest share of the R & D funds is channelled into the metropolitan areas of West Germany.
3. The practice of distributing the funds in this way is intensified by the current development practices of the largest public financier in civil research, i.e. the Ministry for Research and Technology (West Germany—BMFT). Only projects which are considered to have a

Figure 16.2 Financing of R & D Expenses in the Federal Republic of Germany, 1981.

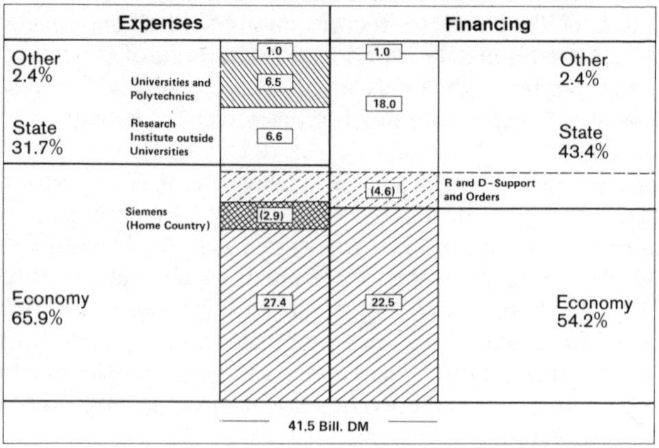

Expenses
Financing
Other 2.4%
Universities and Polytechnics
State 31.7%
Research Institute outside Universities
Siemens (Home Country)
Economy 65.9%
1.0
6.5
6.6
(2.9)
27.4
1.0
18.0
(4.6)
22.5
Other 2.4%
State 43.4%
R and D-Support and Orders
Economy 54.2%
41.5 Bill. DM

good chance of success are chosen for funding by this Ministry (which totalled a sum of DM 2.5 billion in 1983). Basically, this means that funds are not generally made available to regional areas. Projects are only promoted when adequate capacities and initiatives already exist (BMFT, 1979, p. 151). Thus, the partners of the BMFT are 'mainly firms which, on the basis of their organisational facilities and previous achievements, can be expected to pursue a project to its successful conclusion' (Hetzler *et al.*, 1978, p. 59). It is obvious from this that established, large-scale businesses will primarily profit from these funds.

So, from what we have seen so far, it seems clear that the aims of regional policy in West Germany play a minor role in such R & D funding. The degree of co-ordination between regional aims and those of other policies is limited. Thus, innovation-orientated regional policy becomes an appendage of the political system, which, in turn, seems to be incapable of counteracting and balancing out present development in an area's economic structure. The policy tools listed above certainly could be employed as part of innovation-orientated policy in order to exercise an influence on innovative activities, although their actual effect is over-estimated by some. In practice, however, they are only implemented as an object of fiscal policy for general innovation promotion.

Figure 16.3 Location of Research Institutes in the Federal Republic of Germany. **Source:** Atlas zur Raumentwicklung, Bundes-forschungsansalt für Landeskunde u. Raumordnung, Bonn ca. 1981, Karte 2.07.1.

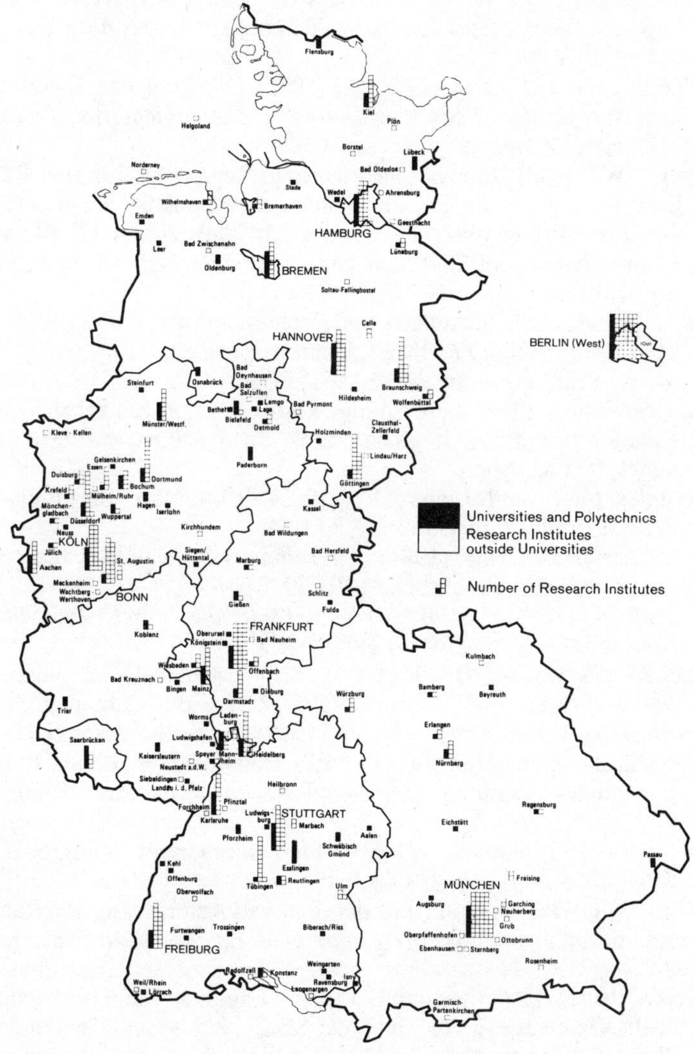

REFERENCES

Blohm, H. (1984) 'Das Personaltransferprogramm der TU Berlin', in *Bundesminister für Bildung and Wissenschaft (Hrsg.)*, Hochschule und Wirtschaft, Bonn.

Bollinger, L., Hope, K. and Utterback, J. M. (1983) 'A Review of Literature and Hypotheses on New Technology-Based Firms', *Research Policy*, 12, 1-14.

Bräunling, G. and Harmsen, D.-M. (1975) *Die Förderungsprinzipien und Instrumente der Forschungs—und Technologiepolitik*, Verlag Otto Schwartz, Göttingen.

Bruder, W. (1980) 'Innovationsorientierte Regionalpolitik und Räumliche Entwicklungspotentiale—zur Raumbedeutsamkeit der Forschungs— und Technologiepolitik des Bundes', in W. Bruder and T. Ellwein (eds) *Raumordnung und Staatliche Steuerungsfähigkeit*, Westdeutscher Verlag, Opladen.

Bundesminister für Forschung und Technologie (BMFT) (1979) *Bundesbericht Forschung VI*, BMFT, Bonn.

——— (1981): *Förderfibel*, BMFT, Bonn.

Bundesminister für Raumordnung, Bauwesen und Städtebau (1980) 'Innovationsorientierte Regionalpolitik' in *Schriftenreihe 06 'Raumordnung'*, 06.042, Bonn.

Commes, M.-T. and Lienert, R. (1983) 'Controlling im F&E-Bereich' *Zeitschrift für Organisation*, 7, 347-54.

Deutscher Bundestag (1984) *Dreizehnter Rahmenplan der Gemeinschaftsaufgabe*, Drucksache 10/1279, Bonn.

Eickhof, N. (1982) *Strukturkrisenbekämpfung durch Innovation und Kooperation*, Verlag Mohr, Tübingen.

Eiff, R. and Wagner, H.-Th. (1984) 'Kann Man in Deutschland "Silicon Valleys" Aufbauen?', *Blick durch die Wirtschaft*, 13 June, p. 5.

Ewers, H.-J. and Wettmann, R. W. (1978) 'Innovationsorientierte Regionalpolitik—Überlegungen zu Einem Regionalstrukturellen Politik- und Forschungsprogramm', *Informationen zur Raumentwicklung*, 7, 467-83.

——— and ——— (1980) 'Innovation-oriented Regional Policy', *Regional Studies*, 14, 161-79.

Flore, G. (1978) 'Instrumente der Innovationsförderung im Rahmen der Raumordnungspolitik', *Informationen zur Raumentwicklung*, 7, 503-14.

Gatzweiler, H.-P. and Runge, L. (1984) 'Laufende Raumbeobachtung: Aktuelle Daten zur Entwicklung der Städte, Kreise und Gemeinden 1984', *Bundesforschungsanstalt für Landeskunde und Raumordnung*, 17, Bonn.

Giese, E. and Nipper, J. (1984) 'Die Bedeutung von Innovation und Diffusion neuer Technologien für die Regionalpolitik', *Erdkunde*, 38, 202-15.

Glaubitz, J. and Priewe, J. (1976) 'Effizienzanalyse der Gemeinschaftsaufgabe "Verbesserung der Regionalen Wirtschaftsstruktur"', *WSI-Mitteilungen*, 29(12), 732-43.

Hetzler, H. W., Müller, V. and Schienstock, G. (1978) 'Staatliche Innovationspolitik' in *IFO-Institut für Wirtschaftsforschung (Hrsg.) Der Innovationsprozess in Westeuropäischen Industrieländern*, 4, Berlin.

Hort, P. (1985) 'Die Unternehmensbrüter von Atlanta', *FAZ*, 15 July, p. 10.

Knauer, S. and Meffert, G. (1984) 'Pakt mit der Zukunft', *Stern*, 18 April, 108-18.

Lindlar, K. P. (1984) 'Technologiepark—Technologietransfer', *Chip*, 4, 32-5.

Mansfield, E. (1968) *The Economics of Technological Change*, Norton, New York.

Roberts, E. B. and Wainer, H. A. (1968) 'New Enterprises on Route 128', *Science Journal*, 4(12), 78-83.

Rothwell, R. (1982) 'The Role of Technology in Industrial Change: Implications for Regional Policy', *Regional Studies*, 16(5), 361-9.

Rupp, E. (1976) *Technologietransfer als Instrument Staatlicher Innovationsförderung*, Verlag Otto Schwartz, Göttingen.

Steigerwald, H. J. (1980) 'Das RKW als Projektträger im Technologie-Transfer' in E. Staudt (ed.) *Innovationsförderung und Technologie-Transfer*, Verlag Erich Schmidt, Berlin.

Thwaites, A. T. (1978) 'Technological Change, Mobile Plants and Regional Development', *Regional Studies*, 12(4), 445-61.

UNI (1984) 'Gründerzentren—American Dream', *UNI-Berufswahlmagazin*, 9/84, 22-3.

Chapter 17

KNOWLEDGE CENTRES, INFORMATION DIFFUSION AND REGIONAL DEVELOPMENT

P. Nijkamp and A. Mouwen

In recent years, much attention has been focused on the long-term pattern of structural changes and economic fluctuations (Kondratieff cycles, for example), at both the national and the regional or urban levels. Various publications demonstrate that (lack of) innovation is regarded as one of the driving forces behind the mechanism of economic growth and decline. This implies that the economic perspective of a city or region is dependent on the innovative activities within the area concerned. On the other hand, the presence of a highly-qualified production environment in a city or region may favour the development potential of the area concerned through its incubator function (the so-called 'breeding place hypothesis').

It should be added, however, that empirical research in the USA (Malecki, 1979) has demonstrated that many large cities—especially the big metropolises—have lost their innovation potential with respect to large-scale firms, while medium-sized towns exhibit a boom process due to the design and use of new technologies. Consequently, spatial dynamics and innovation are closely interwoven phenomena.

A key element in innovative efforts is creativity. Creativity is the ability to generate and to structure information in such a way that it favours a synergistic process of knowledge and skill in order to create a surplus value of insights (higher-order knowledge). Stress and structural instability are some of the driving forces for creativity (the so-called 'depression-trigger hypothesis'). Such a creativity requires a high level of skill and education, a diversity of disciplines, a wide variety of internal and external communication channels, and a discrepancy between available means and socio-economic needs. In light of the important role of creativity in stimulating innovation and hence regional development, it is no surprise that many regions have adopted a strategy of establishing so-called knowledge centres in order to assist the innovation and growth potential of the region concerned. The present chapter will provide a critical review of this knowledge centre concept by examining its theoretical foundations, its potential as a strategic regional-policy tool, and its practical use in various countries. In the second part of the chapter some empirical evidence regarding the Dutch geographical distribution of knowledge centres and the regional innovation potential will be provided, followed by some policy conclusions.

KNOWLEDGE CENTRES

Knowledge centres make up part of the R & D infrastructure of a country or region. They may be defined as a spatial concentration of scientific and technological communication, information and research centres for both the public and private sector (*inter alia* science parks, innovation centres, transfer centres). Consequently, insofar as innovation policies are focused on an improvement or enhancement of R & D efforts (in both the private and the public sector), the creation of knowledge centres is of utmost importance.

The geographical location of knowledge centres has in recent years become an issue of important scientific research. It is often taken for granted that technological innovations, especially product innovations, are strongly favoured by an urban production environment (e.g., Ewers and Wettmann, 1980). Large cities and metropolitan areas appear to provide a favourable breeding place due to scale and agglomeration economies in such areas. It is conceivable therefore that innovation policies are often oriented towards an improvement of the incubator function of urban areas. In this respect, there is a strong parallel between current innovation policies and former growth-pole policies, which also took for granted the existence of centrifugal growth impulses. It is worth noting, however, that various very large cities are losing their contribution to the regional innovation potential and to the restructuring of the industrial sector in favour of medium-sized cities (see, among others, Cross, 1981; Segal, 1983; and Wever, 1984). Therefore the question arises whether large or medium-sized towns are promising locations for knowledge centres which serve to provide the necessary R & D infrastructure for technological change in a certain area.

The latter question can only be answered if more insight is available into the geographical orientation of innovative firms with respect to external R & D facilities (including knowledge centres). In the context of this chapter we will limit ourselves to the industrial sector (a sector where high-technology firms appear to exhibit a high degree of innovativeness and production growth; Premus, 1982). Following Doody and Muntser (1981) we define the high-technology sector as a cluster of firms which show a high growth rate, a high amount of R & D expenditures, a high value added, a strong export orientation, and a labour-intensive production technology (especially as far as high-skilled labour is concerned).

The question regarding the geographical location of knowledge centres is closely aligned to the economic and technological function in a large spatial system: knowledge centres serve to induce, to encapsulate and to transfer innovations. In the initial phase of innovation generation, a spatial concentration and an orientation toward urban centres may favour innovation, as intensity of information and communication is a necessary condition for innovation potential. In the second phase, where inventions have to be operationalised as feasible and viable innovations warranting a commercial production of new commodities or a commercial introduction of new (production or management) processes, more emphasis is being placed on internal entrepreneurial adjustment processes. In the last phase, however, the position in a communication network is of crucial importance. Whether this desired position leads to a concentration in large cities or in medium-sized cities depends on a trade-off of agglomeration economies *versus* scale diseconomies (for example, congestion). The modern information and communication technology in particular has shaped the conditions for an urban sprawl of innovative firms outside the traditional metropolitan areas (see also Brotchie *et al.*, 1985).

In conclusion, the locational pattern of knowledge centres deserves a careful analysis. There is *a priori* no guarantee that innovative firms are mainly oriented toward knowledge centres in their close vicinity. It is now of course an interesting research question whether the desired geographical pattern of R & D facilities—seen from the viewpoint of innovative entrepreneurs—corresponds to the actual spatial dispersion of knowledge centres.

KNOWLEDGE CENTRES AND REGIONAL DEVELOPMENT

In this section the impact of knowledge centres upon regional development will be discussed in more detail. Particular attention will be paid to the question of whether the availability of a highly-skilled labour pool exerts a positive influence upon locational decisions of firms in the high-technology sector. The results of some noteworthy studies in this context will be presented.

Premus (1982) analysed the locational factors of high-tech firms in the US. He arrived at the conclusion that the quality and availability of labour, the wage level, and local taxes (insofar as they influenced the propensity of technical and scientific personnel to migrate) were major determining factors for locational decisions of high-tech firms. Due to a shortage of technicians, scientists and engineers in some regional labour markets, various high-tech firms appeared to attach a very high priority to the costs and availability of human capital. In this respect, universities played a major role as knowledge centres for the high-tech industry. Surprisingly enough, traditional locational factors (such as transportation and energy costs) appeared to play only a minor role in locational decisions of high-tech firms. It is clear that a joint orientation of high-tech firms toward a scientific and research environment implies a concentration of such firms in regions with a favourable R & D infrastructure (including knowledge centres).

Malecki (1980) also came to the conclusion that the regional availability of a highly-qualified labourforce is an important factor in the locational choice of an R & D division of a multi-plant organisation. As this labourforce attaches a high priority to urban amenities (cultural, social, educational and scientific facilities), the urban environment provides a favourable source for skilled labour. Consequently, there is a strong tendency for R & D activities to locate in urban areas.

Cross (1981) studied the location factors of new high-tech industries in Scotland by means of micro-economic choice analysis. He identified the following determinants: the location of the original entrepreneur, the location of the former employer, the location where the new entrepreneur was educated, the location of the most important sales market, and the local environment as a whole. In particular, the local environment and the historically-derived geographical orientation (*inter alia*, place of residence or education) appeared to provide a statistically significant explanation.

Cnossen and Koerhuis (1982) made an attempt at identifying the locational factors associated with computer and software firms in the Netherlands. It turned out, on the basis of detailed questionnaires, that these high-tech firms attached much importance to accessibility via the transportation network and the availability of parking space, due to the high knowledge intensity and communication orientation of these enterprises. Given the geographical scale of the Netherlands, no specific regional or local sales orientation could be identified. Furthermore, these firms also regarded the availability of knowledge potential of universities, the size of available industrial areas, and the presence of firms in the same sector as highly important and risk-reducing locational conditions. Finally, given the required high skill of most employees in this sector, a pleasant residential climate received a high priority.

The conclusion from the abovementioned studies is that the availability of knowledge centres (especially research divisions of universities) is a locational determinant of major significance for the high-tech industry in many countries. In a recent OECD report it is also concluded that the presence of universities in a region is of crucial importance for the socio-economic development of the region concerned, as universities create a production environment that stimulates the high-tech industry through its high-skilled labourforce, through its potential attraction force exerted on skilled labour from other areas, and indirectly through its facilities such as libraries, etc. (OECD, 1984). Furthermore, universities also have spin-off effects in terms of the creation of highly-qualified production activities directly or indirectly associated with and induced by the universities concerned (see also Cross, 1981; and De Jong, 1983). A good example of such spin-off effects can be found in the Greater Boston area, where the establishment of innovative commercial firms is often generated by academic research. In this respect, the incubator function of a metropolitan area is very much favoured by the presence of highly-qualified academic research institutes.

Therefore, a successful knowledge centre strategy as part of a regional development policy with respect to innovative firms requires the fulfilment of the following conditions (cf. De Jong, 1983):

- presence of research institutes that may act as breeding places,
- presence of a highly-skilled labourforce,
- public support for R & D activities of youthful innovative firms,
- availability of venture capital,
- presence of a stimulating and innovative entrepreneurial climate.

The abovementioned analyses suggest that large metropolitan areas are providing the most favourable breeding place for innovations, but this is not always true. Premus (1982), for example, observes that the sunbelt areas in the USA are exhibiting bottleneck factors (such as high labour costs, high land rents, lack of industrial space, high local taxes and traffic congestion), so that there is a tendency among new high-tech firms to move away from the sunbelt States to the Midwest. Cross (1981) also shows that places around bigger cities attract more high-tech firms than cities themselves. Similarly, Wever (1984) finds for the Netherlands a similar pattern: the net number of new firms in the ring of intermediate areas around the (industrialised) Dutch Randstad (the Rimcity) is higher than in the Randstad itself (and also higher than in the peripheral areas). A similar observation was made by Hoogteijling (1984), who found that computer and software firms had a strong preference for a location in the intermediate zone around the Randstad.

The increased orientation of innovative new firms toward accessible areas outside large agglomerations may mainly be due to congestion effects. It is worth noting, however, that the migration of an existing firm may, in addition to bottleneck factors, also be explained from its phase and position in a product cycle. During the design stage of a new product, an intensive communication with R & D institutes (including knowledge centres and universities) is necessary, leading to a geographical concentration of innovative activities in urban areas; such a location is often risk reducing and cost saving. As soon as the production of new commodities has to be implemented, however, other locational factors (market orientation, for example) may become important, so that in this phase of product development the firm may be willing to move to more peripheral regions offering a cheaper location. Thus the locations of new-product firms are co-determined by their product cycle stage.

KNOWLEDGE CENTRES: A COMPARATIVE ANALYSIS

The concept of knowledge centres assumes a wide variety of forms among different countries, including science parks, innovation poles or transfer centres. In the framework of a concise exploratory comparative analysis, some experiences from the USA, Great Britain, Germany, Sweden and Canada will briefly be described.

USA

The idea of knowledge centres has a long history in the USA, with the establishment in 1951 of the Stanford Industrial Park in Palo Alto, California (based on military defence and R & D contracts). Three successful examples of the knowledge centre concept will be described here, followed by a brief survey of alternative knowledge transfer concepts.

Silicon Valley as an offspring of the Stanford Industrial Park is the classical example of an integrated, rapidly-growing knowledge centre based *inter alia* on advanced computer and telecommunication technology. Its commercial basis, its position as the nucleus of a communication network and its efficient organisational structure shaped the conditions for a successful regional development pole.

Route 128 in the Greater Boston area has many elements in common with Silicon Valley. The rapid development of this area induced by military and aircraft R & D expenditures and later by many high-tech firms, is a clear spin-off effect of the Boston-Cambridge scientific climate. Its continued success has been mainly due to the innovation potential in this area which was caused by its favourable sectoral structure, its diversified labour market and its efficient institutional structure.

The Research Triangle of North Carolina may be regarded as a spin-off result of the universities of Raleigh, Durham and Chapel Hill. It started as a purely academic research area, but later it also encompassed new production activities. Nowadays it contains many R & D divisions of private enterprise (such as IBM), while it continues to attract high-skilled labour. The presence of academic research institutes and the favourable social climate were mainly responsible for the success of this knowledge centre.

In addition to these well-known examples of knowledge centres, various alternative initiatives in the USA may be mentioned:

- University-Industry Co-operative Research Centres (encouraged by the Federal Government), which serve to stimulate co-operation between academic research institutes with a specific know-how and industry.
- Small Business Innovation Programs (supported by the National Science Foundation), which aim at stimulating a commercial operational development of innovative ideas.
- Innovation Centres (supported by the Federal Government and based on co-operation between private industry and universities), which serve as training institutes for entrepreneurs, as breeding places for innovations, and as a consultancy centre for inventors.
- Small Business Development Centres (supported by local governments), which aim at providing co-ordinated assistance to the commercial sector and small-scale industries.
- University Industrial Associates (stimulated by the university), which serve to establish a closer formal link between entrepreneurs and research divisions of universities.
- Venture Capital Funds (supported by universities), which serve to stimulate the application of academic knowledge in private industries, so that the risk of introducing new production techniques may be diminished.

Great Britain

In Great Britain much emphasis has been placed in recent years on the development of science parks. In various cases, however, the integration of science parks with the regional or local institutional structure has been overlooked, so that the success of such knowledge centres has not always been overwhelming. A good example, however, is the Cambridge Science Park which will be discussed first, followed by a survey of alternative forms of knowledge transfer between universities and industries.

The Cambridge Science Park is characterised by the following favourable location factors: extensive space for innovative industries, favourable living conditions, potential accessibility to scientific research institutes, and availability of good entrepreneurial facilities. Surprisingly, however, the occupants of this science park do not have more intensive relationships with Cambridge University than those R & D firms outside the park. Therefore, it may be concluded that it is not the science park itself, but rather the whole Cambridge area which provides a breeding place for innovative firms (Moore and Spires, 1983; Segal, 1983). Thus the Cambridge area as such fulfils an incubator function characterised by various high-quality social and cultural facilities, the presence of prestigious private and public research institutes and of the university, a diversified production structure, the availability of skilled labour, and a positive attitude by the university with respect to new business activities and entrepreneurship. The support of banks in stimulating new entrepreneurs, an initiating role by local government in providing the necessary physical infrastructure, and the pressure exerted by way of budget cuts on young university-based inventors to start their own innovative activities.

Besides the science-park development discussed above (and similar developments taking place in the London-Reading-Bristol corridor), alternative initiatives with respect to small-scale business firms are taking place in the UK. Examples include:

- Industrial Research Groups, which serve to combine and transfer expertise in the area of technology and entrepreneurial design.
- New Enterprise Workshops, which aim at making available university facilities to entrepreneurs.
- Industrial Science Parks, which provide facilities for an industrial high-tech complex near a university.
- University Companies, which serve to stimulate industrial activities within universities by means of government subsidies.
- The National Research Development Corporation, which aims at transferring new ideas from public institutions to the commercial business sector.
- Wolfson Industrial Units, which focus on applied technological contract research by small-scale R & D centres in universities.

Germany

In Germany various activities in the area of knowledge centre policy are also taking place. The most promising is the Kernforschungszentrum (Core Research Centre) in Karlsruhe. This is a national research institute which aims to develop new technological solutions for large-scale, complex and interdisciplinary problems requiring a large injection of finance, a high-skilled labourforce and long-term planning. In addition, a system of advisory transfer centres for small-scale business activities has been established (so-called Beratungsstellen), as well as an information transfer centre (in Bochum) in order to provide up-to-date insight into new developments in information technology (for a further discussion, see Giese, this volume).

Sweden

Knowledge centre concepts in Sweden are mainly represented in so-called university transfer bureaus, which serve to stimulate the application of new technological inventions in the business sector. Ways to arrive at this goal are through the distribution of professional information, the design of co-operative projects between researchers and firms, facilitating access to university facilities, the introduction of joint R & D projects, etc.

Canada

Canada is also establishing various knowledge centres around universities (in Waterloo, for example). Other initiatives include:

- contracting out, by which means Federal departments spend R & D money in private enterprises,
- the Industrial Research Assistance Program, which serves to cover the salaries of industrial researchers,
- the co-operative projects of the National Research Council with industry.

In summary, it would appear that in the USA, universities appear to provide a fairly direct incubator function for high-tech industry, whereas in Europe, universities appear to provide a more general academic research climate for innovations. (Clearly, it should be kept in mind that the geographical scale of European countries is drastically different from that in the USA.) Furthermore, it is worth noting that a successful knowledge centre also requires a suitable organisational, institutional and physical infrastructure in order to guarantee fruitful co-operation between industry and academic research institutes. It also can be observed that in almost all cases a successful knowledge centre was accompanied by a favourable development of the region at hand. Though cause-effect relationships are difficult to disentangle in this context, it is plausible that a fruitful regional or urban breeding place encompasses additional elements like educational facilities, skilled labour, venture capital, high-tech infrastructure and appropriate public policies. In this respect, a knowledge centre is only one of the desirable conditions for innovative regional development efforts.

It has to be added that, despite the limited role of universities as direct knowledge-transfer centres, universities may provide teaching programs focusing on high-tech industry, so that by means of appropriate investments in human capital the high-tech sector can be stimulated in an indirect manner. Besides, many new innovative firms appear to be more technology-driven than market-oriented, so that university initiatives focusing on technical, financial, marketing and organisational support are an extremely important complement to R & D efforts.

THE SPATIAL DISPERSION OF KNOWLEDGE CENTRES: A CASE STUDY OF THE NETHERLANDS

Knowledge and information are necessary ingredients for innovation. As a consequence, the spatial dispersion of knowledge can be linked to that of innovation. If a direct link would exist between the knowledge potential in a region and the innovation activity in that same region, it would be possible for the government to use the location of knowledge centres as a strategic tool in regional policy. By influencing the location of the public knowledge centres, which make up an important part of the total knowledge potential, the government is able to reinforce the local and regional development potential in lagging regions.

It is therefore interesting to investigate the regional relationship between knowledge centres and innovation activity. In this study on the Netherlands the following categories of institutes and organisations are regarded as part of the Dutch knowledge centre infrastructure:

- departments of universities and institutes of technology as far as they are involved in R & D,
- research institutes which are government owned,
- research institutes which are more than 50 per cent subsidised by the government, and
- the R & D departments of private industry.

For practical reasons it was decided to restrict the study in some respects; special consideration is given to:

- Knowledge transfer to the industry; this implies that only those knowledge centres which provide knowledge transfer to industry are taken into consideration (especially centres that provide technical scientific knowledge).
- R & D divisions of the five biggest Dutch multinationals in the private sector. Recently the outcome of an in-depth analysis showed that more than 70 per cent of the private R & D is concentrated in these five companies.
- The output of the knowledge centres; this is indirectly measured by taking into account the number of R & D employees working in or providing basic support to R & D sections.

In the Netherlands the regional structure is highly dominated by urban agglomerations, so that (socio-)economic processes are often influenced by agglomeration economies. Therefore, the regional classification used in this study is based on these agglomeration economies and takes into account the important role of inter-urban influences. Besides, such a classification is extremely well suited for describing the process of innovation diffusion, which forms an important part of this chapter. Various types of agglomeration regions have been distinguished. These agglomeration regions are composed of villages and towns in a certain area which are oriented towards a large city (or set of cities). This orientation is measured by means of an agglomeration index number (ranging from 1-9). This index number is computed by analysing the size and neighbourhood of other cities, the quality of the communication network, and the relative importance of the biggest metropolises. Figure 17.1 gives a visual representation

Figure 17.1 Agglomeration Indices for the Netherlands.

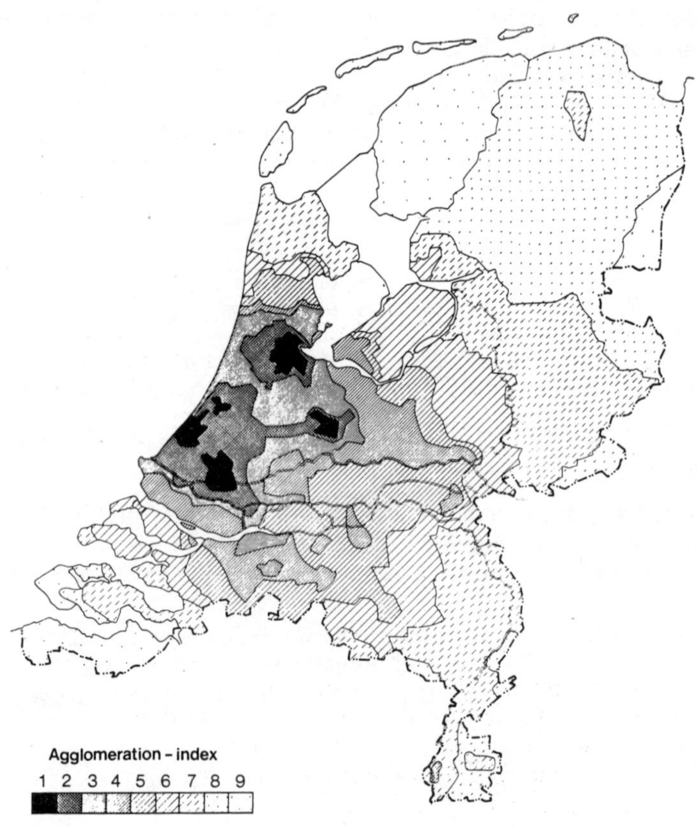

of the results. In the following discussion we will refer to the region numbers which are based on this figure.

Turning now to the actual dispersion of knowledge centres (Table 17.1), it is evident that the knowledge centres of the universities and the technological institutes are primarily concentrated in regions 1 and 2. This is not

Table 17.1: The Regional Dispersion of Knowledge Centres

	Regions									
Knowledge centres (%)	1	2	3	4	5	6	7	8	9	TOTAL
Universities and technology institutes	35.8	27.7	-	15.9	6.4	-	14.2	-	-	(5636)
Non-university research institutes	16.6	31.1	14.8	16.1	0.7	6.7	9.2	4.9	-	(14076)
'Big 5' (multinationals)	28.6	15.8	0.6	11.1	7.6	20.1	3.4	12.8	-	(8847)
TOTAL	24.1	25.7	7.5	14.5	3.9	9.5	8.4	6.4	-	
	(6879)	(7330)	(2132)	(4145)	(1124)	(2724)	(2402)	(1823)	(-)	(28559)

Note: In brackets: absolute figures (number of R & D employees)

surprising because these regions cover a great part of the so-called Randstad, the traditionally highly-urbanised core region of the Netherlands where the four main cities are situated. The regional dispersion of the non-university research institutes shows a diffuse pattern. Although there is also a concentration of these centres in the Rimcity, they are also well represented in the intermediate regions 4 and 7. One finds this kind of research concentrated especially in the zones in the neighbourhood of the big cities but not in these cities.

The regional spread of the R & D of the big five shows a clear concentration in regions 1 and 6. It is remarkable that the regional dispersion of knowledge centres of the big five on the one hand, and those of universities and technology institutes at the other, seem to compensate each other. The geographical pattern of the knowledge centres of the big five multinationals does not exhibit a clear orientation to the regional presence of research departments of universities. The local ties with the parent company seems of far greater importance for the location of most R & D divisions. As a whole, knowledge centres are primarily concentrated in the Rimcity, although the position of the intermediate region is also strong.

In conclusion, knowledge centres are mainly situated on the nodal points of the communication networks (namely the big cities). There is a strong correlation between the ranking of the regions measured in decreasing agglomeration economies and the decreasing presence of knowledge centres. (Kendall's concordance coefficient for these two variables is 0.94).

With the aid of a survey (Kleinknecht, 1985) among industrial entrepreneurs it was possible to measure the use of certain government instruments aimed at stimulating technological innovation (including knowledge centres) with a view to exploring the extent to which knowledge transfer is regionally restricted within the boundaries of the district in which the knowledge centre is situated. The main conclusions concerning the geographical orientation and the impacts of regional boundaries are, that—with the exclusion of TNO, a national research institute with a long tradition of applied natural-science research with a strong regional dispersion—there is little region-specific orientation in the knowledge transfer, so that it is highly improbable that regional boundaries act as a barrier to knowledge transfer. This outcome for an urbanised country such as the Netherlands is not particularly striking because for most of the industrial entrepreneurs in Holland it is possible to reach a specialised knowledge centre within a travel time of approximately two hours. Compared to the benefits of knowledge transfer, such communication costs have only a minor impact. Due to the compact and highly formalised communication structure in the Netherlands (compared to, for instance, the American situation), geographical distance seems to play no significant role in restricting knowledge transfer to industrial companies.

KNOWLEDGE CENTRES AND INNOVATION

Knowledge and information are necessary ingredients for R & D and innovation (e.g. Pred, 1977, and Hägerstrand, 1967). The pattern of the spatial dispersion of knowledge could therefore well be related to the spatial dispersion of R & D and innovation. Two indicators are used as a measure of technological innovation, namely R & D (percentage of companies per region) and innovation (number of realised innovations per company per region). In our view, the R & D indicators are more clearly interpretable and therefore more reliable than innovation indicators.

As can be seen in Table 17.2, only a few companies have exclusively external R & D. Because of the small cell frequencies, conclusions concerning external R & D can only be drawn cautiously. Nonetheless, it is significant that regions 4 and 7 show relatively more companies that spent all of their R & D capital on external contract research. As a whole the R & D pattern shows that the entrepreneurs in region 1 seem not so innovative as would be expected on the basis of their location in the big cities where the innovative climate is usually assumed to be favourable. In

Table 17.2: R & D Indicators

	Regions								
Indicator	1	2	3	4	5	6	7	8	9
Companies with exclusively internal R & D	35.9	37.0	33.0	33.1	44.1	36.6	34.8	37.9	45.9
Companies with exclusively external R & D	3.9	2.2	1.0	4.3	2.9	1.8	6.2	3.1	1.6
Companies with both internal and external R & D	24.8	30.4	31.5	33.9	22.8	28.2	33.5	27.1	26.2
Companies without R & D	35.3	30.4	34.5	28.8	30.1	33.5	25.6	31.9	26.2

all R & D categories, region 1 scores below the average; this results in the highest percentage of companies that have no R & D at all. This means that the core regions are losing the innovation impetus to the intermediate regions 4 and 7, a process that can be observed in many industrialised western countries.

If we compare the regional dispersion of R & D (Table 17.2) with that of the regional dispersion of knowledge centres (Table 17.1), it appears that there is little or no concordance between the two patterns. Concerning external R & D, there is an indication that the few companies per region that exclusively use external R & D have a slight regional preference for using the knowledge centres which are situated in their own region.

Because the abovementioned results seem not totally unambiguous, the analysis has been repeated with a R & D indicator that is corrected for the size of the companies per region. Among others, Malecki (1980) has shown that, on average, big companies invest more in R & D than small companies do. Because the core regions possess relatively larger companies than the peripheral regions, it is possible that the uncorrected R & D indicators show a bias. Corrected for this bias in company size, the results show a clear indication that there exists no positive relation between the regional dispersion of knowledge centres and that of companies who invest in R & D. Relying on the assumption that knowledge is necessary for R & D, the conclusion just reached implies that either the entrepreneurs use knowledge which they get from knowledge centres situated in regions other than where their company is located, or the entrepreneurs perceive the knowledge from the knowledge centres to be unsuitable for their needs, thus forcing them to obtain the desired knowledge from other knowledge sources.

While innovation indicators are not seen to be as strong an indicator as R & D figures, an analysis of innovation indicators may clarify the tentative conclusions reached above. A reasonable innovation indicator that is corrected for the size of the company is the number of innovations (process or product) realised per employee per region. As becomes clear from Table 17.3, the companies in region 4 realise a number of product innovations far above the sample mean. On the other hand, the entrepreneurs

Table 17.3: Innovation Indicators

	Regions									
Indicators (%)	1	2	3	4	5	6	7	8	9	TOTAL
Realised product innovations/ employee	5.10	3.55	4.90	8.39	3.57	3.81	4.73	3.86	5.31	4.80
Realised process innovations/ employee	3.04	2.29	2.45	3.20	4.09	3.28	3.27	3.15	2.72	3.05

in regions 1 and 2 are not as innovative as could be expected. (A similar result was reached by analysing the R & D data.) As far as process innovation data are concerned, the pattern is far less clear. The values of that variable across each region tend to fluctuate more or less around the sample mean. The same distinction between product and process innovations holds true, if the relationship with the regional dispersion of knowledge centres is analysed. The resulting concordance coefficients for product innovations are all approximately 0.5, which implies that the variables 'knowledge centres' and 'product innovations' show no regional correlation. The outcome for process innovations is far less clear because the calculated coefficients vary considerably.

CONCLUSIONS

In the previous sections, which describe the results of a case study for the Netherlands, it has been shown that there is a tendency for knowledge centres to concentrate in regions 1, 2 and 3 (the Rimcity). Despite this result, it should be noted that knowledge centres are also dominant in the intermediate regions (regions 4 and 7).

The main conclusion which has been reached by analysing the regional dispersion of innovative activity in the Netherlands (measured in R & D as well as in innovation terms) is that type 1 regions are not as innovative as might be expected. This result can be seen as a first indication that regional boundaries form no barrier to knowledge transfer. This tentative conclusion concerning the regional direction of the knowledge flows was further elaborated by investigating the spread of innovation activity. Both R & D and innovation indicators showed a regional pattern bearing no relation to that of the dispersion of knowledge centres.

Probably due to the compact and highly-structured communication infrastructure in the Netherlands, a policy aimed at reinforcing the regional potential of lagging regions, by creating new knowledge transfer centres in these regions, is likely to have no substantial impact. Such a policy does not affect the innovative creativity of the entrepreneurs located in these regions. A second and related policy issue is that there seems to be a tendency for the innovative potential of big cities to decline (notwithstanding the fact that the knowledge centre potential in the big cities is enormous). If big cities are to retain a role as a breeding place for new activity, this decline should somehow be stopped. However, it has been indicated that reinforcing the knowledge infrastructure in these cities alone may not be a good policy in attempting to arrest this process of a declining innovation potential.

REFERENCES

Brotchie, J. F., Newton, P. W., Hall, P. and Nijkamp, P. (1985) *The Future of Urban Form. The Impact of New Technology*, Croom Helm, London.

Cnossen, W. and Koerhuis, H. (1982) *De Software-en Computer-servicebedrijven*, Geographical Institute, State University, Groningen.

Cross, M. (1981) *New Firm Formation and Regional Development*, Gower, Farnborough.

De Jong, M. (1983) 'Regionale Condities voor Nieuwe Hoogwaardige Bedrijvigheid: Industriële Vernieuwing in Greater Boston', *Economisch*

Statistische Berichten, 16 November, 1059-63.

Doody, F. J. and Muntser, H. B. (1981) *High Technology Employment in Massachusetts and Selected States*, Massachusetts Division of Employment Security, Boston.

Ewers, H. J. and Wettmann, R. W. (1980) 'Innovation-Oriented Regional Policy', *Regional Studies*, 14, 161-79.

Hägerstrand, T. (1967) *Innovation Diffusion as a Spatial Process*, University of Chicago Press, Chicago.

Hoogteijling, E. M. J. (1984) *Innovatie en Arbeidsmarkt*, Economisch en Sociaal Instituut, Vrije Universiteit, Amsterdam.

Kleinknecht, A. (1985) 'Industriële Innovatie in Nederland', Research Report, State University, Maastricht.

Malecki, E. J. (1979) 'Locational Trends in R & D by Large US Corporations, 1965-1977', *Economic Geography*, 55, 309-23.

——— (1980) 'Corporate Organization of R & D and the Location of Technological Activities', *Regional Studies*, 14(3), 219-34.

Moore, B. and Spires, R. (1983) 'The Experience of the Cambridge Science Park', OECD-Workshop on Research Technology and Regional Policy, Paris, October.

OECD (1984) *Report of the Results of the Workshop on Research, Technology and Regional Policy*, OECD, Paris.

Pred, A. R. (1977) *City-Systems in Advanced Economies*, Hutchinson, London.

Premus, R. (1982) *Location of High Technology Firms and Regional Development*, Study Joint Economic Committee, US Congress, Washington.

Segal, N. S. (1983) 'Universities and Advanced Technology New Firms in Great Britain', International Workshop in the Future of Industrial Liaison, Technical University, Berlin.

Wever, E. (1984) *Nieuwe Bedrijven in Nederland*, Van Gorcum, Assen.

Chapter 18

THE LOCATION OF HIGH-TECHNOLOGY INDUSTRY AND THE TECHNOPOLIS PLAN IN JAPAN

T. Toda

This chapter describes recent trends in Japanese industrial structure and location of high-technology industry and outlines the Technopolis (high-tech city) Plan which has been promoted by the Ministry of International Trade and Industry (MITI, 1985) as a new industrial relocation policy.

The leading industries for regional development in Japan, including steel, chemical and petroleum products, have been capital-intensive and resource-intensive. In the 1960s and 1970s these export-orientated industries developed rapidly; the Japanese economy grew at an annual rate of 10 per cent over this period. About 70 per cent of all industry is concentrated along the Pacific coast in an area which includes Tokyo, Osaka and Kita-Kyushu. This concentration has produced an over-centralisation of the population and urban problems such as a rapid rise in land prices, traffic congestion and housing shortages, especially in the three large metropolitan areas of Tokyo, Nagoya and Osaka. There has been an outflow of the young, male labourforce from smaller local areas to these urbanised areas and this exodus has effectively destroyed the local communities.

Various policies to improve the balance of regional development have been instituted from the viewpoint of effective utilisation of limited land and resources. However, since the mid-1970s the rate of growth of the Japanese economy has declined and a period of low growth has set in. Moreover, there has been a shift from capital-intensive and resource-intensive industries to technology-intensive and processing and assembling industries such as transportation machinery and electrical appliances. The economic differential between metropolitan and local areas which was narrowing in the 1970s is now widening again.

RECENT TRENDS IN INDUSTRIAL STRUCTURE

Figure 18.1 explains the changes in employment ratios by industrial structure (that is, primary, secondary and tertiary) from 1970 to 1984. The

Figure 18.1 Changes in Employment Ratio by Industrial Sector.

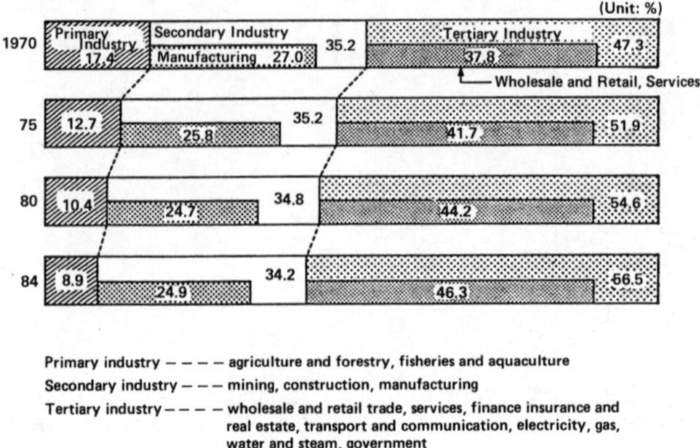

number of people employed in primary industry has almost halved, from 17.4 per cent to 8.9 per cent, in these fifteen years. Secondary industry has also experienced a reduction in the numbers employed, whereas the number of people employed in tertiary industries has greatly increased.

Table 18.1, however, shows that even with the fall in the number employed in secondary industry due to factors such as scale of operation and mechanisation, in terms of real production output the relative importance of manufacturing in total industry has been increasing. This table also indicates that basic material manufacturing such as heavy and chemical industries, which played a crucial part in the era of rapid economic growth, has now been replaced in importance by processing and assembling manufacturing, such as electrical machinery. In general, manufactured goods are now designed to be as thin, light, short and small as possible.

THE DEVELOPMENT OF HIGH-TECHNOLOGY INDUSTRY

High-technology industry is a term used for all industries based on the most advanced technology. This industry includes micro-electronics (computers, word processors), mechatronics (industrial robots, car and medical electronics), fibre optics and communication equipment, bio-

Table 18.1: Changes in Industrial Production and Employment by Sector

	Industrial Production					Employment				
	1970	Share	1980	Share	1970-80 Annual Increase	1970	Share	1980	Share	1970-80 Annual Increase
		100.0		100.0			100.0		100.0	
Manufacturing industry	35,142		69,462		7.1	1,454		1,476		0.2
		(32.7)		(37.7)			(28.8)		(27.0)	
Basic materials		32.4		28.6			22.0		22.3	
heavy & chemical	11,382		19,857		5.7	320		329		0.3
		(10.6)		(10.8)			(6.4)		(6.0)	
		32.0		45.1			32.1		34.3	
Processing & assembling	11,263		31,341		10.8	466		507		0.8
		(10.5)		(17.0)			(9.2)		(9.3)	
		35.6		26.3			45.9		43.4	
Light	12,497		18,264		3.9	668		640		0.4
		(11.6)		(9.9)			(13.2)		(11.7)	

Note: Heavy & Chemical: Chemical, coal, petro, ceramic, metal, metal processing
Assembly: Electric, transportation, micro processing
Light: Food, fibre, pulp and paper
Share: 1975 price million yen, 10,000 persons

technology (drugs and medicine), new materials (fine ceramics, carbon textiles), aerospace (aircraft, artificial satellites), and computer software, social and industrial development systems (medical service, traffic control and information and communication systems). Figure 18.2 shows the actual and expected rates of growth of high-technology industries. There has

Figure 18.2 The Growth of High Technology Industries in the 1980s.

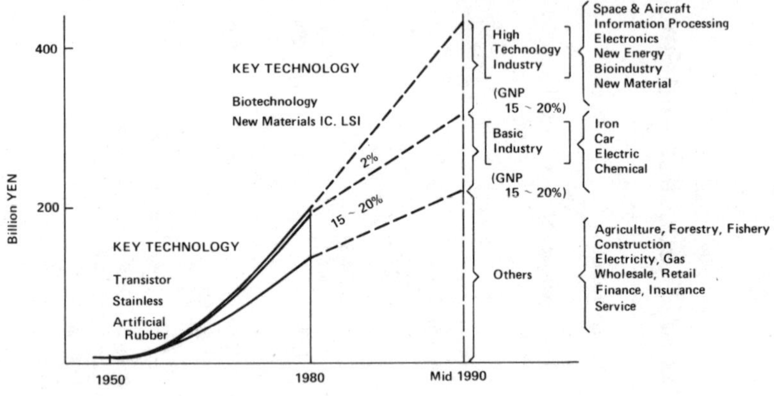

been a remarkable advance in technological innovation centred in the field of electronics, giving expectations for a high level of economic efficiency and creating a demand for industry as a whole. It is also anticipated that there will be significant changes in the fields of information and communications. The frontiers of technical development are likely to be extended in the future, thus the so-called high-technology industry will continue to expand rapidly.

Table 18.2 reveals the allocation of output from high-technology industry to other industrial sectors. The share of micro-electronics is increasing, especially in processing and assembling. The introduction of micro-electronic facilities since 1975 has had a tremendous impact on manufacturing processes and office work. (Table 18.3 shows the rapid increases in R & D expenditure and plant and equipment investment in the integrated circuit (IC) industry.)

The number of high-technology industrial establishments has greatly increased since 1980. Figure 18.3 shows their rate of growth. The rate of establishment of high technology as a proportion of total industry growth has increased from 5 per cent in 1979 to 19.2 per cent in 1984. Table 18.4 provides details of this growth by sector. Together, the number of electronic, communication and parts establishments, combined with communication and related equipment establishments represented more than half of the stock of high-technology industry in the nation.

Locational Patterns

The geographical distribution of factories engaged in high-technology industry is shown in Figure 18.4. In general, most factories are concentrated in and around the Tokyo metropolitan area, which includes Tokyo, Chiba, Saitama and Kanagawa prefectures. There are also relatively minor concentrations in the Osaka and Nagoya metropolitan areas. Except for IC factories, 70-80 per cent of these factories are located not only in the Tokyo metropolitan area, but also in the western island of Kyushu.

Moreover, most R & D activities (78.3 per cent of research institutes are in private companies) are concentrated in the above three metropolitan areas. The location of R & D institutes in Tokyo and Kanagawa prefectures is especially intense, accounting for 37.6 per cent of private research institutes and 38.8 per cent of scientific researchers in Japan.

The Locational Requirements of High-Technology Industry

The locational requirements of high-technology industry are becoming increasingly more difficult to realise. Firstly, accessibility to academic research institutes is important because it is necessary for R & D activities to constantly collect up-to-date information in order to cope with the rapidly evolving technology. Secondly, living conditions are also important and include urban amenities, housing, and a high-quality, natural traditional and cultural environment.

The factors usually influencing the location of industrial activities such

Table 18.2: Plant and Equipment Investment in High Technology Industry

			Plant and Equipment Investment Related to High Technology					
			Micro-electronics		Bio-technology		New materials	
Year	1983	1984	1983	1984	1983	1984	1983	1984
All industrial sectors	18.8	22.4 (31.3)	17.3	20.6	0.2	0.3	0.8	1.0
Manufacturing	23.6	29.6 (46.8)	19.9	25.6	0.4	0.6	1.9	2.1
Basic Materials	14.5	18.2 (35.5)	6.4	8.7	0.7	1.1	4.6	5.5
Textiles	10.7	10.2 (23.1)	-	-	-	-	10.7	10.2
Chemical & Allied products	31.2	30.7 (16.0)	14.3	14.3	1.2	1.1	7.5	7.4
Ceramic, stone and clay products	14.8	19.9 (65.0)	1.6	4.0	-	-	13.2	15.9
Non-ferrous metals and products	14.2	28.3 (127.0)	11.8	23.5	-	-	2.4	4.7
Processing and Assembling	33.1	33.9 (50.6)	32.3	39.2	0.3	0.3	0.0	0.0
Machinery, excl. electric	36.6	43.8 (63.8)	36.6	43.8	-	-	-	-
Electric machinery, equipment and supplies	66.9	71.6 (55.2)	66.9	71.6	-	-	-	-
Transport equipment	12.9	15.2 (28.6)	12.9	15.2	-	-	0.0	0.0
Non-manufacturing	15.4	16.5 (13.5)	15.3	16.5	0.0	0.0	-	0.0

Source: Economic Research Institute of Yamaichi Stock Company 'Monthly Report' (1985)

Note:
1. The numbers refer to the ratio of all plant investment.
2. The numbers in brackets for 1984 refer to the change in rate of investment compared with 1983.

Table 18.3: R & D Expenditure and Plant and Equipment Investment in IC Industry

Year	Sales of IC (A)	Rates against preceding year (%)	R & D Expenditure (B)	Rates against preceding year (%)	B/A (%)	Plant & Equipment Investment (C)	Rates against preceding year (%)	C/A (%)
1975	108,158	129.7	21,524	116.1	19.9	11,379	61.7	10.5
1976	164,924	152.5	24,297	112.9	14.7	35,191	309.3	21.3
1977	155,474	94.3	24,456	100.7	15.7	21,958	62.4	14.1
1978	251,881	162.0	37,997	155.4	15.1	45,932	209.2	18.2
1979	374,910	148.8	54,774	144.2	14.6	84,103	183.1	22.4
1980	547,708	146.1	69,037	126.0	12.6	136,875	162.7	24.9
1981	618,139	112.9	92,179	133.5	14.9	155,449	113.6	25.1
1982	802,616	129.8	124,631	135.2	15.5	223,323	143.7	27.8

Source: Japan Society of Electronic Industry (1984)

Note: Unit: millions yen

as access to markets, transport services, supply of land, water and electricity, security of the labourforce and concentration of related industries also need to be met. However, these basic conditions may be assigned similar weight in the overall location decision-making process as regional differences in these conditions have been shrinking in Japan. Thus the success in attracting high-technology industry to an area depends to a large extent on the two factors, R & D and living conditions.

Figure 18.3 The Growth Rate of High-Technology Industry.

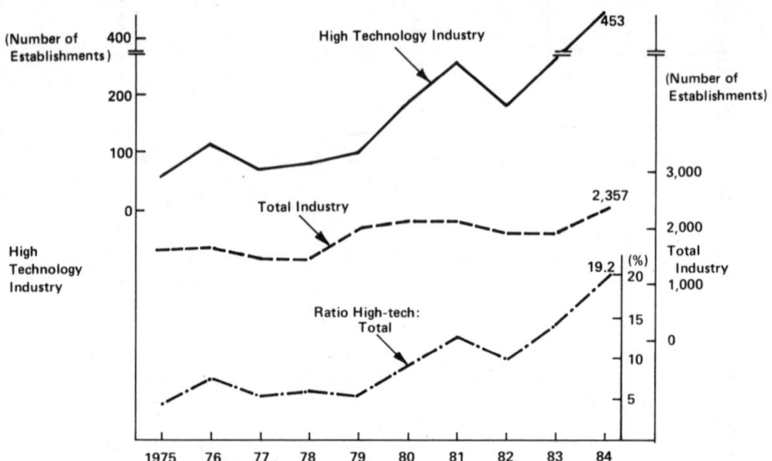

Strictly speaking, the locational needs vary by sector and some companies attach more importance to the arguments put forward by local governments trying to attract particular factories to their areas. However, when the direct relationship between high-technology industry and information and academic research sources is considered, the location of high-technology industry will be largely determined by proximity to large metropolitan areas.

Generally, factories based on high technology can be classified into three types: R & D type, diverse-non-routine/small-lot production type (NRP) and mass production type (MP). The factories of the R & D type normally produce prototype goods and require close contact with head offices and technological and market information in related fields. Therefore, it is more convenient for this type of factory to locate in large metropolitan areas.

NRP-type factories generally produce goods to order which include

Figure 18.4 The Geographical Distribution of High-Technology Factories.

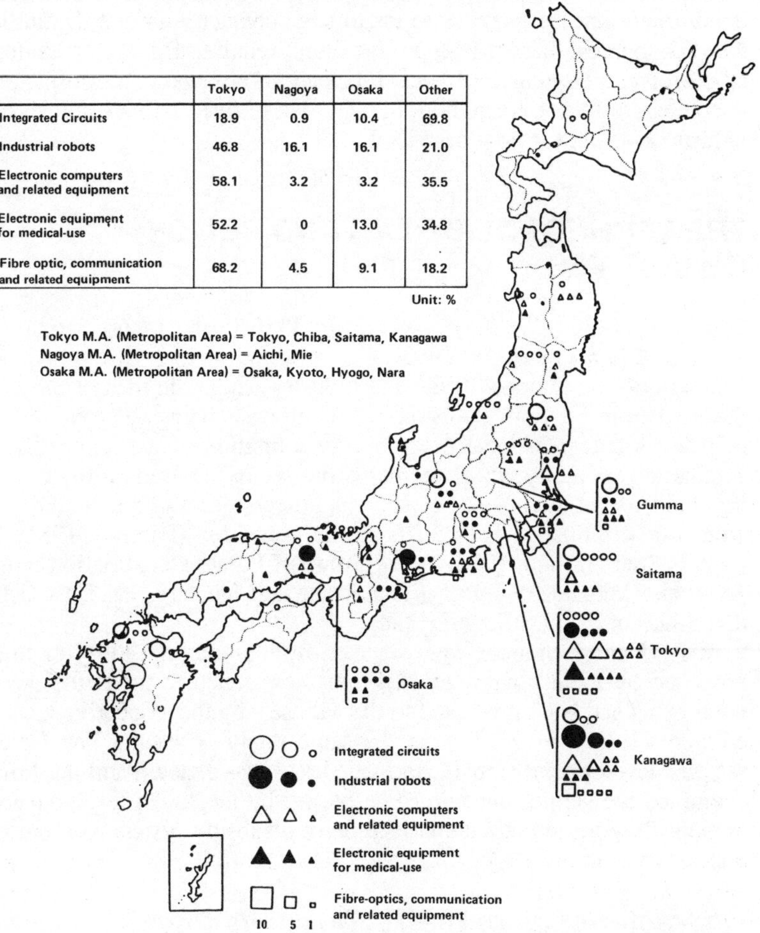

	Tokyo	Nagoya	Osaka	Other
Integrated Circuits	18.9	0.9	10.4	69.8
Industrial robots	46.8	16.1	16.1	21.0
Electronic computers and related equipment	58.1	3.2	3.2	35.5
Electronic equipment for medical-use	52.2	0	13.0	34.8
Fibre optic, communication and related equipment	68.2	4.5	9.1	18.2

Unit: %

such commodities as computers, complex medical electronics, software components and some industrial robots. In the case of production to order, face-to-face information exchange between users and makers is important and the basic research is often carried out by university researchers. The NRP-type factories, therefore, are attracted to large cities where many users are concentrated.

Finally, the MP-type factories produce standard goods including standardised medical electronics, IC, LSI, office computers, word processors, fine chemicals, some industrial robots, fine ceramics and fibre optic equipment. These are mass produced, satisfying the quality standards indicated by the headquarters of each company. Given these requirements, it is not important for these factories to have contact with R & D and information sources. The MP-type factories require a large, high-quality labourforce, a large area of land and clean air and water. These factors induce them to locate in smaller cities. A plan to further focus this high-tech industrial location is now described.

THE TECHNOPOLIS PLAN AND POLICY OBJECTIVES

The Technopolis Plan was proposed in 1980 in the report *International Trade and Industry in the 1980s* after a period of long discussion by the Ministry of International Trade and Industry. The basic idea of this plan is different from previous plans, its main emphasis being to promote economic development in local areas by the formation of a Technopolis. The Technopolis is a city where the industrial sector is based on high-technology industry, an R & D sector such as universities and residential sectors which are established organically in the same area (Hiroino, 1981; Technopolis Study Group, 1983; Study Group of Future Construction Industry, 1984; and Ministry of International Trade and Industry, 1985, for detailed discussion of the Technopolis Plan).

In 1981 nineteen areas were selected from among the thirty-eight areas proposed by local governments. From these nineteen areas the development plans had been presented to the Ministry by the respective local governments before May 1983. In the same month a special law for Technopolis development (the Technopolis Law) was enacted and the Ministry examined the supporting role to be played by the Japan Regional Development Corporation and the utilisation of a subsidy system to promote industrial relocation.

Characteristics of the Technopolis Plan

The Technopolis is developed to combine three functions organically, namely, an industry zone where the production facilities of high-technology industry are built, an academic zone where various facilities for research and education are built, and a habitation zone where living facilities for workers and their families are constructed. The appropriate size of population for a Technopolis is around 50 000 and the minimum land area required is 800-1000 hectares.

A Technopolis should be built close to cities whose population is more than 150,000, which can be reached within thirty minutes, and it should be possible to make day trips to the three large metropolitan areas either by air or by bullet train (Shinkansen). These conditions mean that the Technopolis is not equipped with all urban functions and it must utilise the urban facilities and services in and around neighbouring cities.

Policy Objectives in the Technopolis Plan

There is considerable variety among the Technopolis plans proposed in the nineteen areas. However the industries that were selected to form the basis of these industrial complexes in most plans were footloose industries such as mechatronics, electronics, biotechnology and new material industries which utilise the resources or products of primary industry. Urban-oriented industries such as energy and transport systems were also chosen in many areas.

The characteristics of R & D activities in the Technopolis can be classified into two groups:

1. Transfer type, whose major function is to transfer the products of high-technology industry to various industries in and around the Technopolis. The policy objectives for this activity are reinforcement of R & D functions in factories, differentiation of research functions, enhancement of industrial experiment stations, and utilisation of universities.
2. Frontier type, whose major purpose is to create new technology based on original ideas. This activity is carried out by research institutes of large private companies.

The characteristics of industrial activities can also be classified into two groups:

1. Introducing type, whose major policy objectives are the attraction of high-technology industries, and the establishment of conditions to develop high-technology industries.
2. Fostering type, whose major policy objectives are to make existing industries technology intensive, to promote technical innovation in existing industries, and the fostering of venture businesses.

The target of the Technopolis Plan, therefore, can be classified into four types which have different sets of policy objectives, as outlined in Table 18.5.

Table 18.4: The Rate of Establishment of High-Technology Industries (Number of Establishments per Year)

	1975	1976	1977	1978	1979	1980	1981
All Industrial Sectors	1,487	1,528	1,278	1,353	1,959	2,097	2,091
(1975 = 100.0)	(100.0)	(102.8)	(85.9)	(91.0)	(131.7)	(141.0)	(140.6)
1. Drug and medicines	11	6	19	13	9	20	24
2. Communication and related equipment	20	50	21	23	36	52	76
3. Electronic appliances	4	3	3	6	8	26	27
4. Electric measuring machines	3	3	2	4	3	8	3
5. Electronic and communication equipment and parts	16	38	13	20	30	60	102
6. Medical equipment and parts	2	5	5	6	7	8	5
7. Optical instruments and lenses	3	7	5	8	10	15	23
TOTAL	59	112	68	80	103	189	260
(1975 = 100.0)	(100.0)	(189.8)	(115.3)	(135.6)	(174.6)	(320.3)	(440.7)

CONCLUSION

Recently there have been a number of trends that have made it easier for high technology to locate in local areas. In conclusion, these trends are summarised as follows:

1. The locational constraints on technology-intensive products have been lessened because of a comparative reduction in transport costs. As technology-intensive products have high added value, they can bear high transport costs. For example, in the case of IC products, even if they are transported by air, large profits can still be made because of the time savings, despite the higher costs of transportation. This factor has given rise to new industries located near airports in local areas.
2. The handicap of long distance has been overcome by the development of communications technology. The facsimile, which is now widely used, has made possible the fast transmission of literal information to distant places. In the near future, a comprehensive communications network will be developed which will allow the reference and management of various data and visual information, in addition to the exchange of spoken and written information.
3. The knowledge-intensive and service industries have been growing, and increasing importance has been attached to the role of de-

velopment, information, maintenance and design processes. This tendency will bring about the efficient production of diverse-non-routine/small-lot products which have a special quality of their own. The accessibility, therefore, to a large market in metropolitan areas is not important for the profitability of productive activity.

Table 18.5: The Characteristics of Technopolis

Industry R & D	A. Introducing type	B. Fostering type	Policy Objectives
I. Transfer type	Introducing & transfer type	Fostering & transfer type	a. Reinforcement of R & D function in factories. b. Differentiation of research functions c. Enhancement of industrial experimental stations. d. Utilisation of universities. e. Establishment of corporate development centre.
II. Frontier type	Introducing & frontier type	Fostering & frontier type	a. Establishment of research institutes of private companies. b. Establishment of universities.
Policy Objectives	a. Attraction of high-tech industries. b. Establishment of conditions to introduce high-tech industries.	a. To make existing industry technology-intensive. b. Technical innovation of existing industries. c. Fostering of venture business.	

REFERENCES

Higashi, M. (1985) *White Papers of Japan 1983-84*, The Japan Institute of International Affairs, Editorial Section, Tokyo.

Hiroino, M. (1981) 'Technopolis and Industry; Industrial Location', *Journal of Industrial Location*, 7, 6-17 (in Japanese).

Japan Society of Electronic Industry (ed.) (1984) *Guide Book of IC Circuit Devices*, JSEI, Tokyo.

Ministry of International Trade and Industry (ed.) (1985) *Vision of Industrial Location in the 21st Century*, Survey Group of Ministerial Materials, Tokyo (in Japanese).

Miyazaki, I. and Usui, T. (eds) (1985) *High Technology Industry and the Japanese Economy*, Japan Editorial Company, Tokyo (in Japanese).

Study Group of Future Construction Industry (1984) *Technopolis*, Urban Culture Company, Tokyo (in Japanese).

Technopolis Study Group (1983) *Study Report on Technopolis Industrial Relocation*, Japan Location Centre, Tokyo (in Japanese).

Chapter 19

THE LOCATION OF HIGH-TECHNOLOGY INDUSTRY: AN AUSTRALIAN PERSPECTIVE

P. W. Newton and K. O'Connor

Debate concerning the role of high-technology industry within the Australian economy has been gathering force against the background of an unsatisfactory performance of its economy during the late 1970s and early 1980s and the still-fragile recovery currently underway. Of particular concern has been the relatively weak export prices and stagnant world demand for mineral ores (which represent approximately 30 per cent of current exports by value) and an erosion of the competitive position of Australian manufactured exports (50 per cent of exports). Both trends have weakened the national economy and fuelled unemployment which currently stands at 8.0 per cent of the labourforce. At a sub-national level, prevailing economic conditions combined with technological change have also acted to undermine many local or regional economies while promoting others. Initiatives have been sought by government to revitalise national and local economies by creating an environment conducive to the growth of a high-technology sector (seen as having higher than average prospects as a generator of wealth and employment; Espie, 1983). Government intervention is being actively pursued in all States and the Commonwealth (see Macdonald, 1983 and this volume for a review) in an effort to enhance economic development within their particular jurisdictions.

The hiatus which continues to surround this area is related in part to a lack of consensus as to what constitutes high-technology industry and a lack of knowledge about the developmental and locational requirements of different industries within the high-tech sector. Accordingly this chapter has two principal objectives:

1. To examine the level of high-technology industry in Australia. This calls for a clear definition of what constitutes high-technology industry and a consideration of those nation-specific factors which

may either constrain or facilitate development of this sector within this country.

2. To identify those factors which influence the location of high-technology industry within Australia, and in particular, Melbourne. Comparison will be made with findings from similar studies in the USA and UK.

DEFINING HIGH-TECHNOLOGY INDUSTRY

The problem of deriving indicators of high-technology industry is a now well-recognised one (Breheny and McQuaid, 1984, p.4; Wiewel *et al.*, 1984, p.292). In short this problem involves accommodating a sound conceptualisation of what constitutes high technology with the practicalities of measurement (a problem similar in scope to that faced by the factor ecological studies of urban structure in the 1970s). The numerous attempts which have been made so far to define high-technology industry can be discussed according to whether their concern was principally related to input factors (e.g. worker skills), output factors (e.g. nature of product) or a combination of both.

Technological Sophistication of Products Produced by an Industry

Problems in the use of this definition relate to absence of a criterion for selection (Hall, this volume) and perhaps as a result of this drawback, such data are not collected routinely (Wiewel *et al.*, 1984). In some cases (Economic Council of Canada, 1985) indicators have included industries using high-technology products in their production processes, although this approach has generally been rejected in most empirical studies to date (Breheny and McQuaid, 1984; Glasmeier and Hall, 1983; a principal drawback being its possible inclusion of rejuvenated traditional industries, that is, those adopting new technology). Instead, Breheny and McQuaid (1984) employ a multivariable indicator comprising nature of product and occupational structure. Other multivariable indicators are those which examine R & D expenditure in relation either to sales (Glasmeier, Markusen and Hall, 1983), total investments (SCC, 1981) or by value added (Wiewel *et al.*, 1984).

The Occupational Profiles of Industry

Growth in employment has largely been discredited as a useful indicator of high-technology industry by virtue of the fact that 'the sources of job growth are many and varied, and high-technology content is only one among them' (Hall, this volume). Perhaps the most widely accepted measure to date is one based on 'the degree of sophistication and competence embodied in the technical occupations within the industry' (Glasmeier, Markusen and Hall, 1983). Selection of this indicator appears to have rested more on its technical superiority (satisfying conditions of precision, since disaggregation is possible; comparability, as all industries are assessed according to a set criterion; and objectivity) than its conceptual strengths—other than statements to the effect that '. . . it [a high percentage of scientific personnel] clearly implies that innovation is occurring.' (Hall, this volume). In the section which follows we will attempt to establish a stronger conceptual basis for this particular indicator than has been offered to date.

Conceptualising High-Technology Industry

What distinguishes high-technology from other types of technology and high-technology industry from other types of industry? Macdonald (1983, p.331) suggests that 'high technology is high . . . because of the relatively high risk involved, the possibility of high return, its high pace of change and its high information intensity.' While these may be considered necessary descriptors of high technology they are not, of themselves, sufficient. While successfully describing its future course of development they fail to recognise the origin of the technology, which is at the essence of the entire issue. In this we follow Wild (1984, p.47) who defines high technology as:

> That which originates from science and scientific research through what we call the scientific method. Its findings are not dependent on the common sense inspirations of practical men, but on cumulative advances of scientific investigation of the processes of the natural world . . . Thus, solid state devices—transistors, microprocessors and others in the silicon chip family—are high-technology. They arose from excellent and persistent scientific research conducted over many years in the Bell laboratories.

Hence the origin of the technology (that is, the inventions) is to be found within the cluster of highly trained scientists and technologists, and it is from this base that the definition should spring.

The links between these origins and other stages in the development of a product are explored in Figure 19.1, which begins by distinguishing some stages in the product development process, and does so in such a way that the potential role for science and technological activity can be delineated. The latter are indicated in level 2 of the diagram, and in the definition used here technology will be 'high' in that period where expen-

Figure 19.1 Technological Change and Spatial Impacts: a Schema.

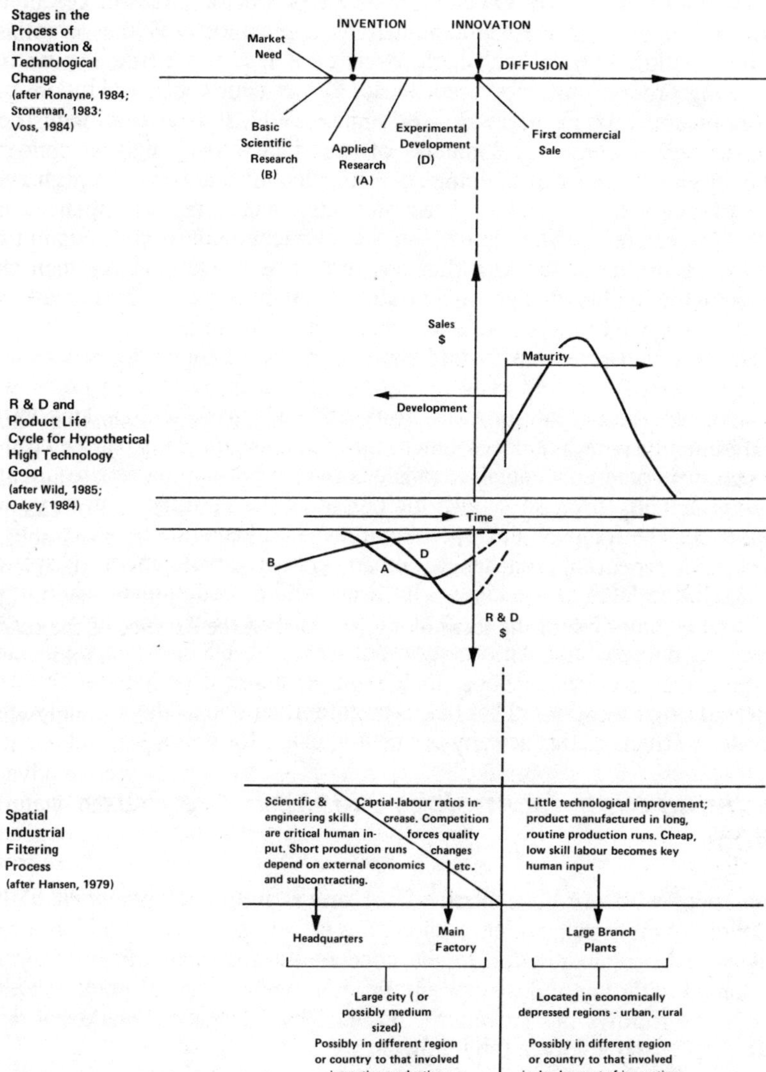

diture on R & D is a significant component in activities A and B. This is naturally followed by experimental development (activity D) which becomes the main focus of expenditure as the product approaches commercial sale. Steps to develop commercial applications of the scientific findings follow rapidly, with high levels of R & D input (roughly in the following proportions: 3 per cent basic, 22 per cent applied, 75 per cent development; Stoneman, 1983—see Figure 19.1). It is at this stage (i.e. progressing toward development of an innovatory high-technology product or process) that the employee mix includes a greater representation of engineers, technicians, accountants, managers, etc., relative to high-level scientists. The 'density' of scientist-technologist positions in the workforce profile of the firm declines further as production commences, although the quality changes in a product wrought by competition requires the maintenance of links to research units within the firm.

This study has set out to find the geography of industries where activities A and B are important parts of overall employment. In doing so, the chapter is concentrating on firms and industries with a long-term commitment to product development through research. This particular approach has some conceptual advantages over other approaches to definition, principally because it links the output of the industry to the type of employee. The focus on the type of employee builds a bridge between this work and recent approaches to urban system development (Noyelle, 1983). In the latter work, cities with firms with a predominance in a type of employment (Noyelle called them control functions) dealing with erratic and unpredictable information dominate the US urban system, and through that dominance boost their own economies, perhaps at the expense of other locations. That link is an important one as the inter-city and intra-city effects of this activity is a central cause for policy concern.

Defining High-Technology Industry: The Present Study

Consistent with our conception of the origin and development of high-technology industry as science based, the present study defined high-technology industry in relation to the concentration of scientifically-trained personnel within an industry group. From a methodological point of view this study follows the procedures outlined by Glasmeier, Markusen and Hall (1983) and involves the following steps:

1. Designation of those occupations considered necessary for identification of industry with capability for high-technology research and/or development. Sixteen occupations were selected from the list employed by the Australian Bureau of Statistics (ABS) in their national census and surveys: civil engineers, electrical and com-

munication engineers, mechanical engineers, chemical engineers, metallurgists, professional engineers n.e.c., chemists, physicists, geologists and geophysicists, physical scientists n.e.c., medical scientists, biological and animal scientists, senior university academics, statisticians and mathematicians, computer programmers and computer systems analysts (ABS, 1981).

2. Identification of those industries (classified at 4-digit ASIC level; ABS, 1985a) which have a representation among any or all of the set of sixteen scientific and engineering occupations (as outlined above) employed within their workforce.

RESULTS

Analyses revealed a low representation of high-technology employees within Australian industry: approximately 88,500 people, equivalent to 1.4 per cent of the national workforce, across the full spectrum of private and public-sector organisations (significantly below the level of 5.8 per cent identified in the study by Glasmeier, Markusen and Hall, 1983, for manufacturing industry in the USA). The data suggest that a principal role of people in high-technology occupations in Australia may be one of translating (or adapting) the technology which is imported into Australia (see later discussion on Australia's ratio of imports to exports in high-technology products). In other words, a level of 1.4 per cent may be necessary simply to successfully facilitate this technology-transfer process. However, much higher concentrations (at least approaching the USA level) are probably necessary in order to mount an export-oriented high-technology industry.

To explore this dimension, and to move away from the lower level activity (probably related largely to activity D in Figure 19.1 rather than activities A and B), it was decided to set a figure of 6 per cent as a cut-off value and exclude those industrial activities where the share of the selected occupations was less than that level. This more rigid definition would identify industries in Australia with many times the national contribution to scientific work and provide a sharp measure of the potential for new product development. This tightly circumscribed group included twenty-four industries, which accounted for a high-tech workforce of 32,250 professional scientists, engineers, etc. (0.5 per cent of the total Australian workforce). These industries are shown in Table 19.1. There is commonality with USA studies (as reported in Hall, this volume) in relation to several industries, namely, organic and inorganic chemicals, synthetic resins, petroleum refining and electronic equipment. The contrasts are more significant, however. Unlike studies published to date where the tertiary sector (which represents about three-quarters of a modern economy) is virtually unrepresented (Macdonald, 1985, p. 3) high-tech employees in

Table 19.1: High Technology Industries in Australia*

Description	High Tech Employees (%)	(N)
Mining - uranium ores	7.7	61
Mining - oil and gas	11.6	186
Services to mining NEC (undefined)	7.9	53
Mining exploration (own account, undefined)	36.7	26
Services to mining NEC (petrol exploration)	18.5	385
Services to mining NEC (mineral exploration)	24.4	287
Services to mining NEC (mining exploration)	17.4	1021
Manufacturing - synthetic resins and rubber	8.0	339
Manufacturing - organic industrial chemicals	10.8	227
Manufacturing - inorganic industrial chemicals	9.2	440
Manufacturing - pesticides	6.1	66
Manufacturing - petroleum refining	6.5	337
Manufacturing - alumina	6.9	338
Manufacturing - electronic equipment	6.5	920
Electricity - generation, transmission, distribution	6.5	4898
Water supply - storage, purification, distribution	6.8	2155
Office and business machines wholesalers	12.1	2347
Finance, property and business services - superannuation funds	7.0	69
Finance, property and business services - surveying services	6.8	604
Finance, property and business services - technical services	24.8	7396
Finance, property and business services - data processing services	36.3	3040
Universities	6.6	2426
Research and scientific institutions	22.9	4470
Meteorological services	8.9	160
TOTAL		32250

Note: * Only those industries with > 6.0 per cent of their workforce in high technology categories of employment are listed (high tech employment in the above listed high technology industries is 32,250, representing 0.5 per cent of the total Australian workforce).

finance and business services account for around one third of the total. The public utilities (water, electricity, etc.) are strongly represented. The importance of public institutionalised research (with approximately 20 per cent of total employment) is also apparent. Unlike most other OECD countries, government is a leading rather than a supporting actor in the development of Australian science and technology (OECD, 1985). Within manufacturing the emphasis is on the capital intensive resource base elements of the sector, so that chemical, oil and resource processing are the most prominent. That perspective is sharpened by the part played by mining in seven separate groups within the total of twenty-four.

LOCATIONAL CONTEXT FOR HIGH-TECH INDUSTRY

To date, most studies of high-technology industry have been conducted in the USA, the UK and Western European countries such as the Federal Republic of Germany—core societies within the world economic system. From the data so far revealed concerning Australia's low number of high-tech employees in high-tech industry relative to the UK (32,250 vs 640,874) and USA (32,250 vs 5,475,200; Hall, this volume) and the somewhat distinctive set of industries in which they are distributed, it is clear that Australia's state of high-tech development is divergent from the aforementioned countries. It is possible to advance several reasons why this is so.

Australia's Position Within the World Economic System

The core-periphery concepts contained in the work of Wallerstein (1978) and Steiber (1979) have emerged as a useful organising framework for assessing the relative economic position of different regions or countries (e.g. with respect to economic or social performance measures) and the nature of the interactions or exchanges which occur between and within particular nations. The model of exchange which derives from this body of theory suggests that within the world market, nations are stratified according to their economic strength. For core societies, their economic diversity and wealth is reflected in the export of large quantities of both raw materials and finished products, with greater emphasis on the latter. Peripheral regions are in a persistent state of underdevelopment, based on an early inability to compete in the world market and subsequent reliance on specialisation in the form of primary industries.

Based on an analysis of international trade statistics, Australia's position in the world economic system could be classed as semi-peripheral; the semi-periphery exports both primary and secondary products, but their trade in secondary products is of modest proportions. Comparing the ratio of imports to exports, I/E, for high-technology products and minerals in 1980 across a group of countries considered to have core status, the ratios for Australia are, respectively, 9.50 and 0.01, which means that for every one dollar of high-technology product Australia exports, it imports comparable products to the value of $9.50 (Newton, 1983). In contrast, for every one dollar of metalliferous product exported, Australia imports an amount to the value of one cent. Increasingly there is the prospect of Australia continuing to pay for high-technology imports by the export of its mineral resources and primary production (Ford, 1983, points to Australia having a trade deficit of $6 billion in high research-component goods in 1980-81; this contrasts with a trade surplus of more than $10 billion for food and crude materials during the same period).

A general explanation for this situation lies in the increasing internationalisation of production whereby most national (and regional) economies are now globally interdependent, locked into a world economic system by (often multinational) enterprises whose objective is to maximise profit on a world scale rather than contribute to the development of particular national or regional economies (see Crough and Wheelwright, 1982). Australia has few world-ranking corporations capable of matching the R & D funding characteristic of their overseas counterparts (e.g. Du Pont's commitment of $120 million for life sciences' R & D in 1981; OTA, 1984). Indeed, across the full spectrum of Australian industry there has been a sharp decline in industry-funded R & D. Nor does Australia match the capacity (and perhaps entrepreneurial spirit) of countries such as USA to raise funds for high-tech firms in the public market (over $500 million was raised by new biotechnology firms in the USA in 1983; OTA, 1984).

Military-Industrial Linkages

Recent empirical research in the USA points to a strong link between high-tech activity and military procurement; for example, the analysis by Glasmeier, Hall and Markusen (1983) indicated that defence spending per capita contributed most (from among nineteen predictors) to explaining high-tech dependence in the economy of a set of 218 metropolitan areas. Placing this in a broader context, Castells (1984, pp. 99-107) suggests that the USA's evolution from the welfare state of the 1960s and early 1970s to the warfare state of the 1980s (reflected in significant shifts in government expenditure away from welfare towards the military, and a share of world arms-trade which has risen from 20 per cent in 1970 to 45 per cent in 1982) is re-shaping the urban and regional structure of that country. For its part, Australia has modest military expenditures, procuring most of its hardware from countries such as the USA, the UK, France and West Germany. As such, the small scale of military and defence research effort in Australia has constrained one of the principal driving forces associated with high-technology activity.

The Role of the State

High-technology industry is becoming increasingly important in State and Federal growth and restructuring strategies. At the State level, many efforts have been made to attract and boost high-tech industry (however defined), something that is common with most countries throughout the world. What is different for Australia in this regard is the oft-repeated problem of technology transfer, where there are difficulties in firms making the critical step between ideas and concepts and production. In part this is due to Australia's peripherality in the world economy and the influence that large corporations can exert on governments' economic and industrial policies. In addition, there is a case put that there are major institutional constraints, like financial resources, acting against new products and new approaches which seriously limits new technology development and marketing. With the deregulation in 1984 of Australia's financial system, the competitive pressures which foreign banks will now be able to place on the indigenous banks may further weaken their incentive to commit their funds to high risk investments, characteristic of high-technology industry. Foreign investment remains overwhelmingly oriented towards the resources sector, with representation in utilities (embracing the power or water projects which service resource development and processing), trade and business services (Table 19.2). A comparison of data in Tables 19.1 and 19.2 indicates some strong links between volume of foreign investment by industrial sector and level of high-technology employment by industrial sector. This suggests that profitability of a particular industry within its national context (e.g. coal mining is not overly profitable in the UK!) determines level of investment (foreign and domestic, but especially the former) and subsequently influences the numbers of highly-trained personnel which can be supported by that industry (and can be employed to add further to its productivity, development of new products, etc.). The inflow of capital to trigger this cycle is based on an evaluation made by the international financial order.

Governments can and do seek to influence the attitudes and decisions of international capital by attempting to underwrite the costs-risks associated with a new industrial venture. Such joint ventures are commonplace in Australia's mining industry and involve provision of some mix of physical infrastructure (roads, rail, ports, telecommunications, etc.), social services (health, education, etc.) and financial incentives (related to tax and royalty

Table 19.2: Level of Foreign Investment in Enterprises in Australia, by Industry (Year ending June 1983)

Industry (ASIC Division/ Subdivision)	Total Investment $A Million	Per Cent of Total Div./Subdiv.	Per Cent of Total Division
A. Agriculture	291	0.8	0.8
B. Mining			26.5
Coal, oil, gas	2970	9.0	
Other mining	3409	10.3	
Services to mining	2396	7.2	
C. Manufacturing			20.5
Food, beverages, tobacco	655	2.0	
Textiles	74	0.2	
Clothing and footwear	98	0.3	
Wood products, furniture	10	0.0	
Paper products, printing, publishing	354	1.1	
Chemical, petroleum and coal products	575	1.7	
Non-metallic mineral products	533	1.6	
Basic metal products	3101	9.4	
Fabricated metal products	193	0.6	
Transport equipment	320	1.0	
Other machinery & equipment	699	2.1	
Miscellaneous manufacturing	202	0.5	
D. Electricity, gas, water	4239	12.8	12.8
E. Construction	371	1.1	1.1
F. Wholesale and retail trade	4154	12.6	12.6
G. Transport and storage	1998	6.0	6.0
H. Communication	179	0.5	0.5
I. Finance, property, business services	5820	17.6	17.6
TOTAL	33040		

Source: ABS (1985b), p. 28

payments). Evidence would suggest that governments are beginning to explore similar packages in the context of high-technology industry; viz: provision of technology parks (Australia's cargo cult of the 1980s?), venture capital, tax incentives, etc. (see Espie, 1983, for a more complete list) That such policies may be merely subsidising the R & D activities of large (overseas) firms through their local branch plants and offices (see McLoughlin, 1984, p. 262) is little different from government assistance to mining activity—except the settlement impacts will be vastly different. The universally applicable caution holds, however: 'Scarce resources are likely to be used inefficiently if the wrong activities are sited in the wrong places for the wrong reasons.' (Macdonald, 1983, p. 335). Australia has witnessed sufficient planning disasters without high-tech industry being added to the list.

These elements of the Australian background bring into focus a perspective that may not have been so relevant to other research conducted on this subject to date and go some way towards explaining the low level of high-tech activity in Australia relative to countries such as the USA.

FACTORS RELEVANT TO THE LOCATION OF HIGH-TECHNOLOGY INDUSTRY IN AUSTRALIA

Locating High-Technology Industry: Methodology

Giese and Nipper (1984) consider Malecki's (1979) findings, which relate to the distribution of R & D in the USA, to have been strongly influenced by the spatial scale at which the analysis was undertaken. The broader the spatial scale (e.g. counties, SMSAs) the more heterogenous are the areas and the potential for confounding associations is increased (witness the spate of unexpected correlations and sign changes in the study by Glasmeier, Hall and Markusen, 1983). The call has been for more disaggregated analyses, focusing on more detailed sub-groups of industry on a finer spatial scale.

In the present study the Australian Bureau of Statistics' Integrated Register System of Australian Industry, a system established in 1983, was employed as a source of 4-digit ASIC industry and employment data at local government area (municipal) level. The spatial unit for which the data were originally tabulated is the location, defined by ABS as 'all the operations carried out under the ownership of one enterprise at a single physical location. Physical location means the place occupied by a single office, shop, factory, plant, farm, mine, depot, hospital etc.'. For reasons of confidentiality, data are not released at levels below the local government area (LGA). Even at this level, low frequency data (less than three entries per cell) are also suppressed, necessitating estimation (in the present case, by allocating an industry average).

A National Perspective

When the 88,430 high-tech jobs are assigned to their broad single digit ASIC groups and allocated to each State, the data shown in Table 19.3 were produced. The comparison of the distribution of high-tech employment against all employment is revealing. Of the larger States, Victoria occupies a distinctive position with 29 per cent of high-tech employment but 26.7 per cent of all employment. That position places Victoria in a somewhat different situation in comparison to New South Wales (NSW) where the proportions are virtually equal, one that is all the more surprising as in the major sectors like manufacturing and finance, NSW seems to have an upper hand (Daly, 1983). Victoria's strengths lie in a broad spread of activity and a predominance in public administration. This information adds to the insight developed from the analysis of competition between Melbourne and Sydney, which in a small range of leading sectors was tending to favour the latter. The present results show, however, that the Victorian economy exerts an effect, somewhat disproportionate to its size, on Australian high-technology activity.

The peripheral role of smaller States, especially Queensland and South Australia, is also apparent from the table. That is consistent with the broad expectations of the role of small regions in this type of activity. However the bias is not as strong as one might expect. This is probably due to a leavening effect of publicly-controlled activity. Universities and CSIRO establishments have been distributed in Australia with some respect for States' rights as well as needs, so that nationally-important research can be found in small States just as much as the larger ones. In addition, the spatial distribution of the mining industry, which favours Western Australia (WA) and Queensland boosts their share in the activity under review here. This quick national survey sets the perspective for a detailed study of the location of this activity within the state of Victoria and its capital, Melbourne.

A State-Level Perspective: Metropolitan-Non Metropolitan Variations

Within the State of Victoria, Melbourne accommodates the vast majority of high-technology establishments (Table 19.4). Excluding utilities establishments (which reflect the State-wide responsibilities for public provision of water and electricity) as well as the State-funded educational and research institutions, the dominance of Melbourne further intensifies with 90 per cent of purely private sector high-tech establishments located within the State capital (similar to the findings by Giese and Nipper, 1984 for West Germany, and Malecki, 1979, for the USA. Hayter and Watts (1983,

Table 19.3: Number of High-Technology Employees in Major Industry Groups, 1981

ASIC Division	NSW	VIC	QLD	SA	WA	TAS	NT	ACT	Nat'l
Agric, forestry, fishing	59	50	53	15	64	14	4	6	265
Mining	792	377	844	345	1388	156	199	35	4136
Manufacturing	7561	6629	1471	1425	1053	380	23	47	18589
Electricity, gas, water	2551	2561	1085	502	636	336	49	40	7760
Construction	1665	978	736	419	660	162	69	57	4746
Wholesale, retail trade	2917	1936	546	371	419	82	25	126	6422
Transport, storage	1200	728	466	195	351	51	8	52	3051
Communication	717	1334	290	170	188	64	10	32	2805
Finance, property, business	6763	4450	2018	798	1768	181	139	247	16364
Public admin, defence	2397	2636	831	854	395	198	195	2084	9590
Health, educ, comm. services	4102	3548	1484	1042	996	256	102	996	12526
Recreation	299	138	102	76	75	62	4	40	796
Industry not stated	546	413	168	85	122	17	7	22	1380
TOTAL	31569	25778	10094	6297	8115	1959	834	3784	88430
Share of High-Tech Employment (%)	35.7	29.1	11.4	7.1	9.2	2.2	0.9	4.3	
Share of National Employment (%)	35.5	26.7	15.1	8.7	8.8	2.7	0.9	1.6	

p. 162) also found that R & D establishments were predominantly located in the more prosperous regions and the larger cities).

A Metropolitan-Level Perspective: Melbourne

Patterns

The focus of the present research has centred on the location of individual establishments rather than employment, as it is the decision to locate a facility that is the means whereby city structures can be changed. Data were subsequently assessed on the number of establishments that had more than 6 per cent of their workforce in the occupational groups discussed earlier. Figure 19.2 represents the aggregate pattern of high-tech establishments across the metropolitan area. What is revealed is a cluster of

Figure 19.2 Total Number of High-Technology Establishments (Excluding Utilities), Melbourne, 1984.

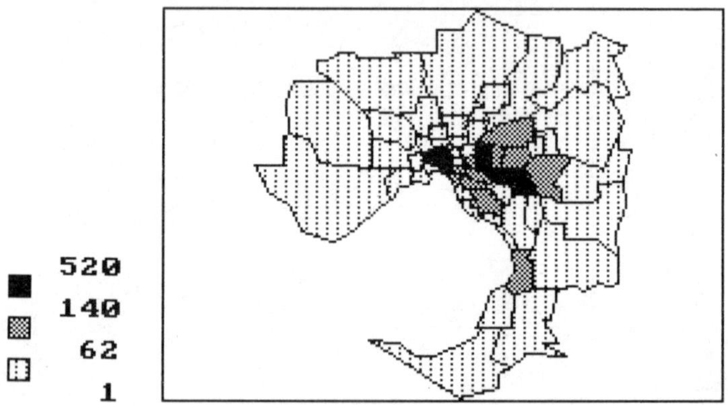

establishments in and south of the central city, together with an area further to the east and south, while the rest of the metropolitan area has few establishments according to the definition developed here. Very broadly, high-technology industry occupies locations within the main industrial and commercial focus of Melbourne, and in simple terms 'on the better side of town'. Inspection of industry sub-categories (Figure 19.3) reveals that although in some sectors (e.g. mining, manufacturing and electronic equipment) there is a greater dispersal than seen in the pattern in Figure 19.2, there are strong common elements in all patterns. These are the centre and selected south and eastern municipalities, though the interstitial suburbs can move about a little.

Figure 19.3 High-Technology Establishments by Industrial Sub-categories, Melbourne, 1984:
a. Mining.
b. Manufacturing.
c. Electronic Equipment.
d. Office and Business Machines Wholesalers.
e. Finance, Property and Business Services.
f. Research and Universities.

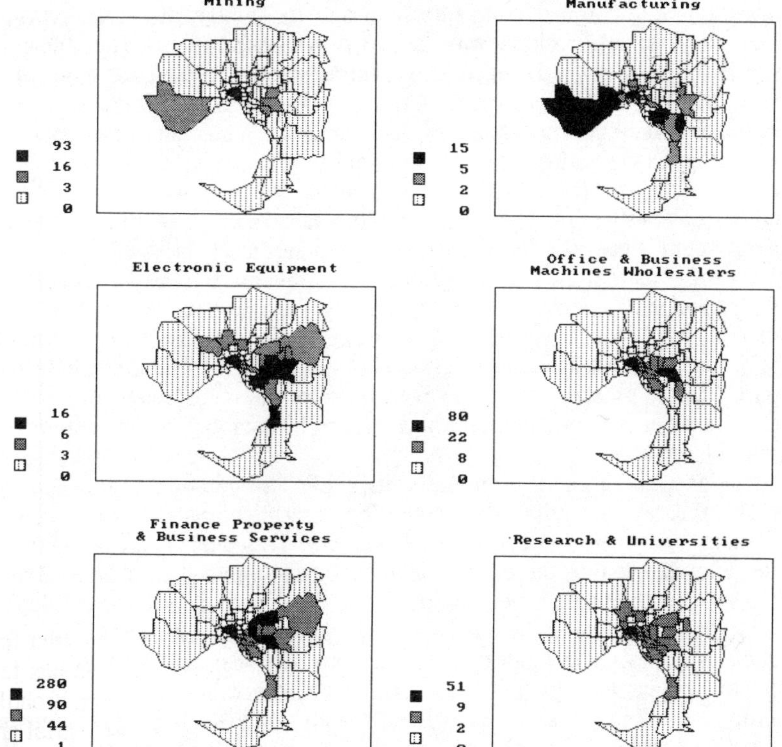

Factors

One of the main emphases of the research was to identify factors that could account for the location of high-technology establishments within a metropolitan area. The approach at this stage was conditioned by two considerations: first by variables shown to be relevant to this type of activity in other countries; and second by a need to involve 'planning' factors, those that could be influenced by government if the latter wanted to encourage high-technology industry in particular locations within a city. These two considerations produced the seven broad factors identified in Table 19.5. The availability of data led to the derivation of twelve associated indices.

Clustering of high-tech firms in close proximity to high quality research establishments has been seen as essential as a source of information about new scientific and industrial developments (the seedbed function) and a source of well qualified labour, in particular, individuals capable of establishing new high-tech enterprises (the spawning function). Breheny and McQuaid (1984) also reflect that the spawning process ensures the continuation of locational concentrations due to residential inertia (unwillingness of locally-resident workers to move house and disrupt lifestyle, social contacts, etc.), the need to maintain established business links and a facilitation of employee recruitment from among a local social and business network. The importance of such informal local (and in many cases social) networks as a catalyst for innovation is receiving growing attention (Mier, 1983, p. 365) and the relevant activity space would typically include universities and research establishments, competitor firms and the local social network. Evidence from case studies (Macdonald, 1983, p. 341) suggests that the informal links are significantly more important than their formal counterparts, notwithstanding the fact that the latter has formed a basis for government policy to facilitate clusters of research-oriented establishments in science parks, and high-technology industrial parks. In the present study, however, only formal measures were possible, e.g. number of universities and research establishments within the municipality.

It has been argued, admittedly in studies of national-regional scope (Hall, this volume), that high-technology establishments tend to locate in or near areas with high residential amenity. This aspect can be linked to the knowledge that these establishments pay higher-than-average wages (at least to the professional cadre of the firm), which provide employees with greater choice in the housing market, the typical effect being, in areas such as Santa Clara County, a bidding up of prices and displacement of lower-income workers (the latter group being a necessary ingredient of the bifurcated labourforce associated with high-technology industry; Saxenian, 1983). This observation may require amendment to allow for the fact that in some cities (e.g. Akron, Ohio) the present high-technology activities have grown up from an earlier base, which could be in older and less attractive areas. To explore this aspect the present study measured both the average price of housing and the socio-economic status of residents in each of Melbourne's municipalities.

A third aspect considered was the level of local rates and charges. This was explored because small, new firms are often sensitive to these costs, and could be encouraged to areas where local government rates were

Table 19.4: High-Technology Industry by Location and Employment, Victoria, 1984

High-Technology Industries	Number of Establishments Vic.	Melb.	Rest of State
Mining - uranium ores	1	1	0
Mining - oil and gas	9	3	6
Services to mining NEC - petroleum exploration own account	42	35	7
Services to mining NEC - mineral exploration	57	50	7
Services to mining NEC - mining and exploration	92	71	21
Manufacturing - synthetic resins and rubber	53	53	0
Manufacturing - organic industrial chemicals	25	25	0
Manufacturing - inorganic industrial chemicals	33	27	6
Manufacturing - petroleum refining	6	5	1
Manufacturing - alumina	1	0	1
Manufacturing - electronic equipment	142	141	1
Electricity - generation, transmission, distribution	234	42	192
Water supply - storage, purification, distribution	278	23	255
Wholesale and retail trade - business machines	532	458	74
Finance, property and business services - superannuation funds	65	63	2
Finance, property and business services - surveying services	213	142	71
Finance, property and business services - technical services	1433	1286	147
Finance, property and business services - data processing services	958	912	46
Universities, libraries, etc.	45	15	30
Research and scientific institutions	161	94	67
Meteorological services	11	4	7
TOTAL (N)	4397	3456	941
(%)	100	79	21
(% minus State funded utilities, educational and research institutions, etc.)	100	90	10

lower. In particular, firms engaged in high-tech activities have often put considerable effort into high standard buildings, providing an attractive work environment. In these circumstances, rating systems that ignored the quality of buildings would be more attractive than systems that built a tax base on the value of the buildings themselves. This aspect was measured according to whether councils rated property on site value only or in terms of the value of improvements. A measure which pertained to the average assessed value of office and factory property within an LGA (i.e. a measure of the investment in infrastructure which has taken place in the past) was employed as an additional pointer to local rate levels as well as quality of industrial-commercial infrastructure.

A fourth aspect considered was the availability of office and factory space. This was considered important because firms engaged in high-technology activity can require space at short notice, to take advantage of shifts in research and/or market opportunities that present themselves in an environment characterised by rapid change. As Hayter and Watts (1983, p. 164) have noted: '. . . at a local scale, space can impose important constraints on acquisition behaviour in certain industries.' For that reason it is likely that more high-technology establishments will be found in areas where factory and office space has been built by the development industry, and can be occupied quickly, rather than in areas where long lead times are necessary to establish premises. This consideration had an urban policy dimension because the availability of space in factories and offices could be influenced by government controls over zoning and building. Empirical measures used in the present study in relation to this factor are the area of office and factory space constructed since 1981, on a local area basis.

Three further aspects that were considered took into account the broad character of locations within the metropolitan area, especially with regard to the type of industry in the area, the standard of the local labourforce and the accessibility to an international airport and to the metropolitan market in general. The first two aspects were included in the study since product cycle theory indicates that high-technology firms may need to draw upon small local firms for components, supplies and sub-contracting and so would be more attracted to municipalities with diverse industrial structures than with specialised activities. As Oakey (1984, p. 157) argues, in relation to the high-technology firm,

> . . . there is little scope for standardization of inputs because products are constantly revised, indicating a need for frequent interaction with local specialist suppliers. Further, the rapidly changing products ensure the need for a constant input of highly skilled research and development and production workers to fill this process of rapid innovation. Hence, both local specialist material inputs and skilled labour inhibits the mobility of high-technology industrial production and contributes to

strong local agglomeration economies in areas where local linkages and labour advantages accumulate.

A measure of the potential for local industrial linkage was obtained by applying a diversity index (Newton and Johnston, 1976) to information on factory activity in municipalities. Similarly, high-tech firms are perceived as more likely to locate in areas with a better-trained labourforce than those with lower-skilled workers. Here the character of the local labour market was assessed according to the level of tertiary qualifications among the resident workforce.

Finally, a range of measures of accessibility were investigated. Here the underlying premise has been that high-technology firms would be located in more accessible sites; in particular they would be interested in locations close to airports (because of a need for interaction with other cities and other countries), close to the centre of a metropolitan market to draw upon its information resources and, related to that, for areas well served with freeways providing intra-regional travel.

Dissatisfaction with the arbitrary geometry associated with the boundaries of LGAs, or any spatial unit for that matter, led to the development of potential measures on seven of the twelve predictors (social status of residents, dwelling prices, office space, factory space, manufacturing diversity, educational qualification of resident population and freeway accessibility). Such a measure represents, for each LGA, its own particular value together with a proportion (in the present study, one-half) of the value present in adjacent LGAs. The resultant measures more appropriately summarise the environment of particular areas as they project themselves to prospective industrial clientele.

Model Results

Correlation and regression analysis afforded one means of attempting to identify that mix of factors which represents an environment conducive to the development of high-technology industry. The dependent variable employed was the total number of high-tech establishments, as listed in Table 19.1, minus the utilities and research establishments (the former in order to confine the study as much as possible to private sector activity; the latter due to the fact that it was to be employed as a predictor). Given the fact that individual sectors of high-tech industry display somewhat different spatial tendencies (a similar finding to that of Glasmeier, Hall and Markusen, 1983), separate regressions were undertaken for six sub-groups of high-tech industry.

Correlation analysis between each of the twelve individual indicators and the total number of high-tech establishments located across Melbourne's fifty-five LGAs revealed significant positive correlations for research potential (.58), new office space potential (.56), average dwelling prices in area (.55), area socio-economic status (.55), academic qualifications of area's residents (.52), manufacturing diversity (.48), and value of area's office and factory infrastructure (.29). All accessibility measures were negatively correlated with total high-tech establishments indicating either that most high-tech establishments in Melbourne are not serving an international market (contrast with Sydney would be useful as that city is being promoted as the nation's international transport and communications gateway), or that road congestion has not grown to the extent that intra-urban proximity to air transport becomes an issue. The performance of the relative accessibility measure suggests that the intra-urban market for high-tech products or services is not uniformly distributed across the city.

Search for multivariate explanation led to the application of multiple regression analysis. The results displayed in Table 19.5 show the order the individual indicators entered the regression and the overall level of explanation (variance) provided. The analyses suggest that there are a large number of factors that are either unimportant, or their statistical influence is accounted for by measures higher up the table (due to the intercorrelation between selected independent variables). The model for total high-tech establishments suggests the importance of a local research environment, high residential amenity and a well-established, high-quality infrastructure for office and factory-based activity, in many ways embracing the standard prescription for high-tech industry.

However, the danger of adopting such a blanket prescription to individual sectors of high-tech industry is also evident from Table 19.5. The city-based, high-tech mining establishments appear the most footloose with new office space as the principal locational determinant (many of Australia's largest mining companies are headquartered in Melbourne's CBD). By way of contrast, high-tech manufacturing appears locationally more constrained with strong associations with established high value office and factory infrastructure (reflected in high net annual value ratings) and need for freeway access. Firms manufacturing electronic equipment appear attracted by localities with a diverse manufacturing base (providing a ready source of components) and a municipal rating system which does not penalise investment in high-quality office-factory complexes (i.e. where rating is undertaken on site value only).

High-tech establishments associated with business machines and data-processing services appear to exhibit a similar configuration of locational requirements to each other, in particular emphasising residential amenity and availability of new office space—the archetypal, clean, post-industrial industries. Inclusion of all high-tech finance, property and business establishments as a single group further emphasises the importance of social amenity and new office space factors (where the latter is concentrated in middle or inner ring localities rather than peripheral green-fields sites), with some reliance on quality office-factory environments and the presence of a suitably-qualified labour market. Taken overall, the present findings suggest that governments could find difficulty in attempting to redirect or relocate the array of high-tech activity outlined above to areas currently unrepresented in this activity, without considerable State intervention and expenditure and detriment to future possibilities for development in areas currently favoured by market forces. After all, there is little enough high-tech industry to go around as it is.

CONCLUSION

At the outset, this chapter established two principal objectives: to examine the level of high-technology industry in Australia, and to identify those factors which appear to influence the location of high technology within this country, and in particular, in Melbourne.

On the first score it is evident that Australia is performing poorly. Less than half of 1 per cent of its workforce could be classified as high-tech employees working in high-technology industry. For every high-tech employee in high-technology industry within Australia, there are twenty in the UK and 170 in the USA (significantly different from the employed persons' ratios of 1:5:20). Export data provides additional verification of this position. Furthermore, high-technology industry is heavily dependent on investment, and up until the present, indigenous investment has been deficient. Most of Australia's high-technology clusters appear to have arisen in response to the injection of foreign capital into those areas where the expectation of profit is relatively high. For Australia this has centred on resource-related industry, utilities and technical services to the finance-property-business sectors. Most high-tech investment in Australia derives from government sources and is reflected in the higher profile (relative to USA and UK studies) adopted by the universities and research institutions.

As governments at both State and Federal levels seek ways of procuring a greater share of high-technology industry within their jurisdictions (as occurred in relation to resource development during the 1960s and 1970s), questions of what industries to support where arise. We have seen that high-tech industry is not uniformly distributed. Concentration occurs in the more populous and prosperous States with an established industrial base (although the resource-based industries and the publicly-funded research accords the peripheral States a larger share of high-tech employment than might otherwise be expected). Within the State of Victoria, the capital, Melbourne, is the focus for 90 per cent of private-sector high-tech industry. Within Melbourne the data reveal that a small, centrally-focused, south-east corridor is where most high-technology establishments have located. There are two important features in that observation. First, the strength of the centre, and second, the selectivity of the suburbs. These features are an important part of the evolution of the metropolitan area.

The significance of the central part of the metropolitan area is somewhat surprising as the international image of high-technology activity is almost exclusively low density and suburban. The re-assertion of the centre in this study is in part due to the actual definition of high technology that was used. The project is dealing with a narrow band of jobs, often within organisations that favour central area locations because of the value placed on day-to-day networks and of the well established infrastructure (measured by net annual value of property and concentration of research establishments) with extremely good freeway and public transport access to the high-quality residential areas in the eastern suburbs. It is also possible that high-tech activity is not perhaps as locationally volatile as was first thought, and is anchored to some currently-established and for the time being efficient institutions that were built in the central part of the metropolitan area, say, fifty or so years ago. The big research hospitals are an example, as are some established research units in medical activity. This is an important finding as it puts a dampener on much of the enthusiastic commentary about the mobility and green fields preference of high-technology firms. The latter locations are significant for some of the high-tech activity within this sample (and more significant for the downstream production, e.g. manufacturing in Figure 19.3, and curve D in Figure 19.1), and the present study's results reinforce the spatial selectivity of those activities.

In general, the suburban areas to emerge as preferred high-tech environments mostly lay across the large middle-class belt of suburbia, generally pulling up short of the working-class industrial areas. In this selectivity, high-technology industry is further strengthening a pattern of well off and less well off areas within the metropolitan area, with positive externalities accruing to middle-class residents (i.e. those best placed to initiate socio-technical innovation). Now there is a broad wedge to one side of the city, circumscribed by the CBD at one end and the better-off suburbs to the east and south. This region takes in most of the good-quality industrial and commercial sites and leaves out large areas of average to low-quality housing and lower-quality industrial development. The correlation between this pattern and the broad pattern of socio-economic development and unemployment, for example, is striking. What becomes clear is that high-technology industry, rather than contributing to the restructuring of a new pattern of development in Melbourne, is more forcibly reinforcing historically established trends.

REFERENCES

Australian Bureau of Statistics (ABS) (1981) *Census 81—Occupation*, Cat. No. 2148.0, ABS, Canberra.

Table 19.5: Factors Associated with Location of High-Technology Establishments, Melbourne, 1984

Predisposing Factor	Type of Industrial Establishment						
	All High Tech Establ.	All High Tech Mining	All High Tech Manuf'g	Electronic Equipment	Business Machines	Data Processing Services	All High Tech Fin., Prop., Bus.
RESEARCH ENVIRONMENT							
• No. research establishments	(1) .58						
RESIDENTIAL AMENITY							
• Price of Housing	(2) .67				(2) .67	(1) .65	
• Area socio-economic status							(1) .63
LOCAL RATES & CHARGES							
• Rating system				(2) .50			
• NAV rating of office & industrial stock	(3) .71		(1) .58				(2) .70
AVAILABILITY OF NEW SPACE							
• New office space		(1) .58			(1) .63	(2) .72	(3) .72
• New factory space							
INDUSTRIAL ENVIRONMENT							
• Manufacturing diversity				(1) .41			
LOCAL LABOUR MARKET							
• Workforce with tertiary qualifications							(4) .76
ACCESSIBILITY							
• Distance to airport							
• Relative accessibility							
• Freeway potential			(2) .64				
Multiple R	.71	.58	.64	.50	.67	.72	.76

Note: Number in brackets relates to step at which factor entered equation ($p<.05$).

——— (1985a) *Australian Standard Industrial Classification. 1983 Edition, Vol. 1: The Classification*, Cat. No. 1201.0, ABS, Canberra.

——— (1985b) *Foreign Investment, Australia, 1982-83*, Cat. No. 5305.0, ABS, Canberra.

Breheny, M. and McQuaid, R. (1984) 'The Genesis of High Technology Industry in the M4 Corridor: A Preliminary View', paper presented at 16th Annual Conference of the British Section of the RSA, University of Kent, Canterbury, September.

Castells, M. (1984) *Towards the Informational City? High Technology, Economic Change, and Spatial Structure: Some Exploratory Hypotheses*, Working Paper No. 430, Institute of Urban and Regional Development, University of California, Berkeley.

Crough, E. and Wheelwright, E. (1982) 'Australia the Client State: A Case Study of the Mineral Industry', in E. Harman and B. Head (eds) *State, Capital and Resources in the North and West of Australia*, University of Western Australia Press, Nedlands.

Daly, M. T. (1983) 'International Forces, Structural Changes in the Australian Economy and the Housing Industry', in Australian Institute of Urban Studies (AIUS) (eds), *Urban Australia: Living in the Next Decade*, Canberra.

Economic Council of Canada (1985) 'Tech Change and the Job Market', *Au Courant*, 6(1), 2-3.

Espie, F. (1983) *Developing High Technology Enterprises for Australia*, Australian Academy of Technological Sciences, Canberra.

Ford, J. (1983) 'Science in Government', *Search*, 14, 180-181.

Giese, E. and Nipper, J. (1984) 'The Impact of Innovation and Diffusion of New Technologies for Regional Policy in the Federal Republic of Germany', *Erdkunde*, 38, 202-216.

Glasmeier, A., Hall, P. and Markusen, A. (1983) *Recent Evidence on High Technology Industries' Spatial Tendencies: a Preliminary Investigation*, Working Paper No. 417, Institute of Urban and Regional Development, University of California, Berkeley.

———, Markusen, A. and Hall, P. (1983) *Defining High Technology Industries*, Working Paper No. 407, Institute of Urban and Regional Development, University of California, Berkeley.

Hansen, N. (1979) 'The New International Division of Labour and Manufacturing Decentralisation in the US', *Review of Regional Studies*, 9, 2-14.

Hayter, R. and Watts, H. D. (1983) 'The Geography of Enterprise: a Reappraisal', *Progress in Human Geography*, 7, 157-81.

Macdonald, S. (1983) 'High Technology Policy and the Silicon Valley Model: an Australian Perspective', *Prometheus*, 1, 330-49.

——— (1985) 'Towards Higher High Technology Policy', Paper prepared for joint OECD-Italian Seminar, Venice (mimeo).

McLoughlin, P. (1984) 'A Note on Job Creation in High Technology Industries and Local Economic Planning', *Prometheus*, 2, 258-66.

Malecki, E. J. (1979) 'Locational Trends in R & D by Large US Corporations, 1965-1977', *Economic Geography*, 55, 309-23.

Mier, R. (1983) 'High Technology Based Development: a Review of Recent Literature', *Journal of the American Planning Association*, 49, 363-70.

Newton, P. W. (1983) 'The Problems and Prospects of Remote Mining Towns: National and Regional Issues', in Australian Institute of Urban Studies (AIUS) (eds), *Urban Australia: Living in the Next Decade*, Canberra.

——— and Johnston, R. J. (1976) 'Residential Area Characteristics and Residential Area Homogeneity: Further Thoughts on Extensions to the Factorial Ecology Method', *Environment and Planning A*, 8, 543-52.

Noyelle, T. J. (1983) 'The Rise of Advanced Services', *American Planning Association Journal*, 49, 280-90.

Oakey, R. (1984) 'High Technology Industry', *Geography*, 16, 157-9.

OECD (1985) OECD's Draft Examiner's Report: Review of Australian Science and Technology, Paris (mimeo).

Office of Technology Assessment (OTA) (1984) *Commercial Biotechnology: an International Analysis*, Congress of the United States, Washington, DC.

Ronayne, J. (1984) *Science in Government*, Edward Arnold, Melbourne.

Saxenian, A. L. (1983) 'The Urban Contradictions of Silicon Valley: Regional Growth and the Restructuring of the Semiconductor Industry', *International Journal of Urban and Regional Research*, 7, 237-55.

Science Council of Canada (SCC) (1981) *Biotechnology in Canada: Promises and Concerns*, Minister of Supply and Services, Ottawa.

Steiber, S. R. (1979) 'The World System and World Trade: an Empirical Exploration of Conceptual Conflicts', *The Sociological Quarterly*, 20, 23-36.

Stoneman, P. (1983) *The Economic Analysis of Technological Change*, Oxford University Press, Oxford.

Voss, C. A. (1984) 'Technology Push and Need Pull: a New Perspective', *R & D Management*, 14, 147-51.

Wallerstein, I. (1978) 'World System Analysis: Theoretical and Interpretive Issues', in B. H. Caplan (ed.) *Social Change in the Capitalist World Economy*, Sage, Beverly Hills.

Wiewel, W., de Bettencourt, J. S. and Mier, R. (1984) 'Planners, Technology and Economic Growth', *Journal of the American Planning Association*, 50, 290-6.

Wild, P. (1984) 'High Technology—Is It the Answer?' *The Australian Director*, June/July, 47-52.

Wild, P. (1985) *CoResearch*, CSIRO, Canberra, June, p. 3.

Chapter 20

HIGH-TECHNOLOGY INDUSTRY LOCATION AND PLANNING POLICY IN THE SYDNEY REGION

B. Hutchinson and G. H. Searle

This chapter discusses the long-term planning strategies being developed for the Sydney region and the role that the advanced technology sector might play in future development. These planning strategies are examined in the context of the changing spatial structure of the Sydney region. A definition of high-technology industries is adopted which includes a core of sunrise industries identified by the Minister for Science and Technology, Mr B. O. Jones, as the focus for Commonwealth Government high-technology policies, and a periphery of other industries which predominantly use the technologies of the sunrise industries. The core industries include computer hardware and software, biotechnology, ceramics, robotics, fibre optics and solar power. The main peripheral high-technology industry considered is electronic equipment. High-technology industries are seen to generate employment faster than other manufacturing industries (Wiewel, de Bettencourt and Mier, 1984) and provide an opportunity for expanding the regional manufacturing employment base.

AGGREGATE ACTIVITY CHANGES

In the Sydney region employment in manufacturing remained roughly constant in absolute terms between 1971 and 1981 but decreased from 39 per cent of total employment in 1971 to 24 per cent in 1981. Rich (1982) has pointed to the deskilling of manufacturing employment over the period and has documented the decline in manufacturing output which reached a peak in 1973-74. Manufacturing output began to increase significantly in 1977-78 but without a concomitant increase in employment. Commercial employment (finance, property, commerce) provided about one-quarter of the jobs in 1961, 32 per cent in 1971 and 34 per cent in 1981. The services sector (public utilities, public authorities, community and business services, entertainment) provided about 25 per cent of the regional jobs in 1961 and 1971, and increased to about 29 per cent in 1981.

Table 20.1: Population and Labourforce, Sydney, 1961-1981

	Population		Labourforce			
	Total	Change	Male	Change	Female	Change
1961	2 278 600		703 819		290 125	
		491 839		126 623		148 201
1971	2 770 439		830 442		438 326	
		162 179		13 000		110 838
1981	2 932 618		843 442		549 164	

Table 20.2: Changes in Labourforce by Occupation Group, Sydney, 1971-1981

Occupation Group	Changes in Labourforce Male	Female	TOTAL
Professional/technical	36 690	39 865	76 555
Administrative	- 9 720	- 465	- 10 185
Clerical	4 509	53 158	57 667
Sales	11 147	15 219	26,366
Trades	2 817	- 10 824	- 8 007
Services	14 105	13 014	27 119
TOTAL	59 548	109 967	

Table 20.1 summarises the population and labourforce characteristics of the region between 1961 and 1981 and illustrates very clearly the sharp reduction in population and labourforce growth during the 1970s. The principal feature illustrated by Table 20.1 is the strong increase in female labourforce participation over the period. The major changes in labourforce by occupation group between 1971 and 1981 are summarised in Table 20.2. Labourforce in the trade and administrative occupations decreased, with the major increases being among females in the professional-technical and clerical occupation groups and among males in the professional-technical occupations.

SPATIAL POLARISATION OF THE LABOURFORCE AND JOBS

Malecki (1985) has argued that technological change will continue to lead to locational polarisation of the labourforce as it had when manufacturing employment dominated the economic bases of many urban areas.

Evidence so far would suggest that the differential changes in job opportunities have had important impacts on the age and sex structure of the labourforce in the different occupational groups. Male and female professional-technical jobs and female clerical jobs grew strongly between 1971 and 1981; this is reflected in the biases towards the younger age cohorts in these labourforce segments as illustrated in Figure 20.1.

Figure 20.2 shows the spatial distributions of job opportunities in the three major occupational groups by broad sub-area in the Sydney region in 1981. The diagram illustrates that job opportunities in the professional-technical occupations dominate the central employment areas along the coast, while jobs in the trade occupations were largest in a broad corridor of suburbs stretching from the City of Sydney to the outer west and south-west, where most of the outer suburbs have developed since the 1960s. The diagram also indicates that white-collar jobs dominate in the outer-northern and southern suburbs.

Figure 20.3 illustrates the spatial changes in job opportunities in each of the three occupational groups between 1971 and 1981. The City of Sydney lost about 16,000 jobs in the trade occupations, about 9,000 clerical jobs and gained about 9000 professional-technical jobs. It is evident that this increase in professional jobs extended only into several of the adjacent, older, industrial areas. This is an important trend and evidence introduced later in the chapter shows that the information-processing industries are beginning to locate in these older industrial areas as industries continue to move out and higher occupation status labourforce continues to settle in these areas. Figure 20.3 also shows that job growth in the trade occupations dominated the changing job opportunities in the outer suburbs,

Figure 20.1 Age and Sex Composition of Major Occupation Groups in 1981.

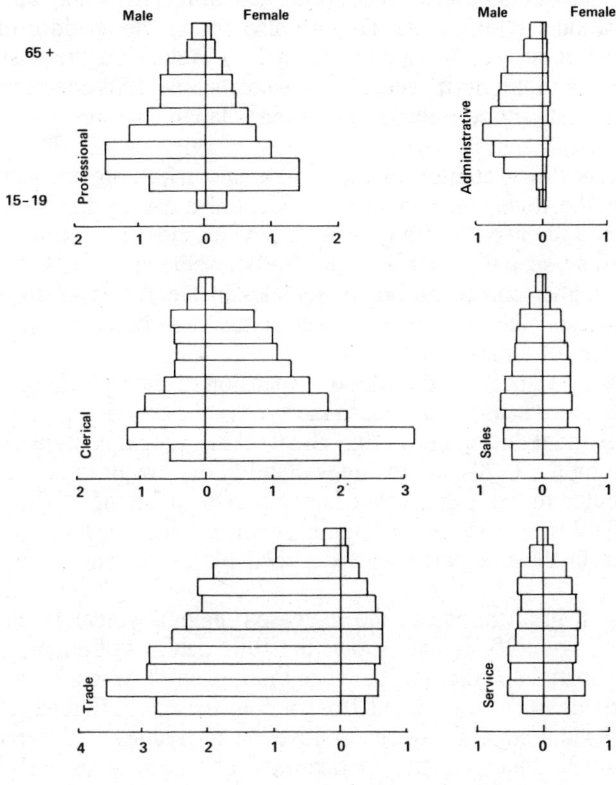

although clerical and professional jobs increased significantly as population-serving activities expanded in these growing fringe areas. The diagram shows that the majority of new job opportunities in a broad range of intermediate suburbs were white collar.

This increasing spatial polarisation of job opportunities was accompanied by increasing polarisation of residential areas by occupation status. Figure 20.4 shows the spatial differentiation of the male labourforce across the local government areas (LGAs) of the Sydney region in 1981. This grouping of the LGAs is based on the divergence of the occupational composition of an LGA from the region-wide average occupational composition. Figure 20.4 illustrates clearly the band of higher occupational-status LGAs running from the eastern suburbs south of the Harbour to the north-west, where these LGAs diverge on the basis of their above average concentrations of labourforce in the professional-technical and administrative occupational groups. The diagram also shows the corridor of lower occupational status LGAs running from Port Botany in the east to the newer suburbs in the outer west. These lower status LGAs diverge on the basis of their strong concentrations of male labourforce in the trade occupations.

The spatial differentiation of the female labourforce is not as extreme because of the dominance of most residential areas by the clerical occupations and the need by many women to work close to home. However, there are some spatial biases with the Harbour-side and northern suburbs containing higher concentrations of females in the professional-technical occupations, and the corridor of lower status suburbs containing higher concentrations of females in the trades.

Rich (1982) argues that the lack of expansion of the manufacturing sector was the major cause of the sharp increase in the traditionally low levels of unemployment during the 1970s. The working age population expanded rapidly during the 1970s and the unavailability of new jobs in manufacturing contributed to the high levels of unemployment among 15-24-year-old males in the lower occupational-status suburbs. Much of the new female labourforce in these suburbs was absorbed by the expansion of clerical jobs.

The spatial implications of the increased unemployment levels are illustrated in Figure 20.5 which shows the 1981 patterns of unemployment. The strong spatial associations between these unemployment rates and occupation status may be detected from a comparison of Figures 20.4 and 20.5. The spatial structure of unemployment in Sydney is different from that observed in many North American and European cities with its high unemployment levels in several residential areas on the developing fringe.

In summary then, manufacturing employment dominated the economic base of the region in the early 1960s and since that time process automation and foreign competition have combined to produce a reduction in and a deskilling of manufacturing employment. The strong expansion of the mineral-extraction industry between the mid 1960s and mid 1970s compensated for the erosion of the manufacturing base as employment in the commercial and services sectors increased sharply. Between 1971 and 1981 there was little net change in full-time employment but part-time employment, particularly for females, increased significantly. The largest increases in job opportunities were in the male and female professional-technical and female clerical occupations, with more modest increases in the sales and services occupations. These changes in employment have strengthened the spatial differentiation that exists between the residential areas of the Sydney region. The Harbour-side and northern suburbs have

Figure 20.2 1981 Distribution of Jobs by Occupation Group.

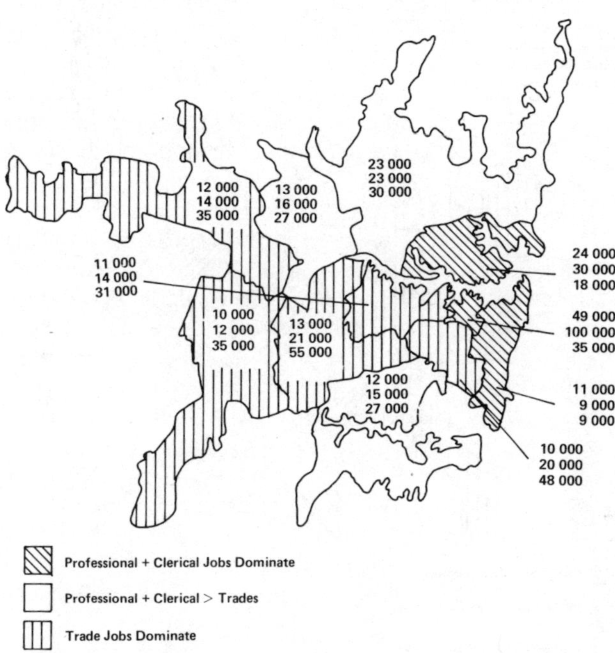

increased in occupational status while the lower-status outer suburbs have decreased still further in status. The issue now becomes one of considering the extent to which development of high-tech industry will reinforce the trends already identified.

THE ROLE OF HIGH-TECHNOLOGY

The case for an indigenous high-technology industry is supported by the degree to which the high-technology sector itself is part of the Australian structural problem. Australia's ratio of technology-related exports to technology-related imports ranks only twentieth among twenty-three OECD nations surveyed (NSW Science and Technology Council, 1985). These imports are strong in the economically and militarily strategic areas of computers, telecommunications and defence. Clearly, Australia has opportunities for significant expansion of its high-technology industries in those areas where there currently exists a large volume of high-technology im-

Figure 20.3 1971-1981 Changes in Jobs by Occupation Group.

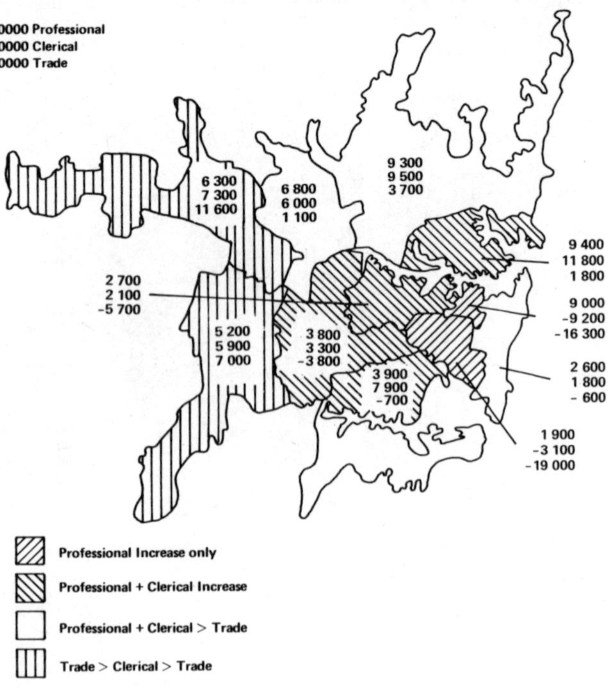

ports (about $A4 billion worth of high-technology equipment is currentlyimported into Australia each year). Further opportunity arises from the present lack of return from scientific research in development of new products: Australia contributes 2 per cent of the world's published scientific papers, but only 0.3 per cent of its technology-intensive exports (Department of Science and Technology, 1983).

The Sydney region offers particular opportunities to capture a large share of any Australian high-technology industry growth. Despite the fact that specific assistance for high technology from State and Commonwealth Governments has only developed in the last three years, there is already a significant high-technology industry in Sydney, even using the relatively narrow definition adopted here. For example, in 1982-83 the Sydney region had 8,987 employees—excluding establishments with under four workers (ABS, 1984) in general electronic equipment manufacturing (i.e., excluding radio, TV and scientific equipment) or 57 per cent of Australia's total.

Figure 20.4 Spatial Differentiation of Male Labourforce Occupation Status in 1981.

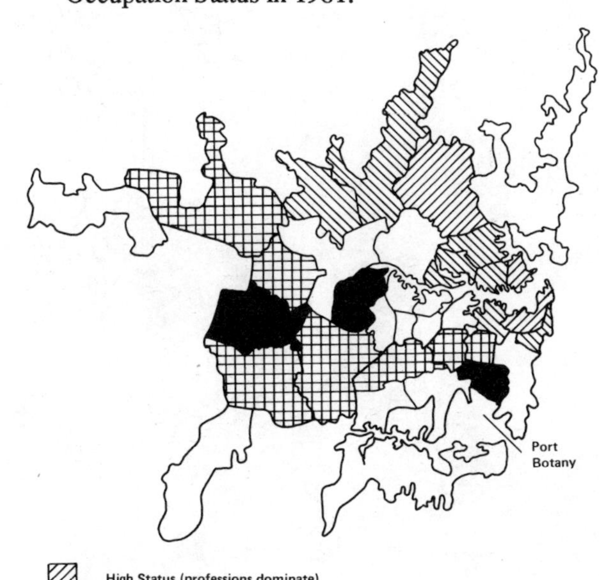

THE LOCATIONAL ENVIRONMENT

The economic and socio-cultural environment of Sydney offers a collective of factors considered highly desirable for high-technology industry, at least at the more stable and innovative headquarters and research and development levels. It is the primary focus of overseas air routes, offers an outstanding natural environment and supports a wide range of cultural activities. These characteristics of Sydney reinforce its commercial leadership, which generates the entrepreneurial, financial and industrial base required to support an innovative, indigenous, high-technology industrial sector and avoid reliance on assembly-line work in branch plants of multinational, high-technology corporations (Peltz and Weiss, 1984). The large and diverse employment base in the region offers the broad range of op-

Figure 20.5 1981 Distribution of Unemployment.

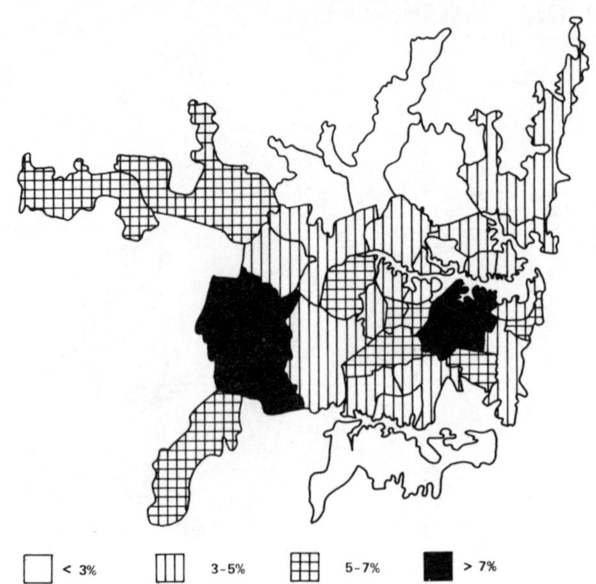

portunities attractive to the skilled personnel required for non-routine high-technology functions (Malecki, 1984).

The region is also well supported by four tertiary educational institutions and a number of laboratories of the CSIRO. While public-sector R & D activities offer an important base for technological development, the R & D activities of the private sector are much more modest. Only 20 per cent of Australia's R & D is funded by the private sector and Australia ranks in the bottom half of OECD countries in terms of the percentage of GDP devoted to R & D (New South Wales Science and Technology Council, 1985). Much of Australia's manufacturing industry has become uncompetitive as a result.

EXISTING HIGH-TECHNOLOGY LOCATION WITHIN SYDNEY

Figures 20.6-20.8 display the location of establishments in three of the principal high-technology sectors, namely, computer software development, electronic-equipment manufacturing and sunrise industry manufacturing. The diagram reveals that the computer software companies are

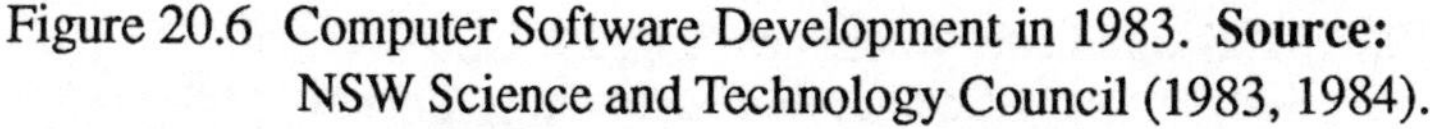

Figure 20.6 Computer Software Development in 1983. **Source:** NSW Science and Technology Council (1983, 1984).

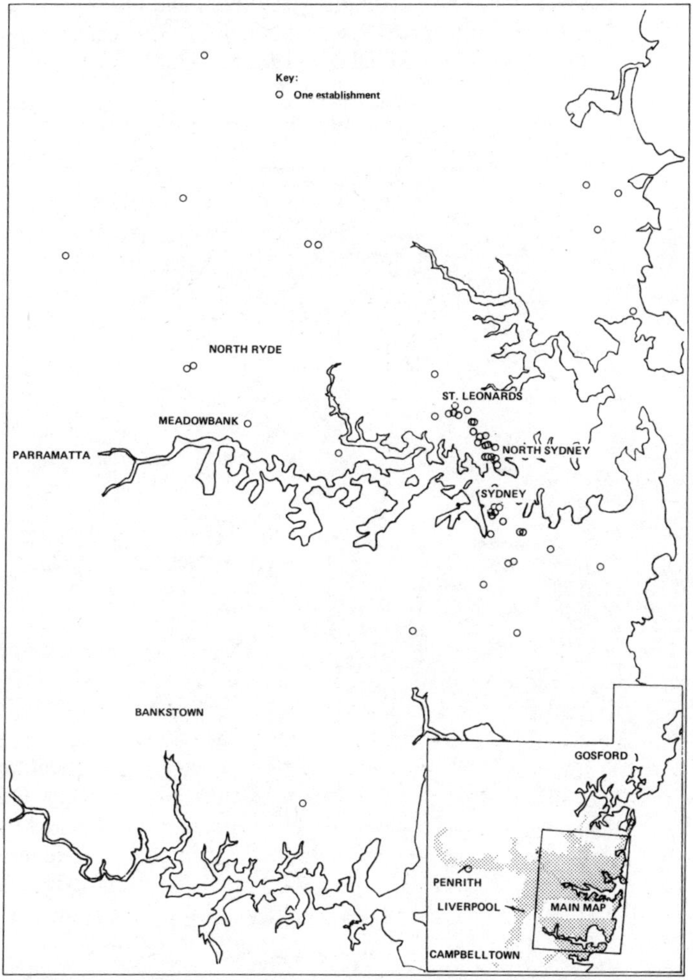

concentrated in a corridor running from St Leonards in the northern suburbs, to the Sydney CBD and its fringes. The main locational factors seem to be proximity to larger corporate headquarters in the CBD and North Sydney as well as to the concentrations of professional firms in North Sydney. It should also be recalled from the evidence introduced earlier that the higher status labourforce tends to be concentrated in northern suburbs.

Figure 20.7 indicates that all but four of the thirty computer-hardware

Figure 20.7 Sunrise Industry Manufacturing, 1983-85.
Source: NSW Science and Technology Council (1984); NSW Science and Technology Council and NSW Computer Industry Advisory Council (1985); Telecom, *Yellow Pages* (1985); DIDD (1983a, 1983b).

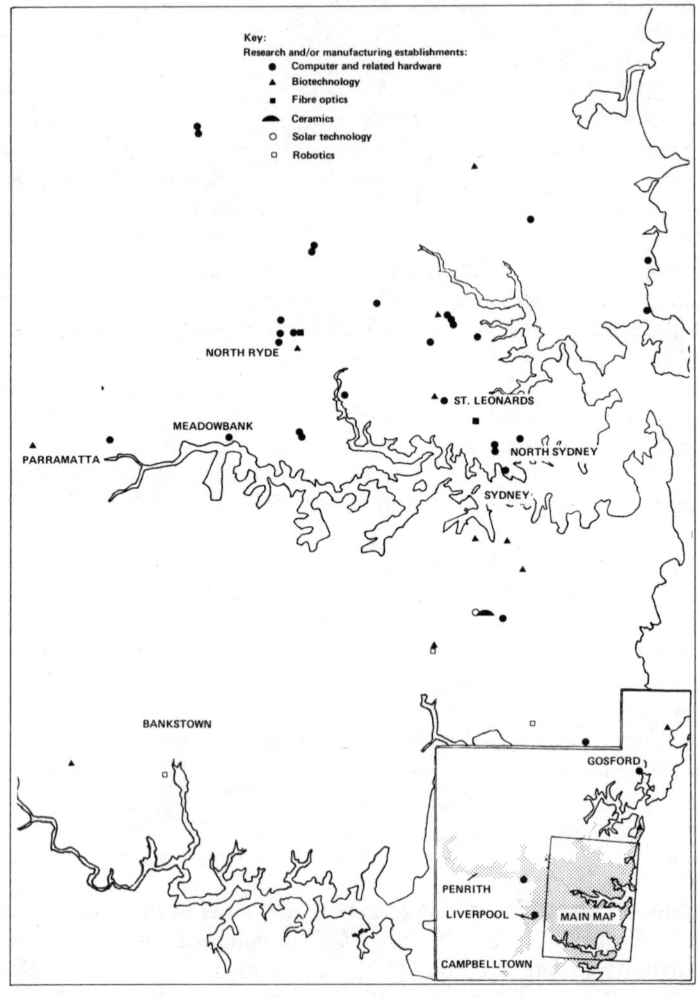

manufacturing companies are located north of the Sydney Harbour. Once again the primary locational factors seem to be the association with the North Sydney-St Leonards concentration of computer-related industries and the higher-status residential areas of the northern suburbs. Figure 20.7 also illustrates that the locations of the wider spectrum of sunrise industries are scattered throughout the region. The robotics, solar technology and ceramics groups are located in the older industrial areas south of the Harbour. These locations may be due to linkages with the manufacturing companies in these areas.

Figure 20.8 illustrates that electronic-equipment manufacturers are located throughout the region and tend to follow the broad patterns of industrial land use, with some bias towards the northern suburbs. There are two significant clusters within this general location pattern and these are Meadowbank-Gladesville-North Ryde in the northern suburbs and the Bankstown-Liverpool area to the south-west. Labour supply considerations appear to be an important reason for the location of some electronics firms in the outer suburbs where large pools of blue collar workers reside. Of the six outermost firms, four are assemblers of printed circuit boards. Areas for urban expansion over the next twenty-five years are mostly at the periphery but would be even more remote from high-technology industry employment opportunities than present outer area residents, if the existing high-technology industry location pattern is retained. Even present outer-area residents have inadequate accessibility to high-technology employment opportunities. The size of Sydney's urban area means that it contains a series of labour markets. The CBD focus of the rail system means that, generally, only CBD jobs have a region-wide labour catchment, and most high-technology jobs are outside the CBD. Changes in employment location in other sectors over the next twenty-five years will provide relatively limited opportunities for outer area residents in the absence of very strong policy measures at all levels of government. High-technology planning policies need to ensure that any employment growth generated by the high-technology industries address this problem of outer-area unemployment, without jeopardising efficiency. High-technology planning policies also need to encourage the transfer of high-technology to the rest of the manufacturing sector.

HIGH-TECHNOLOGY ZONING POLICIES

Zoning regulations are the traditional tool of the planner and still have a potentially important role in selectively encouraging high-technology industry. The experience of the North Ryde industrial area yields valuable guidelines on how innovative zoning policies can be used to develop a high-technology industry zone (Debelle, 1984). The area was zoned for

Figure 20.8 Electronic Equipment Manufacturing, 1985. **Source:** Telecom, *Yellow Pages* (1985); NSW Science and Technology Council (1984).

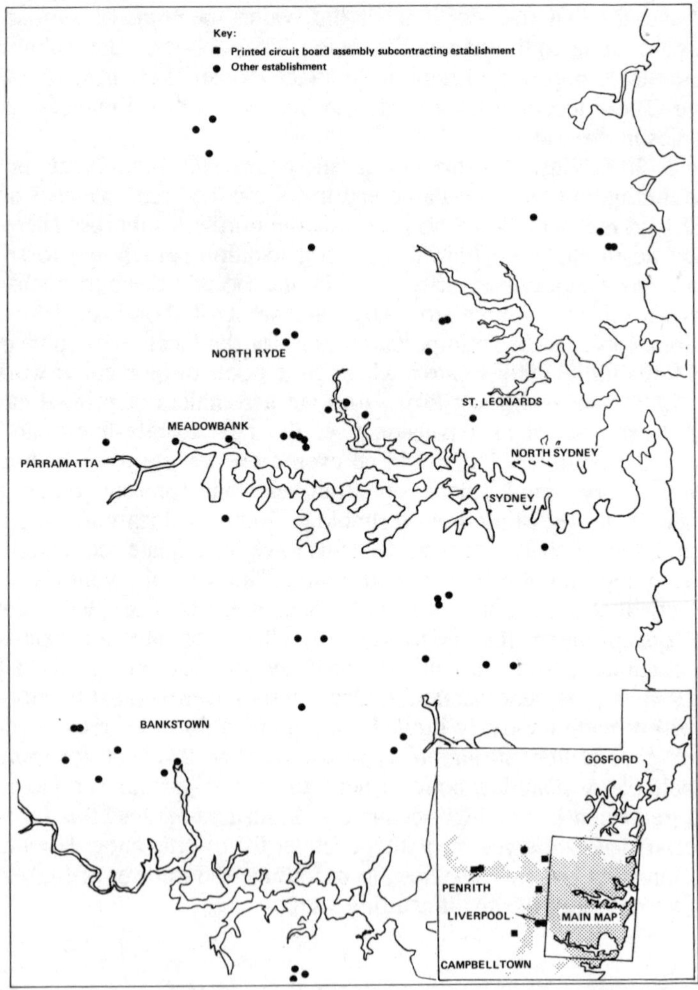

industrial use in 1966, and soon after was designated as a *de facto* technology park, requiring industries there to be compatible with activities of the new Macquarie University, adjacent to the zone, or be substantially dependent on research and scientific development. The local council established development restrictions such as deep setbacks, landscaping, and low-density site coverage, creating a high-quality image and proving a key location factor, along with excellent access to technical and managerial staff. The result has been that the zone has by now been almost fully taken up by companies in high-technology industries defined on a relatively broad basis (including, for example, printing, publishing and cosmetics).

On the other hand, the North Ryde area evidences some policy disadvantages. In the first place, the restrictive development controls have meant only large tracts of land may be developed (Debelle, 1984), discouraging small firms. This has been accentuated by the very success of the zone which now boasts Sydney's highest industrial land costs and rental levels. As a result, the North Ryde high-technology zone today has a pervasive multinational atmosphere. Secondly, little restriction has been placed on the proportion of office space permitted for each company. For some, the proportion of office space is as high as 80 per cent. This flexibility has allowed companies to move their administrative offices from the CBD to their production facilities, creating a business park rather than a true advanced technology park (Joseph, 1985). This has perhaps prevented some potential high-technology plants from locating in the area and from making use of the zone's high-technology information field and skilled-labour pool.

However, planning policies for advanced technology need to take account of the blurring of distinctions between office, R & D and manufacturing activities in high-technology industries (Herring, Son and Daw, 1983). Accordingly, the State Government has recently (1985) advised local councils that upper limits be removed on the proportion of office space in industrial zones, although the office component is still required to be ancillary to factory or warehouse activities. It is intended that the restriction to ancillary office space will meet the needs of genuine high-tech and other industries and prevent squeezing out by CBD-type office activities. Given this modification, the North Ryde zoning innovation could be repeated at other locations, particularly on vacant fringes of the northern suburbs and on attractive sites adjacent to major research establishments.

Another zoning option is to remove restrictions on the growth of small businesses in residential areas where residential amenity is not likely to be disrupted (Edgar, 1985), allowing firms to reduce overheads by using proprietors' residences. A further zoning approach for larger high-technology firms is becoming widespread overseas. This involves the designation of business parks, in which high-quality, low-density environments are provided for commercial firms requiring a prestige image and high-calibre workforce, whether the firms are high technology or not (Worthington, 1982). North Ryde is *de facto* becoming a business park thus defined, as it now also includes head-office employment of a range of service companies.

In planning terms this is generally a less efficient means of achieving high-technology goals than high-technology zones or science-technology parks, since the existence of firms outside the advanced technology sector reduces the opportunities for high-technology clustering and the advantages this brings. It is also more likely to conflict with policies of maximising commercial development in the CBD and other major centres highly accessible by public transport.

Business parks can, however, serve general high-technology objectives for the service sector by allowing the automated or 'intelligent' office to be wired more cheaply and advanced telecommunications facilities, such as cables, satellite dishes and teleconferencing, to be shared more easily. Retrofitting of older buildings with computer and telecommunications equipment, including wiring, is often much more expensive than in new buildings, which can have planned flexibility to accommodate changes in telecommunications technology. Thus advanced business parks can meet more general regional economic goals of promoting corporate efficiency and enhance the attraction for corporate control functions. The range of accommodation for software development in particular, would also be increased.

SCIENCE-TECHNOLOGY PARKS

Zoning policies such as the above are essentially a passive means of encouraging high-technology industries. More active planning policies can be centred on the development of science or technology parks. These comprise high-technology industrial establishments undertaking a high proportion of applied research, located in attractive, well landscaped surroundings with university or major research institutes within a reasonable catchment area (Worthington, 1982; Debenham, Tewson and Chinnocks, 1983). The distinction between science and technology parks rests on the emphasis on research and the degree of interaction between companies and universities or research institutes. In turn, the distinction between technology parks and high-technology zones, of the original North Ryde type discussed earlier, is seen here as hinging on the special facilities and infrastructure that should ideally be provided in technology parks to maximise the development of high-technology research and production.

In general, given the importance of information networks in high-technology development and the extent to which these are fostered by agglomerating information sources and recipients together, physical measures to promote high-technology development should be provided together where possible. This suggests that technology parks in the Sydney region and elsewhere should contain all or most of the following:

- Common facilities, including cafeterias, sporting and exercise facilities, day care and the like, to encourage informal contacts as well as to attract skilled personnel and increase business efficiency; for parks located next to universities, these facilities could include university libraries and computers.
- Nursery or incubator units to encourage small-firm growth, bearing in mind the difficulties which these firms face in finding land or buildings cheap enough and small enough in greenfields high-technology zones.
- A technology transfer centre to raise the technology levels of all firms in the subregion where the park is located, and provide technological advice to firms in the park itself. The technology transfer centre component would allow the technology park to become the focus of technical change in the subregion concerned, maximising the possibilities for local technology information awareness and creating and maintaining informal information networks. This technology park element, in particular, is likely to require public subsidy. But the extent to which technology parks of this kind can promote technology development, and high-technology production in particular, by facilitating information exchange, makes some government subsidy desirable.

One element often included in science parks is an innovation house, generally affiliated with a university, which provides a brokering and evaluative function for people with product ideas, covering technical and market feasibility, developing prototypes, and developing business plans and capabilities (Peltz and Weiss, 1984). It has been claimed, however, that the cost effectiveness of this concept is dubious, with few final endorsements being made (McLoughlin, 1985). The State Government has already set up such an Innovation Centre in the University of New South Wales to promote new products and processes from conception to commercialisation; in view of the uncertainty about its ultimate net benefits, it should not be replicated at this stage in technology parks. The promotion of a small-firm, high-technology zone in the nearby Sydney central industrial area could provide a means of facilitating the development of innovations close to the centre.

Finally, the difficult question of how many technology parks, and where, must be answered. The State Government is already planning an advanced technology park at Homebush Bay, one of the few greenfields sites left in the middle suburbs. The high-technology industry location trends discussed earlier suggest the site will meet the needs of target firms, being on the southern fringe of the northern suburbs and, in particular, close to the Meadowbank-North Ryde cluster of high-technology companies. It will also be accessible to the large population in western Sydney, being adjacent to Sydney's western freeway. The river-front location of the site also offers the opportunity to create an attractive setting for incoming firms. Other potential sites for technology parks in the Sydney region are similar to those suggested above for high-technology zones. In view of the potential public subsidies involved and the risks of failure arising from an oversupply of parks and zones, it may be desirable to leave the development of other technology parks in Sydney until after the Homebush Bay park has begun to attract development.

MULTI-FACETED PLANNING POLICY

Overall, the existence of divergent views on the way in which high-technology development takes place, and on the relative importance of measures to foster this development, suggests a multi-faceted planning approach, such as the one outlined here, should be used in the Sydney region. Indeed, this is the general approach already being pursued by the State Government, enabling a range of policies to be tested. The risks of failure in concentrating on one type of policy, such as technology parks, seem otherwise too great.

REFERENCES

Australian Bureau of Statistics (ABS) (1984) *Census of Manufacturing Establishments 1982-83*, ABS, Canberra.

Debelle, L. (1984) *North Ryde: A High-Tech Industrial Analysis*, Hooker Commercial Industrial Developments, Sydney.

Debenham, Tewson and Chinnocks (1983) *High-Tech: Myths and Realities*, Debenham, Tewson and Chinnocks, London.

Department of Industrial Development and Decentralisation (DIDD) (1983a) *Business Opportunities in Advanced Technology: Biotechnology*, NSW Government Printer, Sydney.

——— (1983b) *Business Opportunities in Advanced Technology: Robotics*, NSW Government Printer, Sydney.

Department of Science and Technology (1983) *National Technology Conference*, Proceedings and Report of the Conference held in Canberra, September, AGPS, Canberra.

Edgar, G. (1985) 'Science Parks: A Town Planner's Perspective', *Urban Policy and Research*, 3, 38-9.

Herring, Son and Daw (1983) *Property and Technology—The Needs of*

Modern Industry, Herring, Son and Daw, London.
Joseph, R. (1985) 'Encouraging High Technology: The Alternatives and the Rationale', Paper presented to Royal Australian Planning Institute's Winter School, University of Queensland, St Lucia, July.
McLoughlin, P. (1985) 'Science Parks—Too Many, Too Late?' *Urban Policy and Research*, 3, 43-4.
Malecki, E. J. (1984) 'High Technology and Local Economic Development', *American Planning Association Journal*, 50, 262-9.
——— (1985) 'North America', in J. Brotchie, P. Newton, P. Hall and P. Nijkamp (eds), *The Future of Urban Form—The Impact of New Technology*, Croom Helm, London, and Nichols, New York.
New South Wales Science and Technology Council (1983) *NSW Information Industry Directory*, NSW Government Printer, Sydney.
——— (1984) *Development of High/New Technology Companies in NSW*, NSW Government Printer, Sydney.
——— (1985) 'New South Wales Industries Disadvantaged in R & D', Press Statement, 25 July.
——— and New South Wales Computer Industry Advisory Council (1985) *Computer Communications and Related Hardware: Directory of Products Manufactured in NSW and ACT*, n.p., Sydney.
Peltz, M. and Weiss, M. A. (1984) 'State and Local Government Roles in Industrial Innovation', *American Planning Association Journal*, 50, 270-9.
Rich, D. C. (1982) 'Structural and Spatial Change in Manufacturing', in R. V. Cardew, J. V. Langdale and D. C. Rich (eds) *Why Cities Change: Development and Economic Change in Sydney*, George Allen and Unwin, Sydney.
Wiewel, W., deBettencourt, J. S. and Mier, R. (1984) 'Planners, Technology and Economic Growth', *American Planning Association Journal*, 50, 290-6.
Worthington, J. (1982) 'Industrial and Science Parks—Accommodating Knowledge Based Industries', in *Planning for Enterprise*, Proceedings of an International Seminar, Swansea, September.

PART 4.
GOVERNMENT INTERVENTION AND POLICY

Chapter 21

TECHNOLOGY, THE MARKET AND GOVERNMENT INTERVENTION

J. Ronayne

This chapter explores in some detail the role of science and technology in economic development and the need or otherwise for government intervention. The arguments that have been used to justify government support will be reviewed together with the difficulties these pose for some policy analysts.

Until relatively recent times economists treated invention and technical change generally as an exogenous variable; it was considered to be quite independent of any economic forces, proceeding according to its own internal logic. Savings and capital accumulation were the prime agents of change in the economy, and economic growth could be understood in terms of the use of more and more physical inputs.

In the 1920s, however, the economist Joseph Schumpeter departed from the mainstream of this so-called neo-classical economic tradition. Though he still accepted the neo-classical form of perfect competition as a suitable description of some types of market over the short term, and regarded technical change and innovation as exogenous, he did suggest that in the long term it was competition based on innovation that was the most important. It was this type of competition that struck at the very survival of rivals, and not just at their output or profit margins. Schumpeter, therefore, placed innovation at the centre of his theory of economic development. His definition of innovation was rather broad and could involve:

a. the introduction of new products or new methods of production,
b. utilisation of new raw materials,
c. opening up a new market, or
d. re-organisation of a sector of the economy.

He stressed the importance of entrepreneurs in bringing together new technical and socio-economic forces leading to innovation; inventions and discoveries have been made throughout history but it was only when dynamic enterprising figures systematically applied new technical ideas in the production and marketing of goods and services that the take-off into sustained economic growth occurred. Importantly, Schumpeter made a sharp distinction between invention and innovation. Invention, the creation of a novel idea, sketch or model for a new or improved product, process or system, need not necessarily lead to innovation, and the application of new combinations by entrepreneurs can take place without technical invention.

So important was innovation to Schumpeter that it was the autonomous factor that caused economic life to go through repeated cycles, which he named the Kondratieff cycles after a Russian economist who first identified long waves of economic growth and recession. These long waves were the result of changing economic conditions for innovation and were not the result of fluctuations in the rate of technical invention or other factors involved in the innovative process; but technical invention is very important in generating them. The Kondratieff cycle theory has recently been revived and refurbished in an attempt to understand the role of technological change in the creation and displacement of job opportunities (Freeman, 1979).

In considering the role of science and technology in economic life, therefore, Schumpeter stands as an important pioneering figure. Although technical invention, which he regarded as exogenous, was only one of a number of factors involved in innovation, he gave it an importance that had been denied to it by his colleagues in the neo-classical economic tradition.

In the post-Schumpeter era, and especially after World War II, economists focused much more sharply on the role of science and technology in the economic system. That war had shown conclusively that both science and technology could be made to respond to needs, in this case military needs; there then seemed to be no reason why they could not respond to non-military economic needs as well. Perhaps they had so responded in the past and economists had ignored their importance. Quantitative economic studies, which were designed to throw some light on the relative importance of the various inputs into the economic system, showed that economic growth could not be adequately understood in terms of the factors labour and capital accumulation alone. The increase in supply of goods and services, measured in terms of GDP per inhabitant and showing a continual rise in productivity and standard of living, could only be understood in terms of learning to use the inputs more effectively.

With this realisation came a renewed interest in technical change as the possible contribution to this third factor of growth. Denison (1962), in a study of US economic growth, estimated that between one-half and three-quarters was due to third factors and he concluded that advances in knowledge (both technical and managerial) were the most important part of the third factor. Most of the post-war economic growth in the US depended, according to Denison, on third-factor growth; technical inventions, either exogenous or induced by economic factors, were an important part of the third factor.

There has been criticism of the ways in which the economic growth modellers arrived at their rather precise percentage contributions of the various inputs to economic growth. The OECD dismissed their calculations as misdirected, and Freeman (1974) was critical of them because their authors failed to come to grips with the actual process of science, invention, innovation and technical change.

However, such studies served to focus the attention of the economists on technical change, on the diffusion of new technology and on the relationship between investment in science and output in terms of new technology. In 1958 Griliches, an American economist, demonstrated in his study of the diffusion of the hybrid-born innovation in agriculture, that the diffusion of innovations could be explained in economic terms, on the basis of profit expectation as shaped by market size. Some years later, with the work of Jacob Schmookler, the economists came full circle in their consideration of the role of technical change in economic growth. Schmookler (1966), by examining patent statistics over a lengthy period of time in the agriculture, paper, petroleum and railroad industries, came to the conclusion that not only could the diffusion of technological innovation be explained in terms of market forces but the pattern of invention itself could be so explained. Demand considerations were the major determinants of variations in inventive effort in specific industries. In the railroad industry, for example, he found that increases in inventive activity, as measured by patents, lagged slightly behind increases in purchases of railroad equipment. Schmookler argued that variations in equipment purchases induced variations in inventive activity. The data he gathered in all four industries indicated to him that inventors (corporate or otherwise) perceive the growth in equipment purchases by an industry as an indication of increased profitability of inventions in that industry and direct their inventive talents and resources accordingly. Technical change was not an exogenous variable after all; it was not an activity having economic consequences but independent of economic forces. Invention was controlled by demand considerations in the economy.

Schmookler acknowledged that there are limitations to the extent that demand can call up the appropriate inventions. Some inventions will always be impossible and others will be impossible until the knowledge upon which they are based is uncovered. Clearly, the growth of scientific knowledge was important since it would influence the specific characteristics of an invention. Inventors will choose the most efficient means of achieving their ends and will choose the area of science that allows them to do this. The characteristics of the invention will be affected by this choice but not the purpose. 'Invention', he says, 'and in all possibility technical change itself generally, is usually not apart from the normal processes of production and consumption, but a part of them. It expresses something not adventitious for a nation's economic life but an inherent part of it.' (Schmookler, 1966, p.207).

ECONOMICS, INVENTION AND THE PUBLIC PURSE

If Schmookler's analysis of the impact of economic demand on technical invention is correct, then the question arises as to whether the development of an activity that is an inherent and vital part of the economic life of nations should be left entirely to the free market and its scientists, engineers and entrepreneurs. If the answer to the question is no, then there is a case for government intervention, which leaves us with further questions. For example, what form should government intervention take? At what stages in the research-invention-innovation sequence should the intervention take place? What is the appropriate governmental financial commitment and how is it to be distributed across the research to innovation spectrum?

Before we examine these issues, however, we must first explore some of the reasons that have been put forward to justify government intervention in the inventive and innovative processes in the first place. By now there is hardly any government that is prepared to leave the production and utilisation of the scientific knowledge that is relevant to economic advancement entirely to the private sector. Most economists accept the view that government intervention in scientific research and development is necessary if nations are to derive the greatest benefit from technological developments. The key papers, setting out the justification for government intervention in research, were written by the economists Richard Nelson (1959) and Kenneth Arrow (1962). They argued that if left to itself the free market will tend to under-invest in research, especially basic research. A number of economic factors cause market failure, by which we mean the failure to invest at the optimal level. Of these, three are especially acute in the case of research and development. Arrow called these indivisibility, inappropriability and uncertainty.

Indivisibility can be of two kinds. First, in conditions of atomistic competition, where many small companies compete in an industry in which social and economic benefits are likely to be increased by the rapid diffusion of technological change amongst all the producers, and where cost-effectiveness would preclude the individual producers from engaging in their own research programs, there is a situation of indivisibility. In this situation the industry as a whole may benefit from a centralised research and development effort and this can be provided by the government, or by the industry itself in the form of a co-operative research association.

At the other end of this spectrum there is the situation in which the R & D costs involved in bringing modern high technologies to the marketplace are so great that it would be beyond the financial capacity of even the largest companies to contemplate without government assistance of some sort or another.

Inappropriability is due to the fact that research results are usually expensive to produce but, in general, are relatively inexpensive to reproduce by commercial rivals. The patent laws allow some protection, of course, but even in the event of successful exploitation of research by the innovating firm it will be unlikely that the firm will be able to fully appropriate to itself all of the benefits of its investment. Patents do eventually run out, and there may be applications of the knowledge produced which the innovating firm does not even know about when the first patent is filed. Since it is generally impossible to obtain patent protection for all future applications of a scientific discovery, the problem of inappropriability may be a disincentive to private firms to invest in research, especially pioneering, long-range, basic research. On this basis there is a case for government support for or performance of this type of research.

Uncertainty is inherent in the research process and companies may be unwilling to risk their profitability or even their survival by embarking on activities that have a high risk of failure. Risk-averse companies, with relatively short perspectives, will, therefore, tend to under-invest in research, especially if they cannot spread the risk over a number of projects, thus increasing the chances of success. Again, long-range basic research will be neglected in the private sector as a result of this uncertainty but so too will applied research and experimental development. There are other reasons for government intervention which are mentioned in passing. In areas of public good—where a service is provided by government—it would be inappropriate to allow technological developments to be completely controlled by the market: the duplication of the market mechanism would, according to economic theory, simply put up the price without necessarily improving the service. We may take defence, meteorology, standards, regulatory and analytical services, public health and public works as examples. The economists' theory of public goods indicates that when goods and services are provided for the general public either wholly or partly independent of each individual's ability to pay for them, then governments should finance R & D related to them.

Now these economic arguments have, in the past, been fairly widely accepted as ample justification for government intervention. For example, in the US the 1972 Report of the President's Council of Economic Advisors said:

> Government has an appropriate role in R & D even when its results will not be incorporated in government purchases, because firms would un-

der-invest in R & D for goods normally purchased by the private sector. Though an investment in R & D may produce benefits exceeding its costs from the viewpoint of society as a whole, a firm considering the investment may not be able to translate enough of these benefits into profits on its own products to justify the investment. This is because the knowledge which is the main product of R & D can usually be readily acquired by others who will compete away at least part of the benefits from the original developer. This is particularly true of basic research.

INTERVENTION IN THE COMMERCIAL SPHERE

Few would argue against government intervention in R & D that underlies the provision of health, defence and so on. Intervention in the commercial area does not command such wide assent: the most vigorous opposition tends to come from the ranks of the economists themselves, especially those employed in government Ministries of Finance or in Treasuries. The statement from the US did not come from the Office of Management and Budget, it will be noted. The arguments against intervention are summarised below.

First, on the matter of externalities, the argument that firms will be unable to appropriate all the benefits of R & D. There is a possible disincentive here for R & D performance in the private sector, but the imperfection is adequately satisfied by the patent system, a system that is, in essence, contrary to the theory of perfect competition in that it allows inventors to charge a monopoly price.

Second, though there is little doubt that there is a clear, strong link between technology and economic growth, the support of any technology, simply because it is a good thing, is not warranted. The issue is not whether technology is a good thing but how much technology do we need and how should it be funded. From a producer's point of view, an investment in R & D is like an investment in capital. Up to a certain point an investment in capital enables a task to be done at a lower cost than if there were less capital and more labour. Beyond a certain level, additional capital, while saving labour, will become uneconomic when its cost exceeds the cost of the labour saved.

Similarly, some investment in research and development can help to achieve worthwhile reductions in costs, but clearly there is a limit. The burden of this message is that while research and development, like capital accumulation, can make major contributions to economic growth, there is a limit. The Concorde program is used to illustrate the 'technology-is-a-good-thing' argument that resulted in an aircraft that the manufacturers, scientists, technologists and government promoted, not the users. Spillover was negligible and, as Robin Maxwell-Hyslop (1971, p.86) has put it,

'To many airlines, Concorde represents an opportunity to buy an aircraft they do not need, with money they have not got, to fly routes which may be forbidden to them. Such a temptation falls short of the irresistible.'

Third, uncertainty is a barrier to all sorts of economic activity. It is not a characteristic that automatically justifies a government subsidy. And the rhetoric that scientists use tends to create the impression that research and development is not all that uncertain. In the view of those who hold this position, research would appear to be a relatively predictable activity. Uncertainty is a fact of life handled best by the amalgam of individual risk-taking decisions we call the market. Individual entrepreneurs will do a better job of picking winners than government bureaucrats.

A variant of the uncertainty argument that receives as little sympathy is the one which states that the risks that deter individuals do not apply to government—that private individuals do not look far enough ahead or discount too heavily the interests of future generations. These views are dismissed because they assume that an imperfectly-functioning market will be put right by a supposedly perfect interventionist, without acknowledging that the interveners may not, and generally are not, free from pressure-group influences and from the temptation to pursue Concorde-type projects.

Fourth, the indivisibility argument as applied to small businesses would seem to be the one area in which a ready rejection of government aid is not forthcoming. This does not mean that such aid is approved by the paymasters. It is suspect because research and development subsidy is given to projects simply because they arise from a particular source, not because there are any compelling reasons for doing so. The counter-argument is that if small businesses have bright ideas that they cannot exploit because of shortage of capital, either exploratory or venture, then they always have the option of selling the idea to a bigger company.

Finally, the argument that other countries who intervene in these commercial R & D systems will create a disadvantage for those who do not is countered by the notion that if such intervention in a particular nation is successful the resultant growth will exert upward pressure on the exchange rate, thus balancing the comparative advantage it holds.

MARKET FAILURE AND THE LINEAR MODEL OF INNOVATION

We turn now to other views that have been put forward concerning the adequacy or otherwise of the market-failure theory as an economic justification for government intervention in the R & D system. The theory is rejected by some because it relies on and is generally supportive of the linear model of innovation (Gannicott, 1980). This model suggests that technological innovation stands at the end of a chain which starts with basic research and proceeds to commercial exploitation via applied research and experimental development. Research, and basic research especially, is considered to be the forerunner of invention and innovation and mechanisms designed to stimulate economic growth must pay due regard to its progenitors. Thus, the market-failure theory points to the research end of this linear sequence for corrective action to be taken.

The linearity or otherwise of the innovation process has been much studied and it is not the intention of this chapter to dwell on the results of the studies. The model is misleading if it is taken as the only interpretation of the innovation process. A great deal of innovation takes place not by the push of a scientific discovery but by the pull of the market. It may take place as a result of adaptation of old technology, and by trial and error. The sources of invention are numerous, scattered and various and any description of the process that assumes linear process starting with basic research is simply wrong (Jewkes *et al.*, 1958). Some innovations can start with basic research and in allowing basic research to be neglected—as it would be if left entirely to the private sector—would mean that governments were taking a conscious decision to forego the sorts of public benefits that have in the past arisen from basic research. The market failure theory cannot be abandoned because of the lack of uniqueness of the inventive process.

Joseph and Johnston (1985) have advocated the abandoning of the market-failure theory for reasons that are different to those that have been discussed so far. Following Eads (1974), they suggest that there are several inherent weaknesses in the theory. First, there is no evidence that, in the absence of measures designed to spread and shift risks, the resources allocated to R & D will be sub-optimal. Second, the costs of administering the corrective measures may be greater than the potential benefits. Third, political and bureaucratic actors who design and implement corrective measures may not themselves be free from bias and self-interest. Finally, there is no mechanism for identifying market failure, or its scale, and there is no way of telling when social and private costs no longer diverge.

However, while acknowledging these difficulties with the market failure theory, these are not the issues that principally concern Joseph and Johnston. Within particular political contexts, they suggest, the theory can be used to oppose government intervention in the commercial R & D - sphere. Because of the potency of the arguments against the theory that can be mounted by Treasuries and Ministries of Finance, the straw man of the market-failure theory can be used to effect to overcome calls for massive intervention. The principal antagonistic arguments are, first, the inability of government bureaucrats to be more knowledgeable than the entrepreneurs of the free market in picking the winners, and second, the rigidities that are introduced into the industrial scene by central direction as to what should, or should not be pursued. Joseph and Johnston, as the result of their deliberations conclude by quoting Mishan (1982, p.144): 'If the concept of market failure has no applicability what is to be gained by developing a formal decision-making apparatus that depends on it?'

But what if we do not use this justification from economic theory? How can government intervention be justified? The Australian Treasury says that government intervention in industry could be justified on certain economic grounds or where government clearly wishes to achieve a specific policy objective. But if the policy objective is economic, and if technology is involved, the market-failure theory is still, implicitly, being invoked. When the Australian Minister for Science announces that the sunrise-industries policy is fulfilling a government objective and market failure does not rate a mention (Joseph and Johnston, 1985) this does not mean that the theory could not be said to fit the actions taken. Whether we conclude that an intervention is designed to correct the failure of the market or not hardly matters. What matters is—do we intervene at all? If we decide to intervene, how do we do it? The fact is that many countries intervene in their commercial R & D systems. Their bureaucrats do try to pick winners all the time—in defence, space, environment and health—and there is little evidence to suggest that they are any better or any worse at it than the private sector (Hill, 1985). Private sector failures do not attract as much attention as government failures and in a number of countries, and especially in the USA, there is an important element of plurality in R & D decisions. Many firms make different choices so that if some work out then it looks as if industry is good at picking winners.

GOVERNMENT INTERVENTION IN COMMERCIAL R & D: AN EXPERIENCE

I should now like to report on the positive experience of one country with its interventionist policy. This experience may not be exportable to other countries but it does at least provide something which is notably absent in the literature on government intervention—an evaluation that says something positive about the government's actions. The country is Israel and in 1967 it decided that, in view of lower labour costs amongst its competitors, it could no longer rely on its exports of primary products and textiles for its economic survival. A decision was made to increase the scientific content of its export products through increased R & D. The country adopted a number of principles and these were judged in retrospect to be designed to overcome market failure (Teubal, 1983). The policies that were adopted included the following. First, there would be a policy of neutrality, where flat subsidies would be given without any explicit preferences for any particular branch of industry, technological area or class of product. There was, in other words, to be no key industry or key technology policy. Second, the firms who were supported had to demonstrate the feasibility of the project, the export potential of the product and the quality of their scientists and engineers. Third, the support scheme was to be located in the Ministry of Commerce and Industry, specifically in the Office of the Chief Scientist and not in the government's semi-autonomous laboratories. The latter decision was reached in order to avoid the danger of an excessive academic approach. The share of government funds allotted to the semi-autonomous laboratories declined as the result of this decision from 27 per cent in the natural sciences and engineering in 1966 to 12 per cent in 1978. Firmly established in the minds of the appropriate authorities at the time was an understanding of the difference between scientific research and industrial technology, and an understanding that the latter does not necessarily result from the former.

The policy of neutrality, though modified slightly later, was chosen because of the difficulty perceived in picking winners across a narrow front and because of the belief that natural selection would in the end weed out the losers. An analysis showed that although the number of technologically-sophisticated firms did not alter very much between 1966 and 1975, the mix altered. Further, though some of the survivors experienced significant difficulties in their early stages they were fast entrepreneurial learners, realising that innovative R & D in itself was insufficient for success but that due regard had to be paid to marketing requirements and the complex relationship between the market and R & D. An indicator of the natural selection process is that the share of subsidies granted to electronics firms increased from 56 per cent to 60 per cent in the period 1967-1975.

Evaluation

An evaluation of the subsidy scheme for the period 1967-1975 showed that it was a success. In 1970 only one technologically-sophisticated branch of industry in Israel, the chemical industry, was in the first five exporters; by 1975 there were three—chemicals, transportation equipment and metal products. Electronics went from ninth to sixth in the export stakes and textiles fell from fourth to eighth. The growth of output and exports in R & D-intensive industries was sufficiently impressive to justify a presumption that the arrangement for supporting industrial technology had been successful. Exports from the industrial sectors utilising sophisticated and intensive R & D—electronics, transport equipment, metal products, machinery, rubber and plastics—which received most of the subsidies, have all increased. Exports from projects supported by the Ministry's scheme rose from $US1.6m in 1967 to $US750m in 1979 and topped $US1 billion shortly thereafter.

SUMMARY AND CONCLUSIONS

In this chapter, the central role of technological innovation in the developed economy is highlighted and the question is posed as to whether an activity that is so vital to the economic survival of nations should be left for its development to the entrepreneurs, engineers and scientists of the market. The arguments in favour of government intervention centre on the probability of under-investment in R & D by the private sector due to the problems of inappropriability, uncertainty and indivisibility. The market, it is suggested, fails to call forth the necessary investments.

The counter-arguments, used by those who see themselves as the custodians of the public purse, attempt to discredit the market-failure theory not because they do not think that there is a problem with externalities but because they believe that existing measures to counter these problems are sufficient and measures over and above these only encourage wastefulness of public funds. More academic arguments against market-failure theory, on which the justification for government intervention rests, focus on the theory's reliance on the discredited linear model of innovation. And finally, some commentators reject use of the market-failure theory because it is an easy target for those who oppose excessive government involvement in an activity that they consider to be the responsibility of the private sector.

Few of the arguments against government intervention or the market failure theory rely on any empirical evidence and it is clear that studies of the sort that were carried out by Teubal (1983) in the case of the Israel's system of governmental promotion of technology should be extended so that the debate can become more informed.

REFERENCES

Arrow, K. J. (1962) 'Economic Welfare and the Allocation of Resources for Invention', in National Bureau of Economic Research (eds), *The Rate and Direction of Inventive Activity*, Princeton, New Jersey.

Denison, E. F. (1962) *The Sources of Economic Growth in the United States and the Alternatives Before Us*, Committee for Economic Development, New York.

Eads,G. (1974) 'US Government Support for Civilian Technology', *Research Policy*, 3, 2-16.

Freeman, C. (1974) *The Economics of Industrial Innovation*, Penguin, Harmondsworth.

——— (1979) 'Technical Change and Unemployment', in S. Encel and

J. Ronayne (eds) *Science, Technology and Public Policy*, Pergamon Press, Sydney.

Gannicott, K. G. (1980) 'R and D Incentives', in *Technological Changes in Australia*, AGPS, Canberra.

Griliches,Z. (1958) 'Research Costs and Social Returns: Hybrid Corn and Related Innovations', *Journal of Political Economy*, 66, 419-31.

Hill, C. (1985) 'Rethinking Our Approach to Science and Technology Policy', *Technology Review*, 88, 14-5.

Jewkes, J., Sawers, D. and Stillerman, R. (1958) *The Sources of Invention*, Macmillan, London.

Joseph, R. and Johnston, R.(1985) 'Market Failure and Government Support for Science and Technology', *Prometheus*, 3, 138-55.

Maxwell-Hyslop, R. (1971) 'But How Many Will Fly In It?', *The Economist*, No. 241, 85-6.

Mishan, E. J. (1982) *Economics and Policy Making: the Tragic Illusion*, Greenwood Press, Westport, Connecticut.

Nelson, R. L. (1959) 'The Simple Economics of Basic Scientific Research', *Journal of Political Economy*, 67, 297-306.

Schmookler, J. (1966) *Invention and Economic Growth*, Harvard University Press, Cambridge, Mass.

Teubal, M. (1983) 'Neutrality in Science Policy: the Promotion of Sophisticated Industrial Technology in Israel', *Minerva*, 21, 172-97.

Chapter 22

FORESIGHTING CANADA'S EMERGING SCIENCE AND TECHNOLOGIES

A. Elzinga

This chapter reviews some of the activities and results of the program of Emerging Science and Technologies (EST program) which was established at the Science Council of Canada in 1984. The program was initiated to identify areas of emerging science and technology of key importance to Canada during the next decade. Experience with forecasting in various countries indicates that there are indeed many emerging technologies worldwide. The issue is to identify those that are strategic to Canada's future development. In this respect opinion varies amongst experts in four different constituencies: business, government, labour and the universities.

To resolve these differences and focus on the opportunities, challenges, constraints and possible recommendations, the Science Council launched its experiment in foresight in 1984. Emphasis was placed on consultation and on the process of eliciting expert opinion, as well as on developing a broadly-based consensus from both producer and end-user points of view. Accuracy of detail in prediction was seen to be less important than the actual process of consensus-building and the raising of public awareness.

The basic rationale for this kind of foresight is that early awareness of new technology will enhance the nation's possibility of participation and raise its level of international competition in the economic sphere. Fundamental research and its mastery are, in many cases, a key to the leading-edge technologies based on modern science. Micro-electronics, biotechnology and advanced industrial materials are three such key areas. Furthermore, in this context it is necessary to know not only one's own strength as a nation, but equally to have a clear and realistic picture of weaknesses, current resource limitations and bottlenecks of various kinds.

FROM FORECASTING TO FORESIGHT

Foresight is relatively new as a technical concept in the forecasting literature. In a less formal sense, of course, it is an activity that has been practised by human beings and institutions all through history. The significance of the concept as it emerges at the present time is twofold: firstly, it reflects the terminological shift from a technocratic reductionist position to an emergent historical perspective in forecasting; secondly, it reflects the urgency and scope of many of society's current problems which are linked to advances in science and technologies. The terminological shift from prediction and forecasting to foresight, outlooks and understanding also reflects a reorientation in futures studies from global to more concrete sectoral, regional or branch studies.

In its contemporary specialist definition, foresight ranges from looking at problems immediately ahead, through trends (5-10 years hence), to long-range emerging issues. In the latter case it may project fifty years into the future, as is done for example in some of the visions projected by the Forecasting and Assessment of Science and Technology (FAST) group in Brussels. Foresight is thus a concept covering an activity and methodologies that are somewhat looser than the forecasting of fifteen years ago. Today foresighting is in vogue in advanced industrial countries, but Third World countries are also showing an interest, as witnessed by the Advanced Technology Alert System (ATAS) at the UN in New York, which does networking amongst researchers, policy people and research administrators of the Third World.

In the context of basic research, foresight is an instrument for directing R & D as a strategic potential for development. Irvine and Martin (1984) find that although nations differ in their approaches, there is nevertheless a striking convergence in their identification of a small number of technology clusters as the most important ESTs. In their estimation the Japanese experience is the most interesting because it systematically involves a bottom-up approach to forecasting rather than the centralised top-down approach often favoured by government. It also seeks to integrate both science-push and market - or demand-pull perspectives by involving large numbers of active researchers from academia, government laboratories and industries. The role of government ministries and agencies, it has been found, is best limited to identifying broad trends only, to provide holistic overviews and raise public awareness around these, rather than intervening and trying to pick winners while sitting in a room at the top. The difficulty is to go beyond the broad overviews to specify actual niches of opportunity or how to build upon existing strong points to serve indigenous needs and interests.

METHODOLOGICAL CONSIDERATIONS

There has been debate over two models of connecting research, policy analysis and decision making. The traditional political model corresponds to a responsive mode of behaviour. In it there is 'a conscious use of research by political decision-makers to strengthen an argument, to justify positions already taken, or to avoid making or having to make unpopular decisions by burying the controversial problem in research' (Trow, 1982). In the newer enlightenment or percolation model, research influences policy indirectly, 'by entering into the consciousness of the actors and shaping the terms of their discussion about policy alternatives'. The framework and ambitions of foresighting, which tries to articulate anticipatory intelligence found in a broad range of constituencies and their visions of the future, lies much closer to the percolation model and its rethinking of the tasks of the policy analyst.

Schwarz *et al.* (1982) point to a relationship between choice of methodologies and organisation. In essence they describe two different ways of bringing together expertise; let us call them two ideal types. The one is the integrating-team model, the other the methodology-expert model. In the first case a core group or team covering different disciplinary and sectoral interests tries to synthesise trend information from various sources. The team is accorded a relatively high degree of autonomy in defining the problem and finding the means to develop a futures study report. By contrast, the methodology-expert model relies heavily on input from panels of experts via Delphi exercises or the like. In a review of three Swedish futures studies the authors find that the second model was mostly applied in a situation where the sponsor of a study has a strong interest in directing or steering the process, while the integrating-team approach tends to be more common in situations where the linkages between decision makers and policy analysts is weaker.

The two foregoing models seem to represent two basic points of departure in foresighting. Panels may be set up on an *ad hoc* basis around specialty areas, while teams may work entirely in-house in an agency, or at other times comprise loose networks of investigators working on contract and located in various different places and intellectual environments. In this way the input to a foresight exercise may be varied and to some degree steered.

Questionnaires are often used in a preparatory round to Delphi panels. In connection with foresight, since the scope and process of consensus-building is more important than the predictive value of formalised methods, it is not uncommon to find results of questionnaires being used directly as input to workshops or symposia organised around a specialty area or theme (Newton and Taylor, 1985). The point of departure may either be the technology-producer perspective (in which case science push and actors from research communities dominate) or the end-user perspective (in which case the workshop will try to facilitate the articulation of market prospects, concerns and identification of possible bottlenecks in funding programs, infrastructure, government policies, and even cultural factors). It may be useful to combine, as we did in the EST program, the two approaches in tandem, starting with a polling of expert opinion and plotting clusters of emerging technologies with anticipated benefit, impact or threat as per sector (manufacturing, forestry, agriculture, environment, transport, each of these sectors being broken down into appropriate sub-sectors). A difficulty is that one does not readily arrive at a fine-grained division of end-user sectors, nor are there always clear-cut boundaries between clusters of technologies. Different respondents may identify the same technology from different points of departure and place it in different end-user sectors. The outcome of this exercise has been a focus on four specific interdisciplinary and intersectoral themes for further development after exploration in a series of workshops with a mix of experts. The four themes are:

- *Expert-based servosystems* (especially important for innovation of industrial processes).
- *Algorithms and processors* (for new computer-systems architecture).
- *Micronics* (including interdisciplinary points of intersection or convergence at the sub-micron level between parts of micro-electronics, biotechnology, materials chemistry, surface physics as well as micromechanics).
- *Humanities* (or the human sciences as a resource for technological development).

The motivating force, as in the Canadian case, is international competition with countries like Japan and the need to develop long-range strategic research and technologies. The foregoing themes point to areas of expertise that need to be built up in preparation for the 1990s.

THE STRATEGY OF THE EST PROGRAM

A four-pronged strategy of national consultation emerged (Figure 22.1) to fulfil the EST program mandate thus covers:

- *telephone interviews* with leading scientists and technologists to provide both an initial oversight and discussion on areas which warrant more detailed attention;
- a *mail survey* of over 11,000 researchers, research administrators and others knowledgeable in S & T, designed to obtain a more holistic and refined overview;

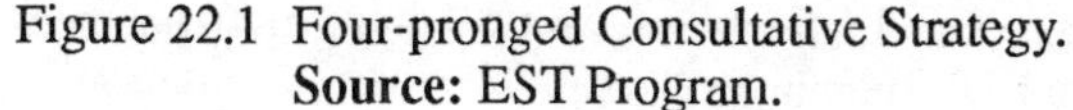

Figure 22.1 Four-pronged Consultative Strategy.
Source: EST Program.

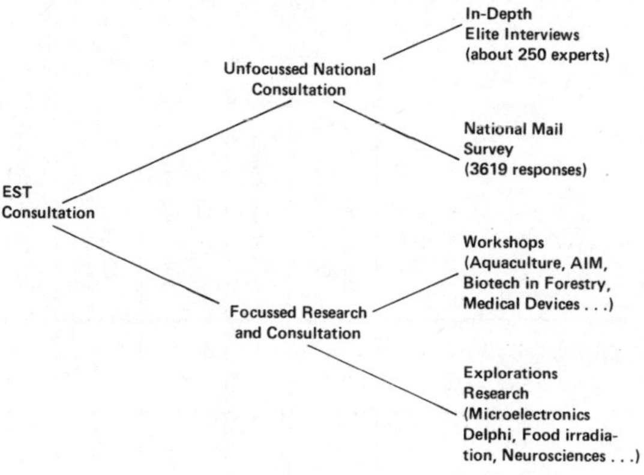

- a *series of workshops*, followed by a report and survey on advanced industrial materials (AIM), the impact of biotechnology on the forestry industry and the future of medical devices and health technologies;
- an *explorations research program*, designed to explore particular areas of opportunity identified by the other three consultative prongs, and to lay the basis for a possible second cycle in the EST program.

The basic design of the program is geared to identifying clusters of ESTs and plotting these against spheres of potential impact in the form of a broad matrix to highlight density points of emerging technologies and sectors of application and impact. The matrix should not, of course, be interpreted as an indication of the potential scale of impact or the priority to be given to a particular technology. Rather, it is an instrument for structuring the EST scan; the density points form points of departure for further consultation and investigation with the help of small panels of experts, background papers and, if need be, larger workshops and conferences where EST producers and potential end-users are brought together. Practically, the procedure may be to use a list (with commentary) of technologies generated in the survey and have it circulated for discussions at a conference. Briefs from such a conference may in turn be used as a basis for further panels or workshops organised, either around a particular technology cluster (e.g. polymers and plastics, or medical devices), or an end-user sector (e.g. forestry, food processing or transportation). In the latter case the point is to determine multi-EST opportunities, potential impacts and possible threats for a given sector.

Figure 22.2 Matrix of Emerging Technologies and Sectors of Impact Derived from EST Telephone Survey.
Source: EST Program—Telephone Survey.

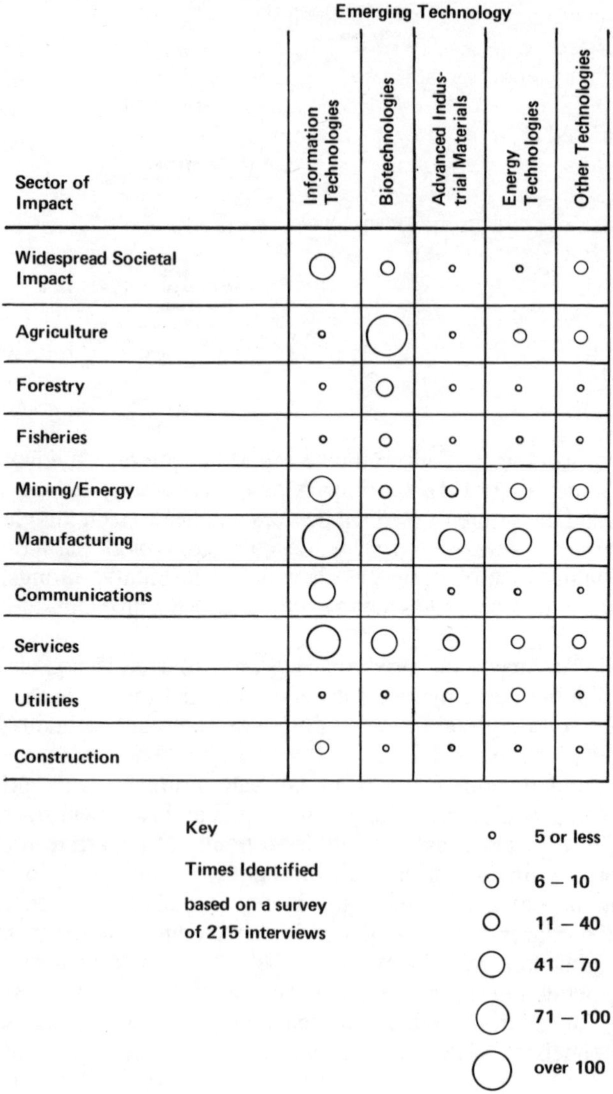

Emerging Technology

Information Technologies
Biotechnologies
Advanced Industrial Materials
Energy Technologies
Other Technologies

Sector of Impact

Widespread Societal Impact
Agriculture
Forestry
Fisheries
Mining/Energy
Manufacturing
Communications
Services
Utilities
Construction

Key

Times Identified based on a survey of 215 interviews

5 or less
6 – 10
11 – 40
41 – 70
71 – 100
over 100

The general idea is that EST clusters (e.g. biotechnology) can be unpacked and their applications plotted in cross-impact *vis-à-vis* agriculture, forestry, fishing, food-processing, health care, etc., sectors. Conversely we can start with a sector (e.g. forestry) and structure the subjective perceptions of future impacts of say silviculture in the biotechnology cluster, sensors or fibre optics in the micro-electronics-telematique cluster, and so on. A schematic representation of the underlying concept is provided in Figure 22.2. In a sense it serves as the conceptual basis for structuring the vast database compiled as a result of the broad mail survey.

In designing the questionnaire we ran into several problems. We attempted various *a priori* approaches to categorising both ESTs and breakdowns of end-user sectors, for example using the Industrial Categories Index or modifications of it. However pilot runs with cross impact formats in the questionnaire were found to be too complex and too constraining for the respondents. After many simplifications we finally ended up using an open-ended questionnaire with five general questions (see Table 22.1).

Table 22.1: Summary of EST Mail Survey Questions

1. Which emerging technology do you consider to be potentially significant for Canadian development during the next decade?

2. Which business or socio-economic sectors will be most affected by this emerging technology?

3. What is your estimate of the probability of commercializing this technology during the next decade?

4. What concern, if any, do you have about Canada's responses to this emerging technology?

5. What are the main areas of your personal science/technology expertise and experience?

The Telephone Surveys

Preliminary to the fully-fledged national consultation-by-mail survey the questionnaire (October and November 1984) served as a basis for structuring a series of telephone interviews with 256 S & T leaders. This was done on contract by four consulting firms. Each was responsible for a particular constituency of experts. At first the consultants covered 205 experts in five groups, and then one of them undertook an additional poll of forty-one experts to complement findings in the industrial domain. The first five groups were as follows:

a. Directors of R & D divisions and research managers in large Canadian *industrial corporations* including some crown corporations.
b. A cream-of-the-crop selection of scientists and engineers in *universities* across the country.
c. Directors of major national laboratories, high level *government* bureaucrats with R & D or S & T policy responsibility, and some high-ranking government scientists.
d. Leaders and policy advisors or planners in *provincial R & D networks* (provincial research organisations, laboratories and special agencies or authorities with specific S & T tasks).
e. A fifth group comprising people in all of the four categories above but within the *francophone community* (mostly concentrated in Quebec).

The group of forty-one added later in a separate telephone survey consisted of:

f. Experts on S & T in either *small* information technology *enterprises* performing R & D or in technology centres.

On the average each consultant dealt with a group of forty interviewees. The questions were sent out beforehand. A time was set up for the telephone interviews, then, on the appointed time, these were carried out. The average time taken was 20-25 minutes per respondent (with extremes of 10 and 70 minutes). Replies were based on the respondents' subjective perceptions and their interpretation of the contextual environment. Some withheld proprietory information, and there sometimes existed vagueness and contradictions concerning the definition of technologies and levels of aggregation. Some included several technologies under one generic name, others combined different ESTs like robots and CAM. Estimates regarding timing of commercialisation revealed a tendency to lack of realism (serious studies often indicate a lead time of anywhere between thirteen and eighteen years for new ideas in laboratories to reach fruitful commercialisation). In the main, however, the respondents' attitudes to the EST interviews were very positive.

At a general level the telephone surveys identified the usual three EST clusters (micro-electronics, biotechnology and AIM). Impacts were perceived across the resource sectors, manufacturing, communications, services, utilities and construction. The category 'widespread' was used to indicate broad multisectoral impacts of a given technology cluster.

At a greater level of detail the first survey gave about thirty specific technology identifications (see Table 22.2). The addition of the sixth group

Table 22.2: Thirty Emerging Technologies and Sectoral Impacts*

Technology	Sector of Impact
Genetic engineering	Agriculture, manufacturing (pharmaceuticals), services (health), forestry, mining
Enhanced chips/ gallium arsenide	Manufacturing (electronic & scientific equipment), communications, defence
Artificial intelligence	Services, manufacturing
Cell & tissue culture	Services (health), manufacturing (pharmaceuticals, food), agriculture, forestry
Microcomputers	Services, manufacturing, defence
CAD/CAM/CAP/CAE	Manufacturing, services, communications
Robotics	Manufacturing, mining
Composite materials	Manufacturing (automobile, aircraft)
Remote sensing	Forestry, agriculture, mining, services, defence
Imaging	Manufacturing (electronics), services, mining communications
Fibre optics	Communications, manufacturing (electronics)
Monoclonal antibodies	Agriculture, manufacturing (pharmaceuticals), services (health)
Computer software	Manufacturing, services, communications, defence
Advanced polymers	Manufacturing
Lasers	Manufacturing (electronics, transportation, medical instruments), services, communications
Synthetic fuels	Manufacturing (refining), energy, services (transportation)
Coal technologies	Mining, manufacturing
Food irradiation	Manufacturing (food, chemicals), agriculture
Telecommunications	Communication, services, construction
Surface chemistry/ plasma technologies	Manufacturing, energy, agriculture, services
Biomass	Manufacturing (chemicals), agriculture, energy, forestry
Hydrogen energy technologies	Manufacturing, utilities, energy
Separation and membrane technologies	Manufacturing (food, chemical)
Fermentation	Manufacturing (food), agriculture
Structural ceramics	Manufacturing (metal, transportation)
Optoelectronic/storage systems	Communications, manufacturing (electronics)
Construction technologies	Construction, mining
Speech recognition	Manufacturing, services
Photovoltaics	Manufacturing (electronics), communications
New alloys	Mining, manufacturing (transportation)

* The emerging technologies are listed in order from the most to least frequently identified.

added many more. Council was provided with the foresight of the 256 interviewees summed into a list of fifty-seven ESTs that the interviewees regarded to be of strategic importance for Canada in the next decade.

The list of fifty-seven ESTs has been circulated to the membership of the Canadian Research Management Association (CRMA), who were asked individually to assign priorities to the list and explicitly state their criteria for choice. In this way the EST program was further assisted in determining a number of focal points for future workshops on specific technologies (several of which have taken place to date; cf. Figure 22.3).

Figure 22.3 Emerging Science and Technology Program 1984-85.
Source: EST Program.

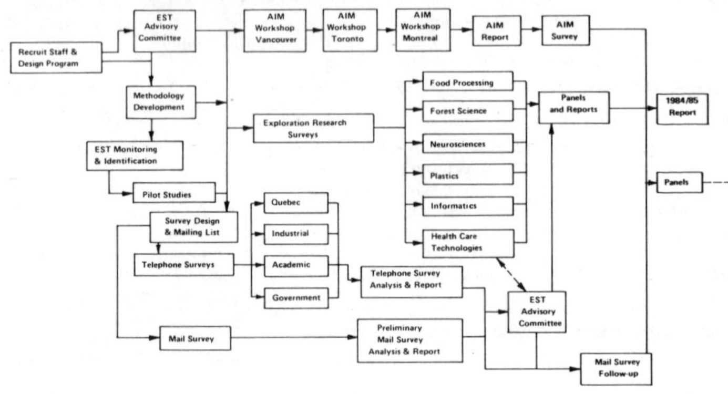

The National Consultation by Mail

The broad, national consultation was finally undertaken in November 1984. This took the form of a mail-out survey whereby the questionnaire was sent to 11,584 Canadians involved with R & D. The population polled included university and government scientists and engineers, R & D managers and chiefs of industry, high and medium-level government people, trade unionists and others who were presumed to be knowledgeable about ESTs. The mailing list was generated with the help of research-grant lists from the Natural Sciences and Engineering Research Council, the Medical Research Council, industrial directories of firms, manufacturing and industrial branch associations, high-tech enterprises, contacts with Federal and Provincial ministries and laboratories, and in-house information from the collective memory of the SCC itself.

In total 3,619 responses to the questionnaire were received and entered into a computer database using a text-recognition retrieval system. The main patterns of responses were tabulated and analysed. Sixteen emerging technology clusters were identified. Apart from the usual top three core technologies (micro-electronics, biotech and AIM) there are four resource clusters (energy, forest, mining and food), two industrially related clusters (manufacturing and transport), and a variety of single issue clusters that may be listed under headings like environment, space, medicine, science and planetary science. Two particularly interesting clusters in the Canadian context are instruments and cold-northern technologies. The clusters are disaggregated in various ways.

In the micro-electronics cluster, gallium arsenide and enhanced chips are seen to be important. Transmission and carrier technologies also received frequent mention. In biotechnology the sectoral impacts were emphasised as most important, especially in the resource-based industries, but also for medicine and pharmaceuticals. Many respondents believe Canada is 5-10 years behind in biotechnology research, and a vigorous plan for an organised industrial thrust is called for. Within the AIM cluster advanced structural materials, electronic materials and the new processes to make them were emphasised, as expected. An interesting point, however, is the attention given to the need to upgrade technologies in traditional materials like wood, paper, concrete, glass and asphalt. Canada has strong points in the plastics industry and in the development of metal alloys.

In the resource clusters, respondents identified surprisingly few ESTs. In many areas foreign domination of these sectors is seen to be a blocking factor. Manufacturing and transportation also evoked surprisingly little enthusiasm, perhaps for the same reasons. Advanced scientific research equipment design and manufacture, robotics, automation and quality control was found to be a potential niche for Canada: sensing and measuring parameters in the environment and sending signals to a decision-making system. Remote sensing systems are also significant in geophysics and environmental geochemistry, areas in which Canada is known to be a world leader in exporting expertise. The space sector, on the other hand, is less well-defined in Canada, despite isolated successes in specific projects like the CANADARM.

Finally cold-northern technologies is a small but strategically important cluster that ought to evoke optimism as to Canadian leadership potential. This cluster includes northern environmental issues and vessels for offshore Arctic petroleum, ice engineering, transport and communications for the Arctic, as well as cold-region agriculture and forestry. Here it is a question of moving rapidly, perhaps in joint ventures with Swedish or Japanese companies.

The most surprising and strongest feature emerging from the analysis of the mail-survey results is the stunning sense of defeatism expressed by Canada's scientific-technological intelligentsia. In the words of one of the consultants, one gets a picture of

> . . . the typical Canadian researcher as an individual with considerable technical expertise, frustrated to the point of despair, unable to mobilise the resources to pursue research or commercialise a product, smarting at unfavourable comparisons of Canada with other countries, lacking the confidence to initiate change, mistrustful of other sectors, and habitually negative toward Canadian endeavours while at the same time obsessed with a "uniquely Canadian" industrial thrust.

'Too late', 'Too little support', 'Behind yet again', 'Opportunity lost to the Americans'—these are the prevalent expressions and attitudes. (The quotation is from a draft of a paper by Steed, 1985, presented to the CRMA.)

This shows that Canada, or at least its scientific intelligentsia, is still suffering from the remnants of a colonial mentality, reinforced by an economic base dominated by US multinationals. We have here an example of the importance of *cultural* factors that must be considered when developing national science and technology policies. Perhaps it might be right to say that Canada needs a cultural revolution so that its economic base may be transformed on the basis of a combination of high-tech *and* traditional resource-based industries.

In addition to this major problem the survey has focused attention on a number of specific issues like funding, S & T infrastructures, manpower needs, etc., where policy instruments must be evolved to help turn frustration into enthusiasm. Two different strategies must be considered: building on existing strengths and/or developing strengths in niches. The first reflects the resource-based industrial interests, the second the interests of venture capital and a new class of high-tech entrepreneurs. The two partially contradictory schools of thought will have to be articulated into a realistic strategy for the future.

CONCLUDING REMARKS

The four-pronged strategy of the EST program has covered considerable ground. The emphasis on the consultative nature of the exercise has shown itself to be fruitful in many respects. The mail and telephone surveys have involved many people. They have given both interesting results and aspects for further investigation, as well as helping to raise the level of foresight and public awareness concerning ESTs.

Follow-up workshops and conferences have already covered a number of areas and end-user sectors. Reports are continually fed back to the different constituencies involved, and the Council in turn receives indications of significant effects of the consultative process it started (Anderson *et al.*,

1985; Elzinga and McCutcheon, 1985; Science Council of Canada, 1985; Cordell, 1985). To cite one example, a major conclusion of the Advanced Industrial Materials workshops was that Canada lags a decade in advanced materials. Participants agreed that advances in materials technology in the United States, Japan and other countries may have serious repercussions on the Canadian economy. New materials, such as composites, ceramics and polymers are replacing traditional ones, such as the metals that Canada exports. The constraints impeding advance in these fields were seen to include the relative lack of corporate R & D funding, the smallness of the domestic market (too small to pull the development of new materials), and the tradition fostered by reliance on the export of raw materials and the foreign ownership domination of a branch-plant economy. The latter two factors have discouraged innovation in materials.

Indeed, there is no lack of ideas, or themes for further activities. The need for the type of foresight provided by the EST program has also shown itself to be vast and crucial. Also the program, after eighteen months along its own learning curve stands well equipped methodologically to generate even better consultation and insight concerning Canada's industrial and technological futures. The scandal is that today the Council is not in a position to match this potential with either budgetary or personnel commitments.

ACKNOWLEDGEMENTS

I want to thank the SCC for permission to use some of the material from the EST program, including a few drafts at the time of writing. In particular I want to acknowledge the support given me by James Gilmour, the Director of Research, and Guy Steed, the Acting Director of the program. Scott Tiffin and Susan Mills have also been helpful in keeping me informed about the EST work. The final responsibility for what appears in this chapter, of course, rests with the author.

REFERENCES

Anderson, F., Bobkowicz, A. and Gerson, F. (1985) *Emerging Plastics Technologies and the Canadian Plastics Industry*, Discussion Paper, EST Program, Science Council of Canada, Ottawa.

Cordell, A. (1985) *The Uneasy Eighties: The Transition to an Information Society*, Background Study 53, Science Council of Canada, Ottawa.

Elzinga, A. and McCutcheon, S. (1985) *Advanced Industrial Materials:*

Canadian Perspectives and Opportunities, Discussion Paper, EST Program, Science Council of Canada, Ottawa.

Irvine, J. and Martin, B. (1984) *Foresight in Science: Picking the Winners*, Frances Pinter, London.

Newton, P. W. and Taylor, M. A. P. (1985) 'Probable Urban Futures', in J. F. Brotchie, P. W. Newton, P. Hall and P. Nijkamp (eds) *The Future of Urban Form. The Impact of New Technology*, Croom Helm, London.

Schwarz, B. Svedin, U. and Wittrock, B. (1982) *Methods in Futures Studies. Problems and Applications*, Westview Press, Boulder, Colorado.

Science Council of Canada (1984) *Aquaculture: A Development Plan for Canada* (Task Force Statement), Ottawa.

——— (1985) *Seeds of Renewal. Biotechnology and the Canadian Resource Industies*, Ottawa.

——— (1986) *The Relentless Sweep: A National Consultation on Emerging Technology*, Discussion Papers, Vol. I by Guy Steed (on the telephone interviews), Vol. II by Scott Tiffin with the assistance of F. Anderson and M. Wallis (on the mail survey results), Science Council of Canada, Ottawa.

Steed, G. P. F. (1985) 'Emerging Technologies—The Science Council's Program', a speech delivered to the Canadian Research Management Association, Jasper, Alberta, September.

Trow, M. (1982) 'Researchers and Policy-Makers in Education: How do They Relate?', Paper presented at Wijk Symposium, Stockholm, June.

Chapter 23

TOWARDS HIGHER HIGH TECHNOLOGY POLICY

S. Macdonald

High technology is widely perceived as an almost cost-free economic nostrum for most of the ills currently afflicting developed economies. This is not a realistic view, but it has nonetheless inspired a proliferation of high-technology policy, much of which seems to be correspondingly unrealistic. It is worthwhile considering both the supposed benefits of high technology as well as some of the possible costs, especially those likely to be associated with inappropriate policy. Much policy, especially that devised by local governments, seems to be based on preconceptions and misconceptions about high technology. Desperate enthusiasm is an inadequate substitute for understanding of high technology, and understanding is not encouraged by the confusing and conflicting findings of much research in the area. A new perspective altogether is required, one that can be gained by focusing on the information requirements and characteristics of high-technology industry.

COSTS AND BENEFITS OF HIGH TECHNOLOGY

High technology, whatever it is (given the problems of definition; see Macdonald, 1983a, 1983d), is a rare and precious economic activity in that it is widely assumed to deliver prodigious benefits while imposing almost no costs at all. With a benefit:cost ratio approaching infinity, high technology has captivated those responsible for the economic management of whole villages, and whole towns, and whole regions, and whole nations. These entranced individuals may well be more numerous now than the managers in high-technology industry subject to their policy. Such rampant political enthusiasm is perhaps surprising: political fingers were just about all that were even singed by the 'white heat of the technological revolution'; it seems only yesterday that technology was firmly cast in the role of despoiler of the workforce. But a week is a long time in politics (as it happens, in high technology, too), and the easy benefits claimed for high technology have proved irresistible. So have the apparently low costs, not just of high-technology industry in action, but of programs to stimulate that action. In fact, the benefits to be gained from high technology have probably been grossly over-estimated, and the costs—especially the indirect and hidden costs—may actually be quite considerable.

Employment

Most prominent by far among the benefits claimed for high technology is its apparent ability to create employment in the depths of prolonged recession. Even here, though, there is some dissent; one study suggests that most high-technology industries in Britain have contributed to decline rather than increase in employment (Breheny and McQuaid, 1984, p. 8). Such sobriety is quite out of place among the usual estimates of employment generated by high technology. According to one American study, 75 per cent of the net increase in manufacturing industry employment in the United States is attributable to employment in high-technology industry (Premus, 1983, p.36). That sounds staggeringly impressive, proof positive of high technology's social and economic value. But just what increase has there been in manufacturing employment? In all developed countries manufacturing-industry employment has been in relative—and sometimes absolute—decline. Three-quarters of very little is even less. Very often the base from which high-technology employment is measured is so tiny that percentage increases are positively misleading. High-technology industry is not a large employer now, and there is no prospect of it becoming one in any future that is even dimly foreseeable (Rumberger and Levin, 1984; Riche *et al.*, 1983). Real job growth has been not in the manufacturing sector at all, but in the tertiary sector and especially in some of the less salubrious service functions. Employment in the fast-food industry is greater than in high technology, and it has grown faster both proportionally and absolutely, all without a hint of policy to encourage the consumption of hamburgers (Windschuttle, 1983). Moreover, it is in such unglamorous sectors that most future employment growth is forecast.

Of course, it could be argued (and is argued) that low-grade service functions create jobs mainly for the semi-skilled and unskilled, and do not make full use of human capital. Unfortunately, the same is true of a lot of high-technology employment. That is why so much of the American industry is located in cheap-labour countries. Highly-qualified professional employees may be more common in high technology than in other industries, but the typical employee is poorly-paid, part-time and female, with the daily prospect of tedious, repetitive work, no union protection and no career structure (Tomaskovic-Devy and Miller, 1983). There is even substantial and illicit back-alley employment associated with some high-technology industry in the United States.

The products of high-technology industry may, of course, help generate employment in other industries altogether. They may equally well reduce that employment by allowing what has become known as 'jobless growth'. It all depends on how the technology is used and on the facility with which organisations adapt to change (Mandeville and Macdonald, 1983). If firms in established industries cannot seize opportunities presented by new technology, or if they adopt that technology with no other intention than to cut labour costs, then high technology may very well reduce rather than increase overall employment. Certainly there is no justification for the bold assumption that high technology necessarily generates employment in the rest of the economy (see Markusen, 1983).

Growth and Wealth Generation

A perpetual dilemma is faced by those whose political life cannot be guaranteed to exceed the time taken for policies to yield their benefits. The problem is thought unlikely to arise with high technology because returns are imagined to be not only huge, but immediate. Certainly the industry has grown, and continues to grow, rapidly. By one calculation, high-technology equipment now accounts for 22 per cent of business capital in the United States compared with just 7.7 per cent in 1952 (Roach, 1983). While it is incontrovertible that some high-technology firms have flourished beyond perhaps even the imaginings of their founders, and have made fortunes for their owners, that is not typical. Many high-technology firms fail, most just linger, and only a very few have instant success. The few, however, are often presented as if their fate were inevitable, as if a major share of the international market were always assured, as if the generation of spin-off firms somehow defies the commercial laws of natural selection. That is most misleading. Moreover, the wealth that is undoubtedly created in some quantity by these few, is likely to be poorly distributed. Whatever the benefits of wealth generation, they should be discounted for the social costs of their maldistribution.

Environmental and Health Advantages

Unlike the smokestack industries with which they are commonly compared, high-technology industries are supposed to be clean and healthy, a boon to their employees and the environment. Just why comparison should always be made with the older parts of manufacturing industry, which now provide relatively little employment, is something of a mystery. Comparison with other growth sectors, with office or even fast-food employment, might be more meaningful, though less impressive. But the smokestack industries offer ample visible evidence of pollution and poor working conditions: in high technology the evidence is simply less visible. In fact, chemical pollution is now a serious problem in the Mecca of high technology, Silicon Valley (LaDou, 1984). So, too, are all those problems traditionally associated with smokestack industries: over-crowding, traffic congestion, smog, poverty and even hunger. It is quite wrong to imagine that thriving high technology is kind to the environment and gentle with humanity (Rogers and Larsen, 1984).

Re-structuring of the Economy

Of all the proclaimed benefits of high technology, its role in the re-structuring of developed economies is certainly the most important and probably the least contentious. The primary and secondary sectors have long ceased to produce enough wealth, or the right sort of wealth, to satisfy the requirements of modern society. The tertiary sector—the service sector—has grown prodigiously and now dwarfs the other two. Its function is not only to organise the production of wealth, but also to produce its own, though often in an intangible form, not always familiar as wealth. Much of this wealth, and a good deal of the wealth produced within the secondary sector and even the primary sector, is in the form of information.

The diffusion of information technology is only an indicator of the information revolution that is taking place; the force of the revolution is information itself. The OECD has performed valuable work in measuring the information sector and has shown that information workers comprise between one-third and one-half of the workforce in developed economies (OECD, 1981; see also Porat, 1978, p.4). The Economics of Information has begun to flourish as a sub-discipline to consider the distinctive properties of information as an economic good (see, for example, Machlup, 1962; Lamberton, 1971; Arrow, 1979). But in proportion to its importance as the major form of wealth in a developed economy, and a prerequisite for the production of all other wealth, far too little is known about information and its organisation. Government policy, in particular, seems to be crippled by an attitude which still regards the manufacturing sector as the only possible engine of growth, which recognises wealth only in tangible form, and which values information as just another input to the production of 'real' wealth (Mandeville *et al.*, 1983). Government high-technology policy is no exception.

High-technology industry is information intensive. High-technology firms survive by their skill in gathering information, using information and selling information (and not necessarily embodied in hardware). In their information orientation they are much more akin to a research consultancy or an advertising agency than to any smokestack firm. But high technology is not a service activity and its growth is less constrained by stagnancy in other sectors of the economy. High-technology industry uses information resources that many other industries have either neglected or used inefficiently (Lamberton *et al.*, 1982). Only those parts of the service sector which have managed to exploit information in human capital have shown much facility for handling information. Other parts of the economy, perhaps all other parts, must acquire the same facility and must undergo the re-assessment and re-organisation resulting from the new utilisation of a major resource. High-technology industry, by example, by its own growth, and by the diffusion of its products through an economy, can play an essential leadership role in what is proving to be a painful re-structuring process.

COSTS OF HIGH TECHNOLOGY POLICY

Contrary to popular opinion, high technology does not produce benefits without any costs. No activity, even the most magical, does. Some of the costs are simply the corollary of benefits unlikely to eventuate—the social and environmental problems emanating from high technology, for example. Some, such as problems associated with inequities in the distribution of high-technology wealth, are costs which accompany the benefits that are realised. Other possible costs are more subtle and are much more the costs of high-technology policy than of the industry itself.

Neglect of Comparative Advantage

Because the market is thought unlikely to work well in such a new and uncertain area as high technology, governments interfere to rectify market imperfections. Governments decide where limited resources may best be applied; they should also consider just where the comparative advantage of their constituencies lies. Advantages that are currently claimed by regional and urban governments are often impossible to take seriously. In Australia, clean water is said to give Canberra an edge, in Tasmania it is pure air, in Perth and Adelaide tertiary education, in Newcastle and Wollongong the existing electronics industry, and in Queensland 'bright people' and an 'established reputation' (detailed references can be found in Macdonald, 1983b). Most governments seem to assume that their countries or regions or cities have an obvious comparative advantage in high technology and, attracted by its benefits, are anxious to direct resources towards that activity. These governments cannot all be right and to the extent that they are wrong, costs are incurred. There is an obvious cost of incentives to encourage companies to locate in the wrong place (Secretary of

State for Trade and Industry, 1983, p.4). In particular, traditional and less glamorous industries are likely to be neglected in order to support high-technology industry (David, 1983). Without any careful assessment of policies, the precise magnitude of these costs is likely to remain hidden. The euphoria for high-technology policy does not encourage such careful assessment.

The location of the Microelectronics Computer and Technology Corporation (MCC) provides just a single example of the frantic, almost mindless, scramble of cities for high technology. Some fifty-seven American cities tried to lure the venture; by May 1983 a short-list had been compiled of Atlanta, the Research Triangle of North Carolina, Austin and San Diego. Austin was selected and San Diego beat its breast, blaming its low education budget and high housing costs, and complaining that Austin was a bad choice because it had poor traffic planning and flood control, and lacked prestigious football and baseball teams. Ultimately the decision was simply dependent on the amount of money each city was willing to pay the MCC. Austin paid $63,500,000 in land, new university facilities and an executive jet (Tempest, 1984).

Inappropriate Policy

If high technology is important, then it is also important that high-technology policy be correct—rather than simply highly visible—if opportunities are not to be missed and resources wasted (Bollinger *et al.*, 1983). There are costs associated with ill-conceived policy, and probably particularly with the general assumption that high technology is a single industry with uniform requirements. It is not; its activities stretch across the whole spectrum of industrial activity and firms are likely to require programs tailored to their particular circumstances. They also require co-ordination of high-technology policy with other policies (Office of Technology Assessment, 1984). For example, national policy to guide high-technology firms to their best competitive location is likely to be frustrated by local policy to attract those firms, however inappropriate the location, or by other national policy persuading firms to go to areas of, say, high unemployment. In Britain, high-technology companies have shown a clear preference for areas without development incentives (Walker, 1982; Thwaites, 1981). Inmos, located in Bristol and South Wales, was both a beneficiary of British high-technology policy and a victim of British regional-development policy. Similarly, there are obvious costs when one region or one city competes for high-technology industry by out-bidding the incentives of others. That is parish-pump mercantilism, and literally a policy for high technology at any cost. But there may be competing policies even within a single government as individual departments seek to exploit their own high-technology policies to further their cause in internecine power struggles. In Australia, where State governments offer incentives to discourage high-technology firms from locating in other States (see LeBlanc, 1984) and thereby negate much Federal high-technology policy, competition certainly descends to departmental level. One unit actually offers as an incentive the ability to circumvent the high-technology regulations of another unit in the same department, but the usual weapon is simply visibility, with each department striving to be seen to be doing most for high technology (Feller, 1984).

Market Distortion

There is but a small step from regarding high technology as a wealth-creating activity increasing the prosperity of the economy, to regarding high technology as a cost to be borne for the prosperity of the economy. Government incentives for high technology are subsidies and, much like tariff protection, help to shelter that industry from competition. The spectre of highly-protected high technology looms large in many countries, created by a formula of infant-industry arguments, examples of government assistance to high technology overseas, defence considerations, protection already afforded to established industries, and acknowledgement of the importance of a high-technology infrastructure to other industries. In fact, a highly-protected, high-technology industry would impose severe costs on the rest of the economy. Not only would high-technology purchases be more expensive, but high technology's leadership role in re-structuring would be substantially weakened, and a new distortion in the market would deprive other activities of resources they would otherwise have attracted. To the extent that policy distorts what is already a highly-imperfect high-technology market, it makes more confusing the market signals on which business decisions must be made. Similarly, rampant speculation in high-technology shares is not altogether a cause for celebration; it confuses the market and makes investment hazardous.

Assessment

It is difficult to know just what high-technology policies have achieved so far. Certainly there is now more high-technology industry than there was, and there are many more governments at all levels with high-technology policies, but there is not necessarily a causal link. There are three problems in any assessment. First, the vast majority of high-technology programs spring from policies which have been operating only a few years and from which little can yet be expected. Second, and notwithstanding, those responsible for high-technology policies tend to make immoderate claims for the success of their programs. Outstanding examples of prosperous firms are presented as typical; failures are conveniently forgotten.

Third, it cannot be known what developments would have taken place in the absence of existing policies. British enterprise zones, for example, are said to have succeeded in attracting new firms, but many of these have taken advantage of the incentives offered by simply re-locating within the same region (*Management Accounting*, 1983). Similarly, many local high-technology policies seem to encourage firms simply to become itinerant.

REQUIREMENTS OF HIGH TECHNOLOGY INDUSTRY

There are now a number of studies, most of them very recent, which have sought to determine which locational factors are most conducive to high-technology industry. Their conclusions just add to the confusion because it would seem that a variety of factors is important and that there is little consistency in the variety. Really this should cause little surprise: high-technology industry is not a single entity, but a wide variety of industries composed of tiny firms and multinationals, firms exploiting different technologies in different ways for different purposes, firms whose characteristics and requirements are apt to change rapidly (Armington *et al.*, 1983). For some firms, the availability of venture capital may be crucial, for others government procurement will be essential, and for yet others a competent supply industry is a prerequisite. Consequently, any high-technology policy based on the simple assumption that there are just two or three universal prerequisites for all high-technology firms—say, a convenient airport, a nearby university and pleasant surroundings—should be treated with considerable suspicion.

If only high-technology industries had conventional locational requirements, if only they would demand nearby markets and a local labour supply and proximity to raw materials, then conventional industry policy would be appropriate. But what conceived and nurtured smokestack industry will not bring forth high-technology industry; on that, agreement is unanimous. Concomitantly, conditions may be thought to be conducive to high technology simply because they are the very opposite of those required by smokestack industry—the green and pleasant high-technology land distant from the dark Satanic mills. Most fascinating of all is the common assumption that, because conventional location criteria are irrelevant, high technology is a footloose and fancy free sort of activity that can be made to thrive anywhere under any conditions, if only there is sufficient political will, if only the policy is right.

What is so commonly forgotten is that high-technology industry's main requirement is information, and that it can thrive only where vast quantities of appropriate information are readily available. Much of this information is not technical at all, but the commercial, production and marketing information necessary to effect successful innovation (Macdonald, 1983c; Morkel, 1984; Bickerstaffe, 1984). While high technology is commonly seen as a scientific activity dependent on basic research, high-technology industry is really almost non-scientific in its desperation to avoid undirected inquiry (Braun and Macdonald, 1982). It is a black-box activity; the object is to make the box work for someone without ever having to open it. The means by which information flows—the information channels—are crucial to all high-technology industries. Of all the ways in which information can flow, personal and informal contact is by far the most useful for high technology. Though formal channels such as licensing and journal articles supplement the informal, only information on the hoof provides the industry with the comprehensive, specific and immediate information it requires. This is why so many surveys of the requirements of high-technology industries reveal the vital importance of qualified manpower, though this is often interpreted by those responsible for policy as an indication of the need for more graduates or better manpower planning. They miss the point altogether: what is in demand is not anyone with good qualifications, but specific information that is embodied in key individuals. High-technology policy must permit these prima donnas to be treated as very special people. Pleasant parkland is unlikely to provide sufficient inducement; stock options and the lures of headhunters might. Similarly, much is made of the need for concentrations of high-technology firms (see Oakey, 1984) without any apparent appreciation that it is largely the means by which information flows in high-technology industry which make agglomeration so conducive to its prosperity.

HIGH TECHNOLOGY POLICY AND INFORMATION

There are just two basic issues which should be considered by those who would shape policy for high-technology industry. The second is whether the policy will satisfy the information requirements of the industry. The first is whether there should be any policy at all. High-technology industry has a strong entrepreneurial tradition in its heartland, the United States, and it could be argued that government interference is more likely to stifle than promote independent vigour. Moreover, governments are not renowned for the intimacy of their relationships with small, new firms rather than with corporate giants, they are not good at picking technological winners, they are not at ease with great uncertainty, and they can move frustratingly slowly when speed is of the essence. Even so, it is naive to think that in a mixed economy there is any area of economic activity independent of government influence. The only way to have no high-technology policy would be to have a policy to have no high-technology policy. Even the United States has never taken that course, and recently has adopted much more overt policy in support of its high-technology industries.

Too often the information requirements of high technology industry are either ignored in policy or are not afforded the central position they deserve. Too often those responsible for policy have only a vague idea what high technology is. High-technology industry is seen to be whatever fills Silicon Valley and therefore the policy task becomes the replication of Silicon Valleys in all sorts of unlikely spots. Already there is said to be a Silicon Gulch in Austin, a Silicon Bayou in Louisiana, Silicon Glen in Scotland, Silicon Alley in Sydney, and Silicon Freeway in Southern England. What is missed in all this energetic innocence is that the original Silicon Valley was planned by no one (neither was Route 128 planned; see Dorfman, 1983), though there are those who would apply retrospective policy and attribute Silicon Valley to the ambitions of Frederick Terman and the presence of Stanford Industrial Park. In fact, the Industrial Park is a product of the Valley's development and not vice versa (Board of Trustees, Stanford University, 1974). It is, of course, possible to mimic the obvious characteristics of Silicon Valley elsewhere, to construct what are really no more than Hollywood film sets of Silicon Valley. Those characteristics can hardly be missed by the thousands of policy makers from foreign governments who are conducted through the Valley each year, though its less desirable attributes—the expensive housing, the uncertain electricity and water supplies, the chemical pollution, the smog, the overcrowding, the traffic congestion, the poverty—do seem to be overlooked (Rogers and Larsen, 1984). What are also missed, because they are intangible, and no one is really looking for them anyway, are the intricate networks of surging information channels which supply high-technology industry with its basic requirement. Key personnel in high technology know other industry experts and mobility is high. For such people, work substitutes for religion and success for heaven. That is attained through a vicious dedication to acquiring and using information. Appropriate and effective high-technology policy must not only allow for such individuals, but also attract and stimulate them. It will be worthwhile looking at typical high-technology policies in the light of this criterion.

Regional and Urban High Technology Policy

At regional and urban level there is little evidence of high-technology policy addressing this matter at all. For example, Bradford, in Yorkshire, advertises itself as an ideal site for micro-electronics firms simply because its workers are 'dextrous, skilled, hard-working, cheerful and friendly' (*Punch*, 5 May 1982, p. 11). Kyushu, in Japan, is supposedly perfect because it has lots of clean water, lots of airports and lots of women (*Japan Quarterly*, 1982). There is no shortage of wondrous high-technology policies, but they are often the product of enthusiasm and parochialism, a dangerous combination which is poor substitute for reason and allows little opportunity for rigorous evaluation. (The Office of Technology Assessment (1983) has identified over 200 States and local authorities with policies directed towards high technology. The variety was enormous, the results variable and assessment almost non-existent.) There can be few urban areas in the developed world now which do not boast some sort of policy to attract and stimulate high-technology industry. The benefits have been assumed, the costs ignored, and high-technology policy, rather than high-technology industry, has flourished. No one seems to realise or care that the world cannot sprout hundreds of new Silicon Valleys, each oblivious of the others and each exploiting a presumed comparative advantage in biotechnology, micro-electronics and information technology to supply the same international market. The situation has become quite preposterous (Macdonald, 1983d).

Among the most common urban high-technology initiatives, perhaps because it is most visible, is the science-technology-research park (the distinction is no more clear than the definition of high technology itself), offering a pretty landscape, space-age buildings, potential agglomeration, insulation from nasty industrial activity, a regulated environment, and—above all—the proximity of a university whence vital information is intended to flow. Technology parks are fascinating; they are high technology as seen by urban planners, developers and architects, who have transmogrified something strange into something they can handle. There are now several hundred technology parks, and hundreds more being planned (South, 1981). They are often established in defiance of any consideration of comparative advantage, of what is known about the requirements of high-technology industry, and of what little experience has been accumulated on the success of such ventures. The last would indicate that many technology parks are really just industrial parks with pretensions, that success is rare and likely to take decades, that high-technology demand is for adaptable and functional buildings rather than for futuristic ones (Debenham, Tewson and Chinnocks, 1983), and that formal, imposed agglomeration is not quite the same thing as informal contact. As for the importance of pleasant surroundings, the work of the Berkeley group suggests that high-technology growth has been greatest in those urban areas which have high pollution, high housing costs, and which lack cultural amenities (Glasmeier *et al.*, 1983, p. 63). As for the mandatory nearby university, it would seem that most high-technology firms have little interest in the sort of information locked in universities and consequently have few links with those bodies (Debenham, Tewson and Chinnocks, 1983, p. 79; Herring, Son and Daw, 1983). The firms on Cambridge Science Park—the oldest in Britain and said to be the most successful—have little contact with Cambridge University (Moore and Spires, 1983; Carter and

Watts, 1984). They do, however, value the commercial advantage bestowed by the name (Willey and Jones, 1982). Nor are local universities necessarily a fruitful source of putative high technology entrepreneurs, even in the United States, where industrial ambitions are not a threat to academic status. Of the 243 high-technology firms founded in Silicon Valley in the 1960s, only six had full-time founders direct from Stanford University (Cooper, 1973).

It is hard to hold out much hope for urban high-technology policy based on a technology park—especially when the whole scheme starts, as it did in Perth and in Adelaide, with the contribution of the landscape architect—but it is harder still to see what else local authorities can do (some suggestions are contained in Lowe, 1984). Innovation centres, providing accommodation, communal facilities and, much more important, referral advice, are popular options, but they usually assume that high-technology is necessarily a small-scale, even amateurish affair and that is patently not always the case (see Udell and Venkatesan, 1980). In Australia, such centres are often expected to cope with both high technology and individual inventors. Local government may also take on the role of information broker, a function of some importance in an industry dominated by new entrants to new markets. It may tackle the problem of existing local planning classifications and regulations which are often totally inappropriate for high technology, combining office, research and manufacturing functions. It may create committees and councils, give grants and scholarships, publish glossy pamphlets and engage in similar activities which give the illusion of action, and which might actually serve some purpose in raising public awareness of high technology. Beyond that there is only the range of incentives normally available to industry—the cheap loans, the tax holidays, the subsidies and concessions. Preferential treatment may be afforded high technology industry, but only if it can be determined exactly what high technology industry is. Under some present arrangements, much high-technology industry does not qualify for existing incentives designed to encourage the production of tangible wealth in the manufacturing sector.

National High Technology Policy

It is difficult to escape the conclusion that high-technology policy is best left to nations rather than to regions or urban areas. National high-technology policies are generally far from impressive, but at least they do avoid the silliness of most local policy. So much of the infrastructure which may be appropriate to high technology—the venture capital, the secondary share market, tax concessions, technological and commercial education, government procurement—can be provided only at the national level, and only at that level can comparative advantage be considered with the seriousness it deserves (Senker, 1983). Moreover, failure is an integral part of high-technology industry (Brophy, 1983), and national governments are best equipped to absorb failure, though they have yet to show much interest in developing high-technology policy specifically designed to cope with failure (Backus, 1984). The assumption is that with the right policy, high-technology industry will always succeed.

That is not to say that local governments should do nothing; they can adapt national policy to their own requirements, not least by helping high-technology industry through the bureaucratic maze associated with central government incentives, by making evident the variety of facilities high-technology industries require, and by lobbying with the strength of local experience for more sensible national policy. There are examples aplenty of misconceived national high-technology policy—the current fad for co-operative R & D exemplified in the MCC project in Austin, or the simple emphasis on research for 'enabling technologies' central to the Alvey Programme in Britain, or the current attempts of the United States to control the export of high-technology information (Macdonald, 1986). Moreover, much that is wrong with the existing environment for high-technology industry is a consequence of market imperfections created by totally separate policy. Much might be gained by exerting pressure to remove some of those imperfections at national level rather than by creating yet more at local level.

To stand any chance of being effective, high-technology policy must be integrated and co-ordinated with other policies, including other high-technology policies. That demands co-ordination at the national level, ideally at international level when variations in national policies bring allegations of unfair competition in high technology (Marcum, 1983; US Department of Commerce, 1983, pp. 26-29). Local authorities seem capable of only cut-throat competition to entice high-technology industry at almost any cost, with little consideration of comparative advantage. Such policy benefits no one, ultimately not even those high-technology firms which would play off one local authority against another (Macdonald, 1982). The end result of what Hugh Stretton (1975, pp.330-3) might identify as a 'biggest and bestest syndrome' is likely to be the wrong activity in the wrong place for the wrong reason. There are already obstacles galore to the realisation of the perceived benefits of high technology; it would be ironic if local high-technology policy were to prove the greatest obstacle of all.

CONCLUSION

It is customary for academics, anxious to emphasise the importance of their topic and, of course, their own importance, to conclude their papers with the observation that further research is required if the problems raised are ever to be solved. In this case, that conclusion is not warranted. By no means has there been too much research on high-technology policy, but what there has been has collectively demonstrated how uncertain are the foundations upon which current policy is built. While that is grist to the academic mill, it is less than helpful to those determined to institute instant high-technology policy. They demand a brief and simple instruction manual, and their demands will never be met. More research will illuminate only more problems and complications and may distance further those who would have some understanding of high technology industry from those who would introduce high-technology policy.

There is, perhaps, a way to avoid this impasse. First, it should be accepted that much local high-technology policy is the product of instinctive enthusiasm rather than careful consideration or even common sense. The laws of probability dictate that some schemes must eventually succeed: the laws of logic insist that most must fail, at least in the sense that total costs will exceed total benefits. There are simply too many players in the game, and most have little idea of the rules. If research offers scant elucidation, then the players would be well advised to pool their knowledge and experience; co-operation is necessary if opportunities are not to be missed and resources wasted. The best way to achieve this would seem to be through integration with national high-technology policy, especially as national policy must impinge on virtually any initiative taken at a local level.

Secondly, a perpetual gulf between research and policy cannot be tolerated. An understanding of high-technology industry is necessary before policy at any level is likely to be effective. That understanding can best be gained by stripping the industry of its glossy reputation and superficial appeal, by laying bare its fundamental parts. There is really only one. High technology industry is basically an engine which runs on information. It follows, then, that appropriate high-technology policy mut be concerned primarily with information, specifically with comparative advantage in information. Present high-technology policies are concerned with an infinity of secondary matters—with land, venture capital, buildings, transportation, labour supply, even pleasant surroundings; but not with information. This alternative is not offered as a solution to the predicament of high-technology policy—there probably is none—but as a means of affording a new perspective. Another view of what it is we are doing with high technology is desperately needed if ever we are to gain the wisdom which dignifies policy and which is so conspicuously lacking in most present high-technology policies.

REFERENCES

Armington, C., Harris, C. and Odle, M. (1983) *Formation and Growth in High Technology Businesses: A Regional Assessment*, Brookings Institution, Washington, DC.

Arrow, K. (1979) 'The Economics of Information', in M. Dertouzos and J. Moses (eds), *The Computer Age: A Twenty Year View*, Harvard University Press, Cambridge, Mass.

Backus, J. (1984) 'In Research, Failure is the Partner of Success', *Research Management*, 27(4), 26-9.

Bollinger, L., Hope, K, and Utterback, J. (1983) 'A Review of Literature and Hypotheses on New Technology-Based Firms', *Research Policy*, 12, 1-14.

Bickerstaffe, J. (1984) 'High-Tech Startups: Searching for that Elusive Management Formula', *International Management*, 39(11), 16-20.

Board of Trustees, Stanford University (1974) *Stanford University Land Use Policies*, Stanford, California, 12 March.

Braun, E. and Macdonald, S. (1982) *Revolution in Miniature. The History and Impact of Semiconductor Electronics*, Cambridge University Press, Cambridge, England.

Breheny, M. and McQuaid, R. (1984) 'The Genesis of High Technology Industry in the M4 Corridor: a Preliminary View', Paper presented to Regional Science Association Conference, Canterbury, September.

Brophy, D. (1983) 'The Venture Capital Investment Market in the US', Paper presented to Hi-Tech 2 seminar, Brisbane, April.

Carter, N. and Watts, C. (1984) *The Cambridge Science Park*, Planning and Development Case Study 4,Royal Institute of Chartered Surveyors, London.

Cooper, A. C. (1973) 'Technical Entrepreneurship. What Do We Know?', *R & D Management*, 3(2), 59-64.

David, E. (1983) 'By 1990 all Industries Must be High Tech', *High Technology*, 3(4), 65-8.

Debenham, Tewson and Chinnocks (1983) *High-Tech: Myths and Realities*, Debenham, Tewson and Chinnocks, London.

Dorfman, N. (1983) 'Route 128: the Development of a Regional High Technology Economy', *Research Policy*, 12, 299-316.

Feller, I. (1984) 'Political and Administrative Aspects of State High Technology Programs', *Policy Studies Review* 3(3-4), 460-6.

Glasmeier, A., Hall, P. and Markusen, A. (1983) *Recent Evidence on High Technology Industries' Spatial Tendencies: A Preliminary Investigation*, Working Paper No. 417, Institute of Urban and Regional Development, University of California, Berkeley.

Herring, Son and Daw (1983) *Property and Technology—The Needs of*

Modern Industry, Herring, Son and Daw, London.

Japan Quarterly (1982) 'Silicon Island—Tomorrow's World Leader?', 29(4), 445-7.

LaDou, J. (1984) 'The Not-So-Clean Business of Making Chips', *Technology Review*, 87(4), 22-36.

Lamberton, D. (ed.) (1971) *Economics of Information and Knowledge*, Penguin, Harmondsworth.

Lamberton, D., Macdonald, S. and Mandeville, T. (1982) 'Productivity and Technological Change: Towards an Alternative to the Myers' Hypothesis', *Canberra Bulletin of Public Administration*, 9(2), 23-30.

LeBlanc, M. (1984) *A Complete Guide to Technology Assistance in Australia and New Zealand*, Rydge Publications, Sydney.

Lowe, J. (1984) *Science Parks, Innovation Centres and Enterprise Development in the U.K.*, University of Bath, Bath.

Macdonald, S. (1982) 'The Relevance of Silicon Valley to Australia', *Australian Electronics Engineering*, 15(3), 38-46.

——— (1983a) 'Towards Higher High Technology Policy', Paper prepared for joint OECD/Italian Seminar, Venice (mimeo).

——— (1983b) 'The Lowdown on High Technology Industry in Australia', in A. Birch (ed.), *Science Research in Australia: Who Benefits?*, Centre for Continuing Education, Australian National University, Canberra.

——— (1983c) 'Technology Beyond Machines', in S. Macdonald, D. Lamberton and T. Mandeville (eds) *The Trouble with Technology*, Frances Pinter, London.

Macdonald, S. (1983d) 'High Technology Policy and the Silicon Valley Model: an Australian Perspective', *Prometheus*, 1(2), 330-49.

——— (1986) 'Controlling the Flow of High Technology Information from the United States to the Soviet Union: A Labour of Sisyphus?', *Minerva*, 24, 75-109.

Machlup, F. (1962) *The Production and Distribution of Knowledge in the United States*, Princeton University Press, Princeton, New Jersey.

Management Accounting (1983) 'Do Enterprise Zones Work?', 61(8), p. 38.

Mandeville, T. Macdonald, S., Thompson, B. and Lamberton, D. (1983) *Technology, Employment and the Queensland Information Economy*, Report to the Queensland Department of Employment and Labour Relations, Brisbane.

——— and ——— (1983) 'Information Technology and Employment Levels', in S. Macdonald, D. Lamberton and T. Mandeville (eds) *The Trouble with Technology*, Frances Pinter, London.

Marcum, J. (1983) 'Trade and High Technology—Results of Work in the OECD', Paper presented to Symposium on Technological Trends and International Trade, Stockholm, June.

Markusen, A. (1983) *High Tech Jobs, Markets and Economic Development Prospects*, Working Paper No. 403,Institute of Urban and Regional Development, University of California, Berkeley.
Moore, B. and Spires, R. (1983) 'The Experience of the Cambridge Science Park', Paper presented to OECD Research, Technology and Regional Policy Workshop, Paris, October.
Morkel, A. (1984) 'Business Strategies in Silicon Valley', *Economic Activity*, 27(1), 12-8.
Oakey, R. (1984), *High Technology Small Firms: Innovation and Regional Development in Britain and the United States*, Frances Pinter, London.
OECD (1981) *Information Activities, Electronics and Telecommunications Technologies*, OECD, Paris.
Office of Technology Assessment (1983) *Technology, Innovation, and Regional Economic Development, Background Paper*, US Government Printing Office, Washington, DC.
——— (1984) *Technology, Innovation and Regional Economic Development*, US Congress, Washington, DC.
Porat, M. (1978) 'Communication Policy in an Information Society', in G. Robinson (ed.), *Communications for Tomorrow—Policy Perspectives for the 1980s*, Praeger, New York.
Premus, R. (1983) *Location of High Technology Firms and Regional Economic Development*, Joint Economic Committee, US Congress, US Government Printing Office , Washington, DC.
Riche, R., Hecker, D. and Burgan, J. (1983) 'High Technology Today and Tomorrow: A Small Slice of the Employment Pie', *Monthly Labor Review*, 106(11), 50-8.
Roach, S. (1983) 'The "New" Capital Spending Spree', *Morgan Stanley Economic Perspectives*, New York, 13 July, 1-13.
Rogers, E. and Larsen, J. (1984) *Silicon Valley Fever*, Basic Books, New York.
Rumberger, R. and Levin, H. (1984) *Forecasting the Impact of New Technologies on the Future Job Market*, Project Report No. 84-A4, School of Education, Stanford University.
Secretary of State for Trade and Industry (1983) *Regional Industrial Development*, Cmnd. 9111, Her Majesty's Stationery Office, London.
Senker, J. (1983) *Small High Technology Firms: Some Regional Implications*, Science Policy Research Unit, University of Sussex (mimeo).
South, H. (1981) 'The Technology Park Scramble: Boon or. . .?', in L. Wisbey (ed.) *Government Assistance to the Scientific Industry*, Australian Scientific Industry Association, Melbourne.
Stretton, H. (1975) *Ideas for Australian Cities*, Georgian House, Melbourne.
Tempest, R. (1984) 'Texas Pays High Price for Luring High-Tech Ven-

ture', *Los Angeles Times*, 21 February.
Thwaites, A. (1981) *Some Evidence of Regional Variations in the Innovation and Diffusion of Industrial Products and Processes within British Manufacturing Industry*, Discussion Paper No. 40, Centre for Urban and Regional Development Studies, University of Newcastle upon Tyne.
Tomaskovic-Devy, D. and Miller, S. (1983) 'Can High-Tech Provide the Jobs?', *Challenge*, 26(2), 57-62.
Udell, G. and Venkatesan, M. (eds) (1980) *The Oregon Innovation Center Experiment: 1973-1978*, Report to the National Science Foundation, University of Oregon, Eugene.
US Department of Commerce (1983) *An Assessment of US Competitiveness in High Technology Industries*, US Government Printing Office, Washington, DC.
Walker, M. (1982) 'The Boom That's By-Passing the New Towns of Britain', *Guardian*, 26 April, p. 15.
Willey, E. and Jones, A. (1982) *Science Parks. The Vision and the Reality*, Technical Change Centre, London.
Windschuttle, K. (1983) 'High Tech and Jobs', *Australian Society*, 3(5), 11-13.

Chapter 24

SPATIAL PLANNING IN THE INFORMATION AGE

M. Wegener

Every generation sees itself as being in a state of transition, but today change is so fast and universal that never before has the future of human civilisation been so uncertain. Indeed, for the first time in history, freed from the constraints of immediate need, humanity has the means to choose from among multiple futures ranging from world peace to the apocalyptic nightmare. Unfortunately, nothing guarantees that the forces at work, market mechanisms or political structures, will make the right choices; rather, the converse is more likely. This makes planning in its broadest sense, rational decision-making on the transnational, national, regional and local scale, more important than ever.

The new freedom is the product of technological progress. Just as the industrial age was brought about by the mechano-electrical revolution, the post-industrial era is introduced by accelerating advances in such fields as laser optics, materials, bio-engineering and electronics. In particular, the latter, by its application to information processing, is rapidly pervading human work from research and engineering to manufacturing, construction, transport and communications, and is quietly transforming non-working life, leading to new leisure, travel and consumption patterns. So the post-industrial era may, until better evidence, be tentatively called the information age.

If planning is so important in the transition to the information age, will it, too, benefit from its potential? After all, planning is concerned with knowledge, so enhanced information-processing technology might greatly improve the comprehensiveness and reliability of its database and the anticipatory power, precision and timeliness of its results, and thus its overall rationality and usefulness. In other words, will the increased complexity of planning in a time of rapid change be matched by a comparable increase in planning competence through information technology?

This question is the topic of the present chapter: based on the limited available evidence, it will speculate about the future of planning, in particular spatial planning (urban and regional), in the computerised society. As there is not one single such future, it starts with three scenarios, each representing one extreme realisation of a particular future option. Because they are purposely presented as extremes, the reader should not accept or reject any of them prematurely in isolation. A synthesis and evaluation will be attempted at the end of the chapter.

THREE SCENARIOS

That knowledge is the prerequisite of good government, is a fundamental belief of Western political culture dating back to Plato who, in his *Republic* (387 B.C.), required that political leaders be philosophers, people 'who know what is'. So the collection of information became central for efficient government; already the Roman Empire conducted regular censuses, but this was not enough, execution of power required intelligence, not statistics. Campanella, an Italian Black Friar, in his utopian dialogue *Civitas Solis* (*The City of the Sun*, 1626) described what was practiced in the ecclesiastical principalities of his time and brought to perfection by the Holy Inquisition: a hierarchical system of obligatory confessions, by which the ruling clergy collected information about non-compliance in their territory. Not much later (1651), Hobbes presented the *Leviathan*, or all-encompassing State, a giant information-processing machine:

> For by Art is created the great LEVIATHAN called a COMMON-WEALTH, or STATE, (in latine CIVITAS) which is but an Artificiall Man; though of greater stature and strength than the Naturall, for whose protection it was intended; and in which, the *Soveraignty* is an Artificiall *Soul*, as giving life and motion to the whole body; The *Magistrates*, and other *Officers* of Judicature and Execution, artificiall *Joynts*; *Reward* and *Punishment* (by which fastned to the seate of the Soveraignty, every joynt and member is moved to perform his duty) are the *Nerves*, that do the same as in the Body Naturall; The *Wealth* and *Riches* of all the particular members, are the *Strength*; *Salus Populi* (the *peoples safety*) its *Businesse*; *Consellors*, by whom all things needful for it to know, are suggested unto it, are the *Memory*; *Equity* and *Lawes*, and artificiall *Reason* and *Will*; *Concord*, *Health*; *Sedition*, *Sicknesse*; and *Civill War*, *Death*. (Hobbes, *Leviathan*, The Introduction, p. 1).

Following this model, in the eighteenth and nineteenth centuries the modern state was constructed which, for its proper functioning depended on three sources of information: external (the Statistical Office), internal (the Bureaucracy), and secret (the Intelligence Service). The notion of government as an information-processing machine is still alive, notwithstanding some substantial reinterpretations, in modern political systems theory and policy science as illustrated by titles such as *The Nerves of Government* (Deutsch, 1963) and *The Intelligence of Democracy* (Lindblom, 1965).

In *spatial* planning, by remarkable contrast, knowledge was for a long time a subordinate category. In particular town planning was long held to be an art performed by architects, and it was only in this century that the importance of comprehensive information (including economic and social as well as physical data) for planning became recognised. 'Survey before plan!' (Geddes, 1915) became imperative, although surveys remained expensive and hence restricted in scope, and so rarely played an important role in the actual plan-making process.

That seemed to change when the advent of the digital computer in the fifties promised to make information storage, retrieval and manipulation much more efficient. The concept of the urban information system emerged as a dynamic, high-resolution representation of the urban-regional system, in a sense a permanent all-purpose survey, from which all knowledge needed for plan-making could be derived. Moreover, computer-based urban models seemed to be able to bridge the gap between knowledge and plan by identifying the optimal plan, or at least by helping to narrow the choice of plan options. Based on these hopes, first in the US and sooner or later in other countries, considerable amounts of money and talent were spent to establish computerised urban information systems and models. The story of the almost complete failure of these early experiments and the disillusion that followed has been told many times and can be skipped here (see, for instance Boyce *et al.*, 1970; Kraemer, 1973; Lee, 1973). Today, thirty years after computers were used for the first time in the Chicago Area Transportation Study, computerised information systems and models for urban and regional planning have in most countries, with few local exceptions, either been completely abandoned or are maintained and applied at a disappointingly low level and practically nowhere play a significant role for policy making and planning (see, among others, Goldberg, 1983; Pack, 1983; Wegener, 1983; Nijkamp, 1983).

That is the situation from which the three scenarios start. The question is, obviously, why the adoption of information technology for day-to-day spatial planning has been so much slower than for other kinds of planning such as corporate or financial planning. Was it for reasons related to the technology itself, such as non-existence, unavailability or overly high costs? Or was it for other reasons? The three scenarios reflect different answers to that question, and accordingly draw different conclusions from the recent past for the likely future.

Scenario 1: The Planning Machine

In this scenario, it is accepted that indeed information processing is a bottleneck factor for successful planning and that with increased availability and affordability of advanced information technology significant changes in the method and process of spatial planning will occur. These changes would be, in the first place, a natural spin-off from the rapid penetration of all departments of local and regional governments by computerisation. Where every administrative transaction is mirrored by a corresponding transaction in a computer, it is easy to organise the records in such a way that they can be made available also for planning purposes. Almost every department of local and regional government produces in its daily routine a continuous stream of data of potential relevance for spatial planning. In addition, there are numerous files maintained by semi-public agencies or by private firms such as utility, transport, telecommunications or housing companies containing detailed client or sales information which, if made available to the public authorities, would provide highly-relevant information on spatial communication and consumption patterns. The same applies to the growing number of private business transactions conducted daily from home terminals through telebanking or teleshopping services or via computerised point-of-sale accounting systems.

All this information is already now assembled or will be assembled in the near future in computerised form. To be sure, most of it has been collected already in manual or even computer files. However, these files, computerised or not, have been maintained separately by each agency or firm. The qualitative new feature of the information age is that it is now possible to link individual computers to local or universal computer networks, and that there are several reasons to do so. Firms install in-house computer networks to enhance the internal communication and consistency of their operations, even at remote plant locations. Computer networks between firms or between firms and private households are the prerequisite for electronic mail, teleordering, telebanking or teleshopping or other kinds of electronic communication making the paperless office a reality.

There is no reason to believe that local or regional government should not in the long run behave like a private corporation. With progressive integration of computers into local-government operations, pressure to integrate the various data sources into one comprehensive databank, or rather a network of integrated departmental databanks, will increase for economic and consistency reasons. As a natural extension, these local databank networks will be linked to similar networks operated by other territorial bodies such as counties, provinces or States, eventually growing together into a national system of spatially organised information systems.

Of course there will be opposition to such schemes, as there has been opposition in the past, first in the mid sixties in the US (cf. Martin and Norman, 1970), later in other countries (cf. Bodelle, 1983) based on the obvious potential of databanks containing personal information to be used for surveillance. But imagine for this scenario that in countries with a reasonably pragmatic political atmosphere and with some safeguards against serious abuse, the legislation necessary to install a national network of spatial databanks will be passed on the grounds that the potential benefits for society are greater than its potential hazards.

The implementation of an urban information system which is not, as earlier urban 'planning' information systems, a collection of data especially assembled for planning purposes, but consists of the operational records of local government themselves and hence is continuously and automatically updated, would indeed dramatically change the information base of spatial planning. With increasing historical depth, that is, more and more transaction records being accumulated over time, ever more sophisticated forecasting and optimising models would become feasible. Forecasting models of hitherto unknown accuracy would probably be of the multistate-cohort type recently developed in demography (Rogers, 1982), either in the aggregate or via microsimulation based on observed transitions taken as revealed preference or on event histories extracted from accumulated records (Hannan, 1982). Even more popular would be optimisation models organised as multi-player games in which players representing real-world actors such as households, firms, travellers, consumers, etc., seek to maximise group-specific objective functions with choice dispersion, while the player representing local government uses policy instruments to maximise the public welfare (for an example, see Roy and Johansson, 1984).

If several of such models could be linked together and connected to the real-time urban database, they could be routinely used for short-term forecasting as early-warning systems. A similar early-warning system plugged into the real-time, local government information system would not only be immensely more sensitive to detect even subtle signals of potential citizen dissatisfaction, but would also be fully automatic. Moreover, the system could be provided with a set of rules concerning how to respond to minor deviances from 'normal' operations it detects, for instance by issuing appropriate advisory messages to the respective agency. Such a system might be called a planning machine, because it would take care of all routine decisions; only in exceptional, high-conflict situations involving controversial political action or large amounts of money would human intervention be required—the planning equivalent to management by exception.

The role of the planner would be, of course, to supervise the planning machine, comparable to the key operator controlling the smooth functioning of a power station or a large automated assembly plant. One imagines planners lolling about in swivel chairs in the Urban Alert Center (looking like a Strategic Air Command situation room) waiting for sensor lights to flash on a floor-to-ceiling city map at places where trouble is likely to emerge. After a few years of successful operation, the planning machine would more and more become immune to criticism; people would believe in its infallibility or even endow it with superhuman power: a veritable Leviathan.

Scenario 2: The Retreat of Planning

The second scenario recognises the same facts (the failure of the first wave of computer applications in urban and regional planning), but draws the opposite conclusions. It is based on the belief that the kind of rationality underlying the planning machine not only misses what is essential for societal decision making, but is also highly dangerous for the existence of human society.

For this scenario, the lack of success of early urban information systems and models is not related to information technology, but to a change in planning paradigms during the last twenty years from blueprint-planning from above to procedural, incremental, small-scale planning from below. In the latter paradigm, planning is seen as a process of mutual adjustment of conflicting interests, as 'social learning' (Friedmann, 1981), the extreme opposite of the centralised rationality of the planning machine. Social learning means that people want to become subjects rather than objects of planning, that they want to actively participate in the decisions about their life and their environment. Therefore, social learning starts at the local, neighbourhood and community level and accordingly focuses on local needs first; only from there does it move on to issues of broader (e.g. national) concern. Hence its primary mode of operation is personal communication.

The kind of information contained in the planning machine is of little value for this kind of planning for two reasons. The first is one of scope. Where the focus is on the problems of a small, local client group, data collection is no real problem, since nobody knows better than the people themselves about their circumstances. More comprehensive information is not only superfluous, it also detracts attention from the problem at hand, in negotiations with other groups it may even weaken their bargaining position and thus prove dysfunctional. The second reason refers to content. As the real-time urban information system is based on transactions, it contains only quantitative or 'objective' information, but no qualitative or 'subjective' information on values, preferences, aspirations, intentions, concerns (this is why the models in Scenario 1 have to rely on revealed preference). But without explicitly addressing values and preference trade-offs, in planning 'from below' conflict resolution between adverse group interests is not possible. For these two reasons, advances in information technology or urban databanks would hardly have any impact on this style of planning.

From the point of view of planning from below, improved information processing and data collection on the part of the State would not only be irrelevant for planning, but would be detrimental, because they would permit the State to extend its power and control over the citizens. The fight against the information monopoly of the state must be seen in a larger context. It is part of the great battle, fought to the end already in a few countries, still going on in some, and not even begun in others, over how society should deal with technology. The critical attitude towards technology is expressed by L. Mumford's 'The Myth of the Machine' (1967), where it is argued that man has already now proceeded beyond the purpose of achieving mastery over nature and is rapidly approaching the point where, like the sorcerer's apprentice, he can no longer control the forces he called upon, with the consequence that he not only destroys the natural environment of the earth, but also deeply alters his own personality. It is what Horkheimer and Adorno (1947) called the 'dialectic of enlightenment': that the achievements of Enlightenment, before they had a chance to bring about, as it was hoped, the great ideas of Enlightenment—freedom, equity, and humanity—turn into destructive forces threatening their very existence and thus pave the way for totalitarianism.

Seen from this angle, the computer is not the magnificent invention capable of extending man's knowledge in hundreds of beneficial ways, but is associated with police information systems keeping track of thousands of innocent people, with secret services' information systems filled wth dossiers on perfectly legal political activities, with industrial personnel information systems designed to tighten the control of workers by management, and with the innumerable other public and commercial information systems encroaching upon one aspect of the private domain of the individual after another. All this has to be fought against, and there is no reason to make an exception for urban planning information systems.

Now, to continue with the scenario, imagine that in one or more countries opposition groups fighting against the implications of information technology, such as civil-rights groups or trade unions opposing the introduction of personnel information systems in industry, form an alliance. Suppose, furthermore, that these groups, including among their members many academics, writers, lawyers, etc., are successful in raising the awareness of the general public for their concerns, then it is not at all inconceivable that they, through court decisions or through legislation, may succeed in effectively preventing the large-scale application of advanced information technology for policy making and planning by the State. This would, of course, also apply to spatial planning on the local or regional scale.

In that case it is likely that planning authorities, never having been particularly innovation oriented, might readily accept this and be content with what they have—and that will not be very much, given the fact that planning departments do not generate their own data, because their access to other agencies' data will be severely restricted, and because special surveys or censuses will become extremely rare because of their lack of acceptance and high costs.

However, if the public authorities will be restrained to engage in data collection for planning, others will not. So it can be anticipated that a commercial market for local and regional spatial-information collection will develop. Customers and/or suppliers in this market will be individuals or firms which, during their business operations, generate and/or use spatial data such as real-estate agents, developers, mortgage banks or other large corporations having interests in the region. Planning authorities wishing to use such privately-collected data will have to pay the market price. Moreover, the privately-collected data will be much less systematic and more *ad hoc* than data from published statistics and normally will be based on samples. This, however, will have implications for the kind of models and forecasting techniques that can be applied. Much effort in applied planning work will be devoted to patching the deficient database by estimation techniques such as biproportional adjustment, contingency table analysis, or analogy methods, and the resulting forecasts will be of questionable credibility. More importantly, the unequal distribution of information between the public authorities and private firms will inevitably shift the centre of gravity of planning from the public to the private domain. With the retreat of public planning from its information base, public authorities will be in a helplessly disadvantaged position in negotiations with industry with its superior 'intelligence'.

A weak public planning authority, however, is a danger for a planning system in which planning from below is the dominant paradigm. Planning from below needs an active counterpart representing the more comprehensive, long-range concerns of the community as a whole to coordinate between the conflicting interests of local groups and the interests of the community, to suggest compromise solutions, and to protect the rights of minorities which might not be able to express themselves otherwise. Without that co-ordination function, planning from below remains fragmented and partisan and is unable to deal with larger strategic issues.

Scenario 3: Computers and Social Learning

This last scenario attempts to sketch a third path between uncritical submission to and total rejection of information technology for urban and regional planning. To do that requires a recapitulation of the basic assumptions that lead the two first scenarios into their specific blind alleys.

The first scenario, which can be associated with planning from above, assumes that information is the bottleneck of planning and so makes use of advanced information technology, but fails to provide for value articulation from below. The second scenario starts at the grass-roots value-conflict level and accordingly rejects any form of centralised information, but has no mechanisms to integrate its disjointed efforts into a coherent whole.

So the two approaches seem to be the total opposite of each other. But on closer inspection they are remarkably similar. In more general terms, both approaches focus on a particular domain of the planning spectrum and seek to optimise its internal information processing capacity at the expense of its attention for external information. External information for the planning machine of Scenario 1 consists of players refusing to obey the rules: consumers not consuming, landlords not maximising their profits, minorities, protest movements, single-issue groups, that is, all irregularities that disturb its smooth functioning. External information for the actors in the second scenario comprises social costs or externalities of selfish behaviour. Both approaches have in common that the exclusion of dysfunctional information is a prerequisite of their specific performance.

Obviously, the clue to solving this dilemma lies in a combination of both approaches that preserves their internal information processing capacity, but enhances their potential for processing information from outside. By this formulation, the central problem of planning becomes one of communication, not information.

There have been numerous proposals in planning theory for intensifying the communication between above and below, between expert and client or between centre and periphery (Dewey, 1927; 1939; Mumford, 1938; Deutsch, 1963; Habermas, 1963; Etzioni, 1968; Fehl, 1971; Friedmann, 1973; Kochen and Deutsch, 1980; Habermas, 1981; Friedmann, 1981). Perhaps the most far reaching, and in some countries most influential, has been the work of Habermas who suggests a scheme of knowledge-informed public debate (discourse) as the basic medium of participatory, democratic decision making, a concept very close to that of the New Humanists' school of American planning theory, for whom planning and social learning in small, cellular, societal organisations are identical (Friedmann). However, these proposals have tended to remain prescriptions without concrete indication of how they might be implemented in practice.

Indeed, the difficulties to be overcome are enormous, as long as communication in planning has to rely solely on pre-industrial forms of face-to-face interaction. Planning by personal communication is characteristic of archaic societies in which interaction system and social system are still identical: the village democracy (Luhmann, 1975). With the increasing complexity of society, public discussion is replaced by more powerful mechanisms to reduce complexity: functional differentiation, representation, and programming by objectives. However, these techniques (as Scenario 1 has shown) cannot be extended beyond certain limits without moving out of democratic control. This is the deeper and absolutely legitimate reason for the re-appearance of the archaic medium of public discussion on the planning scene.

Yet the potential advantages of discursive planning, value-orientation and openness to innovation have to be paid for by serious structural restrictions. Personal communication finds its limits in the scarcity of time and attention: only one theme can be dealt with at a time, arguments have to be processed sequentially, time consumption is high, the complexity that can be handled is small (Luhmann, 1971; 1975). These constraints determine the long-run prospects of participatory planning. If, however, it is possible to extend these limits by some powerful amplifying mechanism, the interaction medium 'discussion' may be reinstated as the constituent vehicle of democratic planning.

The question to be asked in this scenario is whether information technology can provide such amplifying mechanisms. Attempts to use two-way communication media to overcome the fragmentation of local-planning discussions have a long history, from the first phone-ins (cf. de Sola Pool, 1973) to present two-way cable television networks or using various combinations of television and telephone. There have also been experiments to use computers not so much as computing machines, but also as communication media to increase the efficiency, productivity, and substance of group decision making, from computer gaming, interactive modelling and electronic voting systems (for an overview, see Wegener, 1978) to present decision support or expert systems. Before long, the technological potential will be available to combine, first locally, then globally, multi-way television and distributed computing, accessed through nodes in the Integrated Services Digitalised Network (ISDN) providing the infrastructure for networks of electronic town halls from the neighbourhood to the global level.

It still requires speculation to discuss the potential use of this technology. For example, any individual or group would have access to local, regional or national spatial databanks which, for privacy reasons, would not be as comprehensive as those of the planning machine of Scenario 1, but certainly much richer in content than today's published statistics. The system would offer up-to-date tools for manipulating, displaying or analysing data from simple statistics to sophisticated modelling, unless clients wanted to develop their own models and probably come up with different results, which then could be exchanged electronically with other groups, just as today's microcomputer software is exchanged via electronic mailboxes. Groups pursuing a certain issue, alone or pooled with others, might use a public television channel for live discussion, prepared video presentations and on-line, interactive modelling to address a much wider audience than they could ever reach by printed media. Such self-organised events would interact with the audience through remote voting or evaluation techniques providing immediate feedback on the public acceptance of the issue under debate. Another important application of electronic feedback techniques might be forecasting sessions in which scenarios produced by aggregating audience responses are compared and confronted with model forecasts. In a more organised and controlled way, remote voting systems would be used to formulate recommendations for the actual decision making of local, regional or national legislatures.

The technological potential of almost instant feedback between political decision makers and the constituency on all levels from the neighbourhood to the nation, will undoubtedly have deep impacts on the political system of post-industrial democratic societies. It means, in fact, that the historical reasons that forced nations to move away from the grass-roots village democracy to political representation, division of responsibility, hierarchical organisation, etc., in short, the whole complex modern political machinery, are at least partly disappearing, and this will sooner or later be reflected in the rules and procedures of political decision making (cf. Toffler, 1980). That is not to say that government by referendum should become the rule, in particular where emotional or biased judgement is likely. But it implies that the political structures can become less centralised and more responsive to local concerns, minority rights and small-scale innovation and change.

This applies in particular to spatial planning, because it is closer than any other kind of planning to the neighbourhood level, where the first experiments in grass-roots democracy are made. A fast-moving, innovation-oriented planning department might play a vital role in decentralising the local-regional planning machine and stimulating the kind of pluralistic, self-organising planning activity described above, that is, transforming itself from a bureaucracy into an agent of change. An essential task in that transformation would be the implementation of the most advanced public reference spatial information system plus the software tools, education and training necessary for its fullest utilisation by all groups of society (cf. Meier, 1962). Needless to say that this kind of planning would require a different type of planner, more mediator and communicator than, say, engineer or scientist, but at the same time highly competent in the substantial problems of planning as well as in the modalities of the new technology (cf. Batty, 1984).

And certainly it would require a new mix of planning methodologies. In particular, three groups of planning methods seem to be under-developed and under-represented in planning research and education.

Communication with non-experts

The first and most important group comprises methods designed to communicate complex spatial-temporal phenomena, issues and problems to people not trained in planning, that is, to non-experts. The same applies to the design of efficient, user-friendly, man-machine interfaces for planning software (a topic that has received much attention in other planning fields; see, for instance Sage, 1981; Benbasat and Taylor, 1982, but hardly any in spatial planning research) as well as hardware and software development for group decision making such as teleconferencing, local decision networks, decision room design, large visual-display techniques and the like (see, for instance De Sanctis and Gallupe, 1985).

Stated preference techniques
The second group of methods of obvious importance for the new pluralistic-participatory style of planning in the information age deals with the problem of how to incorporate stated (rather than revealed) preference information into information systems and models. There does exist a large body of literature on multi-attribute utility theory and related concepts in mathematical psychology and decision theory (e.g. Keeney and Raiffa, 1976) as well as on multi-objective decision making in regional science (see, for instance Nijkamp, 1979; Voogd, 1983; Nijkamp *et al.*, 1984), however, practical experience with such methods for spatial planning is still limited (see Bauer and Wegener, 1977).

Non-numerical planning information
The third group is concerned with the processing of broader classes of knowledge including, besides numerical, also categorical, causal, etc., information on planning issues extracted from documentary data such as reports, papers and books, as well as from expert knowledge. Despite increasing developments of decision support or expert systems for such planning fields as corporate, financial, marketing and product planning, only very few such systems seem to have been developed for spatial planning purposes (among these are Nakamura *et al.*, 1982; White *et al.*, 1985).

Models in the traditional sense, that is, mathematical planning models, will remain important for research and education and training, and may become more important for policy making, if they are embedded into the communicative environment discussed above. Modelling will benefit vastly, as in Scenario 1, by improved data availability and the possibility of being linked to the real-time urban database as learning monitoring systems (see Brotchie, 1984). However, no model will ever be allowed to gain a dominating influence on policy making and planning (as the planning machine in Scenario 1 had it). This will be achieved first, by requiring that all modelling performed for local and regional government be exposed to unrestricted public review and criticism, and second, by always commissioning more than one model coming from different, competing modelling approaches.

CONCLUSIONS

What are the chances that Scenarios 1, 2 or 3 will become reality? The answer will be different for each country. For a start, the experiences and attitudes of people in different countries toward the computer are vastly different. For instance, in a survey conducted by major newspapers in nine industrialised countries in early 1985 (Ginsburg, 1985), 37 per cent of the adults interviewed in the US indicated that they had already used some kind of computer, whereas the corresponding percentages for other countries were 28 per cent (Great Britain), 26 per cent (France), 14 per cent (Japan), and 11 per cent (West Germany). However, the countries with the greatest personal experience with computers also had the highest proportion of people concerned about the adverse aspects of a computerised society. When asked if computer databanks are likely to be used to intrude into personal privacy rights, 75 per cent of the respondents in the UK answered yes, 68 per cent in the US, and 71 per cent in France, but only 51 per cent in West Germany and 50 per cent in Japan. Does this mean that the deeper societies move into the information age, the more they resign to the inevitable loss of democratic innocence? Martin and Norman (1970, pp. 534-5), in their early preview of the computerised society, may have discovered a sad truth:

> Reluctantly, the public may have come to realise that the machines have information about them. They cannot prevent it. It is regarded as being for the general good. Most of it is reasonably safeguarded. Their relationship with the data networks is in some ways like that with their confessional. At least the data are accurate now. The earlier years, when embarrassing errors appeared on the files, have gone. Most of the public would now be upset if the machines did not have all their personal details. They might miss out on many of the beneficial aspects of the computerised society.

Knowing the hazard, but accepting it as the price to be paid for the good life—if the attitude practised towards nuclear energy, global pollution, traffic accidents and cigarette smoking is also the post-industrial attitude towards information technology—Scenario 1, the planning machine, will be the ultimate prospect. However, there are countries like the Netherlands and West Germany, which are highly industrialised but relatively less advanced in the application of information technology, and hence have had a chance to experiment with participatory planning before the advent of the information age. In such countries opposition against information technology such as the anti-census battles still can occur (see Wegener, 1985). The danger is that such opposition may lead these countries into Scenario 2, the retreat of (public) planning.

Scenario 3 remains the challenge to be accepted. It offers the only visible chance to put information technology into the service of planning instead of letting it destroy planning in one way or the other. To accept the challenge requires the joint efforts of regional and computer scientists, planners and citizens to transform the potential of information technology into a progressive rather than oppressive force.

The problems to be overcome towards achieving this goal are enormous, and there is no guarantee that it can be achieved. Therefore a few caveats are in order. Throughout this chapter there have been three assumptions implicitly made which are so crucial that if only one of them turns out to be wrong, all efforts will be in vain. The three assumptions are:

The information technology will be present
In an unspecified sense, the assumption is likely to be correct. But it is also likely that access to advanced information technology will be unevenly distributed in society. This is especially true for the ISDN technology which will for a long time be expensive and only affordable by firms and more affluent households. But only with technology open to all members of society, can the pluralistic planning market place develop. Seen from this angle, the three scenarios may not be real options to choose from, but phases each country has to pass through—from early technocratic use, while the technology is still rare—through a period of distrust and refusal, to eventually a more mature, constrained, and responsible integration into everyday life, when the diffusion process approaches completion.

There are no limits to artificial intelligence
This assumption underpins the idea of the planning machine of Scenario 1, but also the more civilised notion of the learning monitoring system of Scenario 3. However, the issue is far from being settled. The emerging view is (cf. Dreyfus, 1979) that despite impressive achievements in limited, clearly-defined contexts (microworlds), machines will never be able to understand things outside that context (i.e. extend it), because they lack the genetically-inherited and socially-acquired experience that enables humans to do just that. This limitation may not be critical where planning is concerned with clearly-defined issues, but characteristically in planning from below the context cannot be prespecified from above—it is always changed by the people. If this view is correct, it would not only explain why in the past problems have always outrun models, but it would also suggest that the relationship between people and computers in planning will always remain one of misunderstanding and alienation.

Post-industrial democracy will be grass-roots democracy
Obviously, this assumption is essential for Scenarios 2 and 3, and indeed there is much current evidence of political decentralisation in many countries. But is this evidence sufficient to project it into the future, as many

futurists do, as a megatrend (Naisbitt, 1982), or are these projections merely idealist dreams of their bourgeois authors? Other futures are conceivable. One would be that cities would be the scene of, probably violent, class struggles between the job-possessing class and the jobless victims of computerisation, hardly a place for enlightened discourse. The other, no less depressing vision is much more suggestive: that the freedom of material need and added leisure hours gained through computerisation of work do not lead to increased community involvement and active participation in social life, but, on the contrary, to a loss of solidarity and political interest. A depoliticised public, conditioned by the media to entertainment and consumption, may well be the real client of post-industrial spatial planning.

REFERENCES

Batty, M. (1984) *Information Technology in Planning Education*, Papers in Planning Research 80, Department of Town Planning, University of Wales Institute of Science and Technology, Cardiff.

Bauer, V. and Wegener, M. (1977) 'A Community Information Feedback System with Multiattribute Utilities', in D. E. Bell, R. L. Keeney and H. Raiffa (eds) *Conflicting Objectives in Decisions*, John Wiley and Sons, Chichester, Sussex.

Benbasat, I. and Taylor, R. N. (1982) 'Behavioral Aspects of Information Processing for the Design of Management Information Systems', *IEEE Transactions on Systems, Man, and Cybernetics*, SMC-12, 439-50.

Bodelle, J. (1983) *Computerstaat? Nein danke* (*Computer State? No Thanks*), Arbeitskreis Informationstechnologie und Sozialimplikationen, Berlin.

Boyce, D. E., Day, N. D. and McDonald, C. (1970) *Metropolitan Plan Making*, Monograph No. 4, Regional Science Research Institute, University of Pennsylvania, Philadelphia.

Brotchie, J. F. (1984) 'Technological Change and Urban Form', *Environment and Planning A*, 16, 583-96.

Campanella, T. (1626) *Civitas Solis* (*The City of the Sun*), Translation into German, 1955, Akademie-Verlag, Berlin.

De Sanctis, G. and Gallupe, B. (1985) 'Group Decision Support Systems: A New Frontier', *Data Base*, 16, 3-10.

Deutsch, K. W. (1963) *The Nerves of Government: Models of Political Communication and Control*, The Free Press of Glencoe, New York.

Dewey, J. (1927) *The Public and Its Problems*, Henry Holt, New York.

——— (1939) *Freedom and Culture*, G.P. Putnam's Sons, New York.

Dreyfus, H. L. (1979) *What Computers Can't Do: The Limits of Artificial Intelligence*, Harper and Row, New York.

Etzioni, A. (1968) *The Active Society: A Theory of Societal and Political Processes*, The Free Press, New York.

Fehl, G. (1971) *Informations-Systeme, Verwaltungsrationalisierung und die Stadtplaner* (Information Systems, Administrative Rationalisation, and Urban Planners), Stadtbau Verlag, Bonn.

Friedmann, J. (1973) *Retracking America: A Theory of Transactive Planning*, Anchor Press, New York.

——— (1981) *Planning as Social Learning*, Working Paper No. 343, Institute of Urban and Regional Development, University of California, Berkeley.

Geddes, P. (1915) *Cities in Evolution*, Ernest Benn Ltd, London (1968 edn).

Ginsburg, H.J. (1985) 'Computer, nein danke' ('Computers, No Thanks'), *DIE ZEIT*, 31 May.

Goldberg, M. (1983) 'The Technology of Urban Systems Modelling: Nagging Questions, Sagging Hopes and Reasons for Continued Research', in M. Batty and B. Hutchinson (eds) *Systems Analysis in Urban Policy-Making and Planning*, Plenum Press, New York.

Habermas, J. (1963) 'Verwissenschaftlichte Politik und öffentliche Meinung' (Scientised Politics and Public Opinion), in J. Habermas *Technik und Wissenschaft als 'Ideologie'*, Suhrkamp Verlag, Frankfurt.

——— (1981) *Theorie des kommunikativen Handelns* (*Theory of Communicative Action*), Suhrkamp Verlag, Frankfurt.

Hannan, M. (1982) *Multistate Demography and Event History Analysis*, Working Paper WP-82-50, International Institute for Applied Systems Analysis, Laxenburg, Austria.

Hobbes, T. (1651) *Leviathan*, J. M. Dent and Sons Ltd, London (1962 edn).

Horkheimer, M. and Adorno, T. W. (1947) *Dialektik der Aufklärung* (*Dialectic of Enlightenment*), S. Fischer Verlag, Frankfurt (1969 edn).

Keeney, R. L. and Raiffa, H. (1976) *Decisions with Multiple Objectives: Preferences and Value Tradeoffs*, John Wiley and Sons, New York.

Kochen, M., Deutsch, K. W. (1980) *Decentralisation: Sketches toward a Rational Theory*, Oelgeschlager, Gunn and Hain, Cambridge, Mass.

Kraemer, K. L. (1973) *Policy Analysis in Local Government*, International City Management Association, Washington, DC.

Lee, D. B. Jnr (1973) 'Requiem for Large-Scale Models', *Journal of the American Institute of Planners*, 39, 163-78.

Lindblom, C.E. (1965) *The Intelligence of Democracy: Decision-Making through Mutual Adjustment*, Free Press, New York.

Luhmann, N. (1971) 'Systemtheoretische Argumentationen' ('The Systems Theory View'), in N. Luhmann and J. Habermas *Theorie der*

Gesellschaft oder Sozialtechnologie—Was leistet die Systemforschung?, Suhrkamp Verlag, Frankfurt.
——— (1975) 'Interaktion, Organisation, Gesellschaft' ('Interaction, Organisation, Society'), in N. Luhmann *Soziologische Aufklärung 2*, Westdeutscher Verlag, Opladen.
Martin, J. and Norman, A. R. D. (1970) *The Computerized Society*, Prentice-Hall, Englewood Cliffs, New Jersey.
Meier, R. L. (1962) *A Communications Theory of Urban Growth*, MIT Press, Cambridge, Mass.
Mumford, L. (1938) *The Culture of Cities*, Secker and Warburg, London.
——— (1967) *The Myth of the Machine: Technics and Human Development*, Secker and Warburg, London.
Naisbitt, J. (1982) *Megatrends*, Warner Books, New York.
Nakamura, K., Iwai, S. and Sawaragi, T. (1982) 'Decision Support Using Causation Knowledge Base', *IEEE Transactions on Systems, Man, and Cybernetics*, SMC-12, 765-77.
Nijkamp, P. (1979) *Multidimensional Spatial Data and Decision Analysis*, John Wiley and Sons, Chichester, Sussex.
——— (1983) 'Information Systems for Regional Development Planning: A State-of-the-Art Survey', *Environment and Planning B*, 10, 283-302.
———, Leitner, H. and Wrigley N., (eds) (1984) *Measuring the Unmeasurable*, Martinus Nijhoff, The Hague.
Pack, J. R. (1983) 'The Failure of Model Use for Policy Analysis in Regional Planning', in M. Batty and B. Hutchinson (eds) *Systems Analysis in Urban Policy-Making and Planning*, Plenum Press, New York.
Plato (387 B.C.) *The Republic*, Translation into English, 1941, Oxford University Press, Oxford.
Rogers, A. (1982) *Parametrized Multistate Population Dynamics*, Working Paper WP-82-125, International Institute for Applied Systems Analysis, Laxenburg, Austria.
Roy, J. R. and Johansson, B. (1984) 'On Planning and Forecasting the Location of Retail and Service Activity', *Regional Science and Urban Economics*, 14, 433-52.
Sage, A. P. (1981) 'Behavioral and Organisational Considerations in the Design of Information Systems and Processing for Planning and Support', *IEEE Transactions on Systems, Man, and Cybernetics*, SMC-11, 640-78.
de Sola Pool, I. (ed.) (1973) *Talking Back: Citizen Feedback and Cable Technology*, MIT Press, Cambridge, Mass.
Toffler, A. (1980) *The Third Wave*, William Morrow and Co., New York.
Voogd, H. (1983) *Multicriteria Evaluation for Urban and Regional Planning*, Pion, London.
Wegener, M. (1978) *Mensch-Maschine-Systeme für die Stadtplanung*

(Man-Machine Systems for Urban Planning), Birkhäuser Verlag, Basel.
——— (1983) 'The Impact of Systems Analysis on Urban Planning: The West German Experience', in M. Batty and B. Hutchinson (eds) *Systems Analysis in Urban Policy-Making and Planning*, Plenum Press, New York.
——— (1985) 'Urban Information Systems and Society', in J. Brotchie, P. Newton, P. Hall and P. Nijkamp (eds) *The Future of Urban Form: The Impact of New Technology*, Croom Helm, London.
White, K. P. Jnr, Sage, A. P., Rodammer, F. A. and Peters, C. T. Jnr (1985) 'The Environmental Advisory Service (EASe): A Decision Support System for Comprehensive Screening of Local Land-Use Development Proposals and Comparative Evaluation of Proposed Land-Use Plans', *Environment and Planning B*, 12, 221-34.

Chapter 25

CITIES AND REGIONS IN THE ELECTRONIC AGE

B. Harris

The present era of technical revolution has been designated, incorrectly I think, the information age. The use of the term information to characterise the inner quality of the changes which are in progress is almost certainly too narrow and too limiting, and it seems doubtful that any single substitute of a descriptive nature can be found.

We could of course—and perhaps should—refer to the instruments of the revolution, rather than to its content. It is surely not far off the mark to talk about the electronics revolution. Previous revolutions in technology have been similarly named, for example after steam, automobiles, airplanes and atomic power. In addition, we need to take a look at the confusion of purposes which the electronics revolution serves, and see how we cannot afford to slight some of them.

In the first instance, electronics provides for communication, and had done so for many years before we realised that we were in the midst of a revolution. Of course, communication transmits information among other things, but the recent heightening of the density and intensity of communications suggests that this is not the whole story. A televised account of a Reagan news conference, or of starvation in Ethiopia, conveys more than information—and it is not for informational reasons that in the US the Bell Telephone Company urges us to 'reach out and touch someone.' In brief, electronic communication heightens interactions beyond what might travel by print or by telegraph, and may in some ways be able to substitute for direct interaction. In addition to formal information, it transmits some of the effects and many of the visual and auditory cues of personal contact.

We may thus say that the transmission and storage of information is at best a second purpose served by the electronics revolution, but an important and growing one. Nevertheless, it is not clear that this aspect of electronics is now or will ever be the main strength of the new technology. Partly this is because the effect of an information revolution is substantially misunderstood or misrepresented. The idea that information is power is a particularly culture-bound concept from the West in this century; one need only look at the successes of Ho Chi Minh, Mao Tse Tung, the Ayatollah Khomeini or Corazon Aquino to see that low-information organisation can overthrow high-information technology. Power is force, and information controls force at the level of the population only when social relations are regularised so that information generates wealth which can purchase force.

Even at this level of social coherence, the idea that information itself generates wealth and power is a compression of a process into a phrase. It has been said that the users of information need processes for condensing and filtering it—even more bluntly, it is claimed that information is the enemy of knowledge, while knowledge is the enemy of wisdom. Too much information gets in the way of the thought that goes into acquiring knowledge, while knowledge which is irrelevant, outdated, or false can betray wisdom.

A dynamic society thus needs not only access to information, but the knowledge to organise it and the wisdom to use it. Here we find that there is an important social implication of the electronics revolution which is only now beginning to unfold. We have yet to see what will happen when instruments of communication and computing are in the hands of large segments of the general population, and whether it will be the affective, the informational, or the computational aspects of this dispersed power which will have the greatest importance.

Thus the third, and in my view most important, impact of the electronics revolution has to do with computing, because this power serves to amplify the thought processes which underlie the knowledge and wisdom which are needed to use information. Simultaneously this tends to be the aspect of the electronics revolution which is most badly served by the wrong terminology.

Computing and intelligence have very little to do with information *per se*, and to assume that it characterises their essential qualities is a great mistake. I have long held this view, and it is now apparent that recent work in cognition tends to support it. Research in human cognitive processes currently suggests that there are two broad types of learning: one having to do with facts, and the other having to do with procedures. Knowledge having to do with procedures seems to be more basic and more permanent. It is, of course, easy to see that procedures involve more than facts, since they constitute more general means for dealing with a large number of facts. It seems likely (although I have not seen this idea discussed in relation to cognition), that concepts, analogies and theories are more readily defined as procedures than as factual information.

Computers are many things to many people, but in my view their principal strengths are procedural. Consider one or two simple common problems. If I think I have made a certain type of mistake in writing this chapter, I can ask my computer (through my word processor) to search for some indication of the error, for example a misspelling. At the end of the process I have some new information, but what the computer used was a procedure for processing information. Or suppose we have all the information needed to find the lowest-cost shipment pattern for a collection of warehouses and customers—we still have to submit that information to a procedure for solving the transportation problem of linear programming. Here we can hardly even call the procedure one for processing information: it is one for solving a problem. True, the input may be regarded as information, and the output the same, but the bridge between the two involves something more. Indeed, each case given here involved something which previously could only be done by human intelligence following an extended procedure. In these cases the computer is able to function more rapidly and more accurately than a human editor or calculator, and of course such cases (many more dramatic than these) are the ones which are early candidates for computer implementation.

The essential point is that computers are expected, with increasing - scope and sophistication, to solve problems which people solve procedurally, either consciously or unconsciously. At the same time it seems likely that some human activities will remain beyond the reach of computers. Thus the relation of computing to human thought and other human activities is to extend the power of the individual and social mind, and to permit these to focus principally on activities (creative, insightful and affective) for which they are distinctively suited. This is more than an information revolution.

In a workshop on this subject I dealt mostly with modelling in the present age and perhaps not enough with the substance of the electronic and biological revolutions and how they will affect our lives (Harris, 1985). At that conference also, I failed to listen with an inner ear to the prescient remarks about the future of computer modelling in the age of micros. This time around, I will talk mainly about the real-world phenomena of the electronic revolution and how it will affect living and working in the urban areas of the world. Inevitably I will include some mention of the work of the planner and policy maker in this environment, and thus of the use of models.

The balance of this chapter will deal with three general topics. First, I will provide a general characterisation of the multiple impacts of the electronic revolution, together with some of the worldwide developmental implications which may be drawn from this analysis. Second, I will translate these implications into locational patterns and lifestyles as they may develop in the future, both among and within cities. Finally I will consider very generally how the new technology, and especially the technology of microcomputers, may be expected to impact planning in the near and middle-term future.

BROAD IMPACTS OF THE ELECTRONICS REVOLUTION

The electronics revolution has impacts at many different levels and in many different ways. I will try to sketch these with respect to computer technology and communications in particular, with emphasis in this section on the first of these, up to but not including those applications which fall under the general heading of artificial intelligence.

In the developed countries at least, computer technology is driving the shift in production and employment out of primary and secondary industries and into tertiary or service industries. Some analysts distinguish a class of quaternary services which are largely electronically based. This driving trend has several components.

A movement in the direction of services is almost inevitable under rising incomes, as the income elasticity of demand for services appears to be greater than for other commodities. This trend is intensified when the relative prices of services provided by electronic means falls drastically, as has been the case in the computer and related fields for most of the last twenty-five years. This falling cost has both an income and a substitution effect, and both are increasing over time. This trend is offset in part by the fact that some household consumers of services use the services of purchased equipment, which then appear in national accounts as manufactured goods—TV sets, microwave ovens, home computers and so on. Here prices are falling as well.

The falling cost of all manufactured goods, and especially of electronic equipment, reflects the increased automation of production and in some cases a reduced input of energy and raw materials. Micro-electronic equipment is manufactured largely by environmentally benign processes, its operation is likewise benign, and it facilitates similar effects as it is used in other industries. All of these manufacturing processes use an increasing proportion of service inputs, so that if the services are purchased the volume of service employment is seen to increase, while if they are produced internally to the manufacturing industry the shift is reflected in its occupational structure. Conversely, and once again partially offsetting this trend, there is a heightened use of manufactured inputs (including equipment) in the production of services.

Automated manufacturing and the increased production of services require a more highly-educated workforce and larger amounts of space per employee for any given productive process. The need for more space may be offset at least in part by the compositional shift away from heavy manufactured goods.

Improved communications constitute one of those services which are in increasing demand in this age. In the West, this improved communication has facilitated the decentralisation of population and employment, first from central cities to suburban areas, and more recently from larger metropolises to smaller ones and to non-metropolitan small cities. As a consequence of these improvements, of generally improved education and of economic growth, many of the new centres of production (though smaller than present major metropolitan centres) are in general equal or superior to the metropolitan centres of fifty or more years ago in the availability of services and amenities, both in consumption and production.

This improvement in the conditions of life outside of the metropolis arises in a complicated way out of a long history of development. Initially, it depended to a great extent on mechanical aids to interaction, especially the automobile and trucking, then on improved refrigeration and storage which heightened the availability of goods. But more recently, electronic communications have drastically reduced the relative prices of telephonic voice and, later, data communication. The effect of reduced cost and of enhanced reliability has benefited consumers, but more importantly it has facilitated many forms of decentralised conduct of business. Business services are more readily available, inventories and their carrying costs have been reduced and so on. In effect, many of the agglomeration economies of metropolitan centres have been dissolved, and both the costs of doing business and the relative disadvantages of living in smaller centres have been reduced to the point where these centres are competitive—even though dependent upon—their larger counterparts.

In the next section, we will explore further the implications of declining relative prices for these communications facilities for future locational patterns and spatial organisation. Meanwhile, however, we have still one more line of thought to pursue with respect to the broader implications of the electronic revolution. We have witnessed steadily falling costs and prices for computing power, and greatly expanded capabilities for individual machines. Within the past three years or thereabouts, these trends have been translated into a qualitatively new phenomenon: the availability of personal computers (for home or office work and for play) at prices which make them widely available and with capabilities which rival those of computers of twenty-five years ago which were a thousand times as expensive.

These computers are supported by an outpouring of software, which amounts to generalised procedures for solving a vast number of personal, managerial and intellectual problems. We are approaching the state in which this software is able to meet most of the needs of a large number of users, in a way which requires them to master procedural thinking but which does not require them to understand or use computer programming. While procedural knowledge is already the essence of survival in an industrialised society, microcomputers will open up a vast number of new procedures to their users, and these will become available in a form which vastly reduces the costs of error in the process of learning. Learning with the microcomputer is analogous to learning through game- playing—but it must be emphasised that we are not talking about computer games. We are talking about the situation in which acquiring useful procedures in itself has the nature of a game—in terms of enjoyment, challenge, skill acquisition and reduced risk.

This strong motivation for the use of microcomputers, overlaid on their intrinsically useful nature, gives the system in which they are embedded powerful positive feedback. The use of microcomputers and workstations creates trained personnel who generate new demands for the means of this form of work, and it familiarises a workforce with the essential elements of software and hardware design. These forces working together will speed up the spread of computer applications and increase the productivity of nations in which computer utilisation has started to grow in this format.

At the same time, the wide availability and falling prices of the microcomputer and of all computers will tend to democratise computer use and the scientific skills which accompany it. This trend will be real in spite of the fact that some of the skills of computer users will be superficial and even spurious, like the culture and sophistication of some literate persons. These skills will greatly broaden the base for increased literacy and numeracy, for scientific advance, and for industrial productivity, and they will be achieved at relatively low capital cost per worker. Because microcomputer users, sometimes connected to independent networks, are able to amplify their independent thinking, to generate and disseminate ideas, and even to print articles and books, the wide use of microcomputers may pose a dilemma for authoritarian societies. To employ them as widely as seems possible may be politically destabilising, while to restrict their use may also restrict economic growth.

SPATIAL DEVELOPMENT IN THE ELECTRONIC AGE

The discussion thus far has already begun to suggest that the electronic revolution could have profound effects on the patterns of spatial organisation of modern societies. These spatial patterns will also have important correlates in the lifestyle of the populations they accommodate. Both are worth some exploration.

We are essentially concerned with those spatial urban problems which arise out of population increases and growth of urbanisation, increases in income combined with the persistence of poverty, and technological change associated with both production and consumption. In order to understand, let alone solve, such a broad range of locational problems over time and in the context of dynamic change, we need much more knowledge about the types of interaction, agglomeration and occupancy of space than we have now readily at hand. A sketch of some of these problems of understanding is appropriate. Ultimately we need to know how far many of the present trends will go. To what extent will various lines of manufacturing be automated? What form will future increased consumption of services take? If prosperity and technology bring higher productivity, how will this affect the division between work and leisure? We especially need to know about the locational tendencies of the service industries—both consumer services and business services. To what extent will these industries become specialised in a way which will make large agglomerations unnecessary between specialisations? To what extent will improved communications and computer technology affect this, and will they dissolve the concentrations which might otherwise persist within specialisations? To what extent will communications make it unnecessary for this specialised labour to be physically present at a common place of work?

In recent months I have been reviewing the urban modelling efforts of the last thirty years, and their place in planning. During that same thirty years I have been doing my own share of modelling, and I have found much value in assuming that there is a substantial continuity in people's behavior and system performance from generation to generation, from culture to culture, and from one level of development to the other. In a world that is constantly undergoing drastic changes we must understand how these changes affect the basic needs of consumers and producers; this is essential to making good predictions. We are in the same position that a planner in 1920 might have been, trying to foresee the impact of the automobile by 1970, except that the time scale of change has been compressed and the new technology is not only more stealthy, but already more ubiquitous. We must run very fast even to stay in the same place, and we must not repeat the myopic response which we may have had in gauging matters like the energy crisis purely in terms of known concepts and established technology.

THE ELECTRONIC PLANNER

We now know that planning is a complex and uncertain business, plagued by political problems which arise naturally wherever the vital interests of the population and its organised activities are affected. We may legitimately ask the question of how the electronic revolution will affect planning and analysis of the kind which I have been obliquely viewing here.

At first glance there is a great deal of grounds for optimism, and if we can overcome some difficulties that I foresee in the middle distance, that optimism might prove to be well founded. Even then, however, there will be enduring problems to which our optimism should not blind us.

Electronic data collection and data filtering will become very easy, so that almost anything about the urban scene that we want to know and ought to be able to know will become accessible. We have had a parallel experience with Landsat earth photography from satellites (which is almost entirely electronic). This has taught us that literally trillions of bits of information may be too much, and is certainly not useful, without a clear idea as to its employment and well-conceived means for that employment. In brief, therefore, we will need to curb our appetites for data and form ever more clearly-defined ideas about how and why we want to use it.

Making substantial use of data and living comfortably with it, while at the same time preserving a creative and independent attitude toward planning, calls for new types of planners and analysts. In the twenty-five years between 1960 and 1985, I have been continually hopeful that the work of planning agencies would reflect the enormous potential of the computer revolution; to a limited extent that has happened. More often, however, the most sensitive planning students have continued to rebel against the use of data and computing, and agencies themselves found that the state of the art together with political and fiscal constraints severely limited their ability to use innovative methods.

Today, my own planning department—for the first time among graduate departments in our university—and other planning departments throughout the world, requires entering students to provide themselves with a personal computer, while making additional computers and software available in the school. I take this not as a great advance in itself, but as an indication that the wind of change is rising. Soon, not only most planners, but many bureaucrats and elected officials, as well as lawyers, doctors and teachers will be at home with and competent users of computers.

Through this means, the power of critical and creative thought on matters of public interest may be greatly expanded. As the volume of publicly-available data and software expands, more and more people will be in a position to analyse problems, present plans and (yes) invent models. Of course this kind of activity will be most evident among the professionally concerned, but the capability of interest groups to engage in counter-planning will increase, and the availability of well-prepared consultants will no longer be an obstacle.

All of this might be limited by the inability of the average citizen to secure good software and to understand the state of the art of planning and modelling; indeed, this inability might extend to the offices of planning agencies themselves. It has been said of the perils of trying to optimise, that the best is the enemy of the good. But in terms of the acceptance of ideas and procedures, we might very well find that the good is the enemy if not of the best at least of the better. For example, there is no doubt that Visicalc and Lotus 1-2-3 are superb aids to management and planning—but we would not like to have a generation of planners whose good models would only use spreadsheets and never do better.

For these reasons, it is easy to hypothesise that by the year 2010 every planner could be a modeller, every citizen a planner. More realistically, it will certainly be true that the scope of scientific analysis and professional insight will extend more deeply into the population. If this occurs, and I hope that it will, then we will have to deal with the as-yet-unstudied problems of a democratisation of science. These problems will display in a new form the problems of democracy itself, of leadership and followership, of the diversity of participation and the ultimately appropriate unity of action, and of education and self-discipline as a need but not as a prerequisite for membership in the polity.

There are two persisting and related problems that the electronic revolution will not solve—and which, indeed, I hope can never be solved in their entireties. The first of these has been discussed by example throughout this chapter. In an era of technological change, the techniques of understanding social systems and planning their development must also constantly change. It seems unlikely that with widening participation in scientific activity and with increasing available resources, the rush of technology can or should be hobbled. But it is also clear that adaptive responses by society are not automatic and may, in extreme cases, not yet be adequate. It seems unlikely that an inventive technology can ever be mastered and controlled by an uninventive and automatic social response, so that the need for inventive, humane and competent social action will always persist.

It might, however, be argued (as I have done in some contexts) that these problems of social guidance are in essence problems of finding optimal policies or plans. By extension, it might then be believed that enormous increases of computer capabilities would bring these optimal solutions within more or less automatic reach. I now believe that this is a wholly mistaken idea, fortunately or unfortunately, and that computers can never assume the creative role which is inherent in solving large problems. If this were not true, the prospect that computers could take over endless other human creative activities might indeed be very dispiriting.

The middle-term dangers in these two cases should not be overlooked. The less likely event is that the continued development of new technology might be brought to a world-wide halt, through political means based on a kind of apocalyptic view of progress. But given the power and fascination of computational innovations, it is not improbable that the idea will be put forward (and, worse, accepted) that planning and governance can be entrusted largely to computers. Government by software seems a likely front for more human but malicious schemes.

On this front, therefore, we can conclude that planning and social action will be very different in (say) 2010 than they are today, partly in recoil from the electronic and other revolutions, and partly with their assistance. Planning education, the profession, and its institutions will have to change in corresponding ways. In all of this, we will look forward to the amplification but not the replacement of the social intelligence by electronic means, but the electronic planner will remain—fortunately—unachievable.

CONCLUSION

To a degree, I regret having written in a speculative vein about what I do not fully know, and what our various professions do not yet understand. But this has seemed important to me because the changes that are being generated by the electronic revolution have great potential both for good and for ill, and insofar as we can guess the technological future or tease it out from the present, we will be able to help turn its impacts in the right direction. The future, if we reach it, can be a good one.

REFERENCE

Harris, B. (1985) 'The Urban Modelling Implications of New Technology', in J. Brotchie, P. Newton, P. Hall and P. Nijkamp (eds) *The Future of Urban Form: The Impact of New Technology*, Croom Helm, London.

PART 5.
THE DIMENSIONS OF CHANGE

Chapter 26

THINKING BEYOND POST-INDUSTRIAL: THE SOCIAL PERSPECTIVES

R. Meier

The purpose of this exercise is to portray a stable state of society toward which the world is moving. It is primarily a developed society that provides basic needs for all but a few who fall through the safety nets that support unfortunate individuals. Pressing behind, trying to reach this stable state, is a much more populous developing world. The developed world is influenced by the institutions and traditions that we know today, so the themes and pictographs that communicate about the expected future are images selected from the present—they derive from various blends of socialism and capitalism.

Currently-developing regions, known as the Third World, are able to apprehend the evolutionary stages of the societies preceding them, but their capacity to imitate is greatly restricted by resource shortages, both internal and global in scope. The stable state that is open to these societies will be based on social inventions of this generation, and perhaps the next, that are adopted, disseminated and institutionalised thereafter. Thus the social structures toward which the bulk of the world is moving are more inchoate and indescribable, but even there some leading trends can be detected (Hawley, 1981).

No standard, accepted format has come into being for communicating seriously about futures. This chapter draws on clusters of appropriate images that can be fitted into a quantitative framework, so that the degree of truth can be confirmed by subsequent observations and accounting. It begins with a few paragraphs on the concepts employed. How do they generate features of future social organisation? What are the key definitions, devices and frameworks? Thereafter the discussion rises to a more general level of abstraction. It appears possible to characterise large-scale social orders available to the generation now being born. Implications for long-range planning follow from these observations, so I can reach some worthwhile conclusions.

PRELIMINARIES

The term 'post-industrial' was a catchy phrase when it was introduced. It was picked up quickly by intellectuals and spread beyond America very soon thereafter. An immunity was left behind, because the term developed no content. It lacks image (Bell, 1974). No set of agreed-upon expectations for the future is offered, and that is precisely what we would like to draw upon when deploying a time-dependent metaphor. As the future marches into the past we would like to distinguish what is surprising from what is to be expected.

This is not quite the same thing as forecasting. There may be two or three mutually-exclusive outcomes that can be envisioned quite clearly, but only one of them can possibly come to pass. Whichever happens, we are not surprised, but if something else happens . . . ! Expectations are the shapes and forms of possible futures.

Each of these outcomes may have a value and, simultaneously, a likelihood of coming into being. Both value and likelihood are based on historical sequences of events. An expected value, the decision theorists say, is calculated to be the discounted future value multiplied by the likelihood. Decision between investments, for example, will depend heavily upon these appraisals of expected value.

Whom would I trust to produce a decent set of expectations? It would depend upon the methods that person employed in reaching them. The rigor in the method is much more dependent on commonsensical soft science than pure logic. It is based on a hypothetical game where the other person senses the points that could be made by adversaries and builds a plausible response. The image of the future so generated has embodied in it a recognisable pattern of relationships. It would include location, scale, range of participants' activities, timing and recognisable qualities such as colour, style, pace, etc. Thus the statements in the outcome remain plausible after surviving a set of severe tests.

What does such a person do in order to produce an acceptable collection of expectations? Initially I see him bemused by a sharp trend, probably constructive, that is happening to or within a population. It might also be affecting a community or its environment. This is the phenomenon that has gained his interest. To understand it better he would pick a prototypical example or a site, or a small sample from a large, partially characterised class. He would then try to distinguish it from its context, perturb it, design for it, and otherwise discover what transitions seem to be forbidden and which are common.

Then there are also questions about the immediate neighbours to be answered. What influences do they have? Are there competitors, predators, dependants or catalytic agents? What transactions are involved? Are the important responses delayed? How will all this affect the aggregation of situations similar to the example? What happens when converted into the dimensions of large-scale phenomena?

All this sounds as if I expect the person to be a competent field investigator, but if he is to get dependable expectations, even more is needed. That person should be doing this repeatedly, with other new trends, so he is obliged to be a generalist, or a member of a generalising profession like geography, management, design, planning, etc. He is able to write a persuasive scenario that shows how trends interact. He will also achieve a fair sense of the 'betting odds' between alternative outcomes. By this time he has reached a frontier; perhaps no other person has surveyed the cluster of phenomena that he has. This represents a challenge. Is there a trend of trends, a meta trend, that can be deduced? (See, for example, the prescient forecast by Webber, 1968, produced about the time post-industrial was conceived).

THE RANGE OF THIS ENQUIRY

What factors will be the most important instigators of visible, experienced change? When historians review this period of entry into the twenty-first century they will most likely find that the pre-industrial forces, held over from many generations back, remained the most significant triggers for change. For example, a pre-industrial Israel booted Arab Palestinians out onto the world; it modernised its own society while creating peripheral instabilities. The chaos in Lebanon has already spread to Iraq, Iran and Kuwait, and it seems very likely to inflame the frictions between Sunni and Shia throughout Islam.

The impulse exerted by many new university graduates in their espousal of Koranic law and institutions without knowing, or even exploring, the implications is an important component in the eruptions. Accumulated, cartel-derived wealth, amounting to hundreds of billions of dollars, provides huge stocks of weapons for the forthcoming chaotic conflicts, converting them into spreading firestorms. Millions more refugees will diffuse away from Islamic societies—more numerous by far than those who fled fascism and communism four to five decades earlier and enriched the Americas with the science and culture they brought with them as mental baggage.

Diplomats, reporters and scholarly specialists sense the multiplication of tragedies that will almost certainly envelop most of Africa and much of Asia, but they can offer no strategies for defusing the ideological dynamite. As a consequence, the future of Europe, still heavily dependent on Middle East petroleum, is vulnerable to major deprivations, and energy price changes will again shock Third World countries.

Once holy wars are mentioned, we can do very little to integrate them into the many-stranded, technology-driven images of the future created by scores of individual attempts to understand where civilisation is going. We will address here the islands of stability that show promise of surviving into the period when urbanisation has reached its climax, and as many people are running away to some attractive niche in the hinterland as are coming to the city to seek their fortunes (Table 26.1).

The completion of the build-up of urban population is expected about three generations hence, if the trends emerging after World War II are allowed to work themselves out. We will take up the more novel transformations, such as are envisaged in the insights introduced earlier. We can also explore some of the more modest transitions that lie between these extremes.

Labels will be assigned to the patterns of stability that are identified. This naming procedure enables them to be discussed, sometimes with greater specificity than can ever exist. It also gives the impression of offering a choice in lifestyle. Our children and grandchildren can vote with their feet and their participation, determining the milieu within which they prefer to circulate, thereby contributing to the further evolution of those communities.

THE STRUCTURAL COMPONENTS OF THE CITY OF THE FUTURE

The Fundamentalists

The largest single body of people in the cities of the future are those who started from the bottom in the countryside. They arrive in the urban slums, barely literate at best, then seek stable niches in the ghettos and the working-class districts. Their children frequently rebelled against the folkways of the parents, and a few of them burst out to make spectacular careers for themselves, but the predominant majority found work in the industries and in commercial services. The original urban settlers and their children were buffeted by strikes, plant shutdowns, illness, crime and family dissolution, but they and their communities consistently invested in their children to an extent not paralleled in the countryside.

The third generation in the city, often endowed with ten to fourteen years of education, frequently elected to join reborn fundamentalism. The

Table 26.1: Sites That Should be Important When Flows Between City and Countryside Approach Steady State

Pace-setting Agglomerations Forming the Ecumenopolis

Tokyo	Rome
New York	Sydney
Mexico City	Osaka/Kobe
Seoul	Toronto
Sao Paulo	Houston
Beijing	Frankfurt
Shanghai	Amsterdam/Rotterdam
Los Angeles	Zurich
Paris	Geneva
London	Stockholm
Delhi	Singapore
Bombay	Tel Aviv/Jerusalem
Washington/Baltimore	Boston
San Francisco	Vancouver
Hong Kong/Canton	Seattle/Tacoma

Mainstream Urban Development Supporting Added Participation

Buenos Aires	Barcelona
Rio de Janeiro	Bangalore
Bangkok	Melbourne
Caracas	Vienna
Pusan (Korea)	Taipei
Bogota	Madrid
Madras	Santiago
Brussels	Athens
Lisbon	Colombo

Struggling to Follow

Jakarta	Lima
Kathmandu	Addis Ababa
Manila	Abidjan

Cities Surviving Islamic Turmoil and Adapting to Surroundings

Cairo	Casablanca
Karachi	Beirut
Istanbul	Tunis
Teheran	Damascus
Alexandria	Lahore
Ankara	Kuala Lumpur
Baghdad	Tashkent

Parallel Evolution in Various Socialist Societies

Moscow	Warsaw
Leningrad	Berlin
Prague	Budapest
Havana	Ho Chi Minh City
Tianjin	Wuhan
Chungking	Mukden
Fushun	Qingdao
Chengdu	Kinming

Highly Problematic Futures

Calcutta	Lagos
Johannesburg	Dakha
Nairobi	Kinshasa
Kabul	Rangoon

phenomena have been much studied for Protestant, Catholic and Jewish households, but an even larger number of Hindu, Buddhist and Confucian families are now undergoing these kinds of succession phenomena in outlook on life.

The third and succeeding generations are the salt of the earth—warmly human, often stubbornly so, and relatively unambitious. They prefer to live in the periphery of the inner city, and they often take over inner-ring suburbs that were settled two to three generations earlier. Except for the few that served in the military—and disliked it—they know no other physical environment than that of the urban community, and their language is the argot that has evolved around them but is not spoken by the professionals. As factories become automated, or move to the hinterland, these people become service and transport workers for the city—police, firemen, clerks in public offices, small businessmen, private-sector office workers and building maintenance staff. They fulfil religious obligations zealously at times and backslide quite often. They are captivated by fads, symbols in the popular culture and slightly differentiated brands in the marketplace, but strongly resist basic changes in their environment or in urban society.

The social networks of the fundamentalists are moderate in size and often involve multiple mutual relationships. For instance, they may be co-workers, members of the same parish, neighbours, alumni of the same high school. These bonds tie them down so completely that they refuse to move when the principal industry dies or trade languishes. Fundamentalists are budged only by catastrophes.

Buildings may be reconstructed, families get smaller, diets may be commercialised and the sources of livelihood shift, but the essential threads of life do not change for this segment of the urban population. If national political forces impose a revolution, the fundamentalists make up a large part of the inertia that restores most of the relationships, when the revolution wanes, to what they were in the earlier urban environment.

Ecumenopolitan Bourgeoisie

While the fundamentalists have retained and revived the old myths, selected and modified in practice to fit the local circumstances, another urban population bubbles up new enthusiasms, creates many added services, promotes exchange, fosters knowledge accumulation and re-organises increasing numbers of urban institutions. This stratum introduces new myth to replace the declining strength of nationalist and religious orthodoxies.

These people are world-servers, as distinguished from nation-servers that predominate in public service today and for several generations into the future, or the community-servers among the reborn fundamentalists. The messages the world-servers put out bounce around the urban centres as new ideas, new technologies, shortcuts in administration, interesting styles, novel images and works of popular art, such as books and films.

They are called ecumenopolitan bourgeoisie because their intellectual commitments are not just to the society that grew up within the walls of an old city, but to people like themselves who are scattered throughout the ecumenopolis—viewed by Doxiadis, the enterprising Greek architect-planner who reached his zenith in the 1960s and early 1970s, as the necessary city of the Future (Doxiadis and Papaioannou, 1974). He foresaw parts of metropolises knit together by satellite communications, jet aircraft, multinational firms, global voluntary organisations, United Nations' protocols, world science and a mutual respect for a biosphere that remains poorly understood. These very noticeable developments induce the urbanisation of a population that takes advantage of the new connectivity. For the mid 1980s I estimate that at least one million adults belong to the ecumenopolitan stratum; several times as many are in the educational stream with ambitions to join them. I find that this kind of urban settler is most heavily concentrated in headquarters staffs of the Boston, New York and Washington megalopolis, but there are strong concentrations in London, Paris, the Ruhr, Zurich, Geneva, Rome, San Francisco and Hong Kong.

In what part of the city do they live? Where do they hang out? They increasingly tend to occupy condominiums with access to downtown and airport; they cluster around universities and in suburbs with the reputation of providing good schools at the secondary level; they invade the better-built slums with a wave of gentrification. Obviously, the reciprocal is not true, since only a minor fraction of the condo-dwellers will be ecumenopolitan, and the same may be said of the gentrifiers. I speak of residential compatibility only.

The intellectual homes of these people are to be found in research institutes, modern offices, studios and international assemblages. They transact as readily with counterparts overseas as in the cities in which they were born and educated.

It was the rise of this segment of urban population to positions of influence that created the Common Market and made it function despite the pressures from nationalists. Only a few tens of thousands of them overcame the chauvinistic and ethnic animosities that previously made war seem inevitable, but that small number was still enough to make an important difference.

Nowadays, parallel attempts are continuously being made to pierce the Iron Curtain. Progress is demonstrated by the exchange statistics, so someday they may succeed. As the Second World of socialist nations becomes increasingly dependent on world trade and modern science, the ideologies of nationalism and communism would no longer be in the driver's seat (Pipes, 1984). If present trends continue, several cities on the east coast of China will attempt to belong to both worlds so as to modernise the whole of China more rapidly.

After World War II the Japanese produced a considerable number of professionals with the competences needed to connect East with West, but in large part they were assigned to overseas market development. The initial mission was that of company-serving, but many were swept into the global integration of textiles, autos, electronics, chemicals, steel, shipbuilding, finance, etc. Many thousands of Indians, Koreans, Taiwanese and

Iranians were graduated from North American universities with motivations that are virtually indistinguishable from their classmates, but their command of Asian languages and their entrepreneurship generates a backflow of ecumenopolitan commitments to Asia.

In the arts, the strongest world-serving commitment is found in the cinema and closely associated television features. Music synthesised without the aid of musical instruments has hardly begun, although recorded popular music, such as rock'n'roll and reggae, undermines all prohibitions at national boundaries aimed at repressing the flow across national borders. The black market in popular culture is capable of elevating a performing group to global stature. Their themes are chosen for general appeal to youth, regardless of culture. In the field of literature, decades are required for translations to diffuse major works, therefore growth rates are slow. However, international journalism is generating an increasing flow of inexpensive paperbacks that explore popular themes with thrillers, psychotherapy and cuisine which have gone a long way toward global integration already.

The 'Dark Proletariat'

Population explosions currently under way in Africa and other tribal parts of the world will have widespread repercussion for well over a century. Epidemic control has vastly improved, as has famine relief, but the birth rate (even among the literate) has declined very little from the seven to eight births per woman during the reproductive years. This fecundity results in an extraordinarily young population; if birth control is adopted, the numbers must be expected to expand significantly for a generation afterward. The present population doubling rate is seventeen to eighteen years and at the moment no significant indicators can be identified by the experts as to where fertility reduction will become common or how fast the practice will spread through the society.

In the early twenty-first century about 80 per cent of the urbanised tribal population will have been born and raised in the countryside. Almost all of the remainder will have grown up in the urban slums. Migrants continue to arrive at ages 17-25 in search of work. Employment in manufacturing and construction seems to be preferred, but the jobs will be too few, so the informal sector of the services will be oversaturated. The habits, customs and attitudes of recently-urbanised immigrants remain basically rural, but street smartness is added as a gloss. The circumstances of life they experience in the cities will be similar to those of the lumpenproletariat in the years of Malthus and early Karl Marx. Many will be rootless, flowing from city to city according to rumours of opportunity, living by hustling and panhandling. Meanwhile the ruling classes in these societies remain the lineages of the tribal leadership who get formal education in èlite schools joined by the missionary-educated and the early-urbanised households. They are expected to be able to function as bureaucrats in societies that become as densely settled as the Asian.

Added to the ethnic and class divisions are the racial differences. The people slowest to urbanise are very often black or brown. Moreover they will be five to ten times as numerous as at present. Although the nation-serving bourgeoisie tend to develop a conscientious minority that fights racial prejudice, this splinter faction is, so far, rarely dominant in China, Japan, Arab, or even African countries. Thus we see the prototypical setting for race riot episodes that propagate as an epidemic of social disorder. The media can amplify the effects if not kept under control. Therefore the channels of communication cannot be free, censorship may allow a fair press, but that is only from the point of view of those in power.

The apparent inevitability of communal violence makes one feel very depressed. This is a twist to the Malthusian crisis that has not been widely discussed. Future influxes of poorly-educated immigrants and refugees assure bad feeling and turbulence in the cities for generations to come. Europe and America have found already that they are not completely immune. Misunderstandings that once caused petty wars will appear to be urban disorders in the future.

Much of the future strategy in African and Asian city planning will be devoted to finding short-term resolutions for these internal ethnic conflicts. Sometimes the policy of separation will be used; at other times carefully managed integration will be attempted. The long-term strategy requires education beyond ten years per capita, and acculturation in institutions like the military, athletics, churches, drama, musical performance, etc.

Impacts on North American cities will be far from negligible, particularly along the boundaries, because the numbers of refugees and the opportunities for illegal entry are larger than ever before. Latin Americans are entering each year by the millions, with a major fraction staying for the season and then returning to their homelands. The Vietnamese 'boat people' and other Asians arrive by the hundreds of thousands, but exceedingly few ever return. The United Kingdom slammed the door on immigrants and refugees after its government closed out the leftovers of the British Empire and concentrated on home affairs. The waning of nationalistic fervour—referred to earlier as an important feature of the rise of the Ecumenopolis—renders boundaries more permeable to the flow of peoples.

The typical West European Gastarbeiter wants to hold on to his urban job, resisting pressures to return. Rural Italian, Spanish and Greek workers have been largely superseded by Turks and Arabs. Even larger numbers were attracted to the Middle East from all countries in South and East Asia, more than 90 per cent villagers, so that considerable sums of money have flowed into the back country. A result is an increased demand for land and an increase in price. In my recent contacts throughout the region I can detect no constructive groundswell, only a postponement of crises.

The Chinese remain a big question mark. The same Chinese who are building intellectual bridges between East and West recognise that China has a history that goes in cycles, but does not return to the old state. As the present policy gains momentum at least several millions of them will be anxious to join the ecumenopolitans. However, even if they find a way to emigrate, they often find themselves in districts of the metropolis occupied by the forerunners of the dark proletariat. Despite prejudice, they have used education and business to become highly successful in South-East Asia and North America. When the Chinese start moving up in the world, many metropolitan projections will be invalidated by the consequences.

The Patriots: Nation-Servers All the Way

Here we see a dying breed that merges into the fundamentalists mentioned earlier. Nation-serving has a quick resurgence among a minority when the fatherland (or mother country) is believed to be under attack. Leaders gain instant support when they uphold the honour or security of the nation should it be challenged, but it evaporates within months unless a still greater challenge must be faced. Patriotic sacrifices are less and less common as nations grow older, but are frequent in the newest States.

At the moment the urban patriots constitute an intervening stratum represented by middle-level to upper-level bureaucrats and political leaders and their followers. They are required to behave like patriots when in public. The national interest legitimises their actions. Quite a few contractors to the government must play similar roles. Many are virtuous in this behaviour, since it appears to be untainted by personal interest, but an increasing number use their influence to enhance the status of communities to which they belong. Some endeavour to empower their social networks, while not a few are intent upon personal prestige or wealth, and are thus regarded as hypocrites.

Yet all are nation-servers when at work. They accept hierarchical discipline in the name of the nation, and their careers are linked to its continued sovereignty. The world still has tens of millions of them. The total is probably not yet declining because new nations are still recruiting for additional posts.

As international obligations are assumed and education becomes professionalised, urban residents in public and private bureaucracies play roles in the Ecumenopolis. Lower levels tend to be absorbed in the ranks of the fundamentalists. Therefore, although the patriots play an historic role in nation-building, the increasing permeability of national boundaries and the coincident rise of the Ecumenopolis render it vestigial. We pay our compliments to the true patriots and move on to identify some relatively new actors in the city.

This structure envisioned for urban society is based on social roles. For the same period social psychologists have been viewing value orientations in the population that respond to role changing. They employed measurements made in the United Kingdom—the most mature of the modern societies—over the last twenty-five years as a basis for projection into the future (McNulty, 1985). Their subdivisions labeled 'survivors' and 'aimless' belong mostly to the proletariat, while 'belongers' and 'conspicuous consumers', accompanied by some 'social resisters', are mostly fundamentalists. Their 'experimentalists' and 'self-explorers' have social values associated with the ranks of the Ecumenopolitans.

Other Niches: Specialists, Vehicles, Automata, Robots

Coastal waters constitute a major new environment for urbanisation. The Arctic environment is theoretically possible, but so much less desirable it does not seem to be significant. The desert regions have greater attractions, because saline water can be desalinated with the aid of local energy resources, but sites away from the sea are not at all promising economically.

During the 1960s studies were made which surveyed the implications of a nuclear-reactor design that could desalinate large quantities of sea water while producing enough electric power for associated industries and settlements (Oak Ridge National Laboratories, 1968; Meier, 1969). The most economic design for producing thermal energy could be placed on coastal deserts; it would mix its own manufactured fertiliser with the distilled water to produce a series of crops that required the least water per unit of output (wheat, maize, beans, potatoes, etc.).

Calculations suggested that the typical nuclear-powered agro-industrial complex (or nuplex) might need a quarter of a million people to produce enough staple foods for 5-8 million people in the urban area receiving the trans-shipped foodstuffs (Meier, 1974). Although North African nuplexes would be economic the earliest, Australia has the largest resource. It could feed the whole coastal area of China when food becomes truly scarce in the world. These foodstuffs are not as highly regarded by Chinese urbanites as rice, but they are acceptable.

As food production volumes in all parts of the world advance, the growers will resort to sprinkler irrigation, and then to drip irrigation, thereby obtaining increasing yields per unit of land, sunlight, water and labour. Simultaneously, the processes become capital intensive. Such food producers will live as if they were city dwellers and their jobs will resemble factory work, even though it is a post-industrial society. They would be complementary to the growers of perishable vegetables and minor fruits in the interstices of the city itself. Food growers, processors, distributors and retailers are then likely to make up 30 per cent or more of the urban workforce. They are most likely to fall within the categories of fundamentalists and members of the proletariat. More importantly, however, we see new species and cultivars of flora becoming common in the cities of the future—a kind of edible landscaping. The fauna are likely to be enriched by species of fish occupying ponds in urban open spaces (Meier, 1984).

After food is assured, cities are most dependent upon imports of fuel for transport. How will the vehicles be operated after the inexorable decline of the production of petroleum has begun about one generation hence? The standard assumption has been that the vehicular population will be adapted to conserve fuel. Once the limits for conservation are approached, it is expected that the populations of autos, airplanes and powered boats would decline drastically. Because the price of liquid fuel would then be high, the casualties among the vehicles, it should be possible to exploit tar sands and shales. Diesel-powered vehicles will become more prevalent; they displace the high octane-requiring sub-species. Transport energised by electricity will continue to be promoted, but probably not used to any important degree. Most speculations about the future of urban environments have overlooked a dark horse in the supply picture—methanol. For personal vehicles and for gas turbines methanol is about 70 per cent as energetic as hydrocarbon fuels. Methanol is also cleaner burning. For the next five decades the principal source of methanol would be low-priced natural gas from places such as Arabia, Mexico and Siberia.

A major long-term source of methanol is biomass. By-product biomass is the cheapest, but the supply is insufficient to meet the anticipated demand. Tree plantations, particularly of the fast-growing, nitrogen-fixing varieties, can add capacity sufficient for millions of vehicles. After that, algal plantations, drawing equally efficiently upon sunlight but costing more to process, could eventually support millions more vehicles than at present. A century or more hence, the city dwellers will have almost completed the switch from fossil fuels to many varieties of solar and some nuclear sources. Capital and labour requirements will almost certainly, however, be much higher.

Once all the conveniently-located fossil fuels have been exploited, the competition between humans, animals and vehicles for the biomass that can be produced will greatly intensify. We must expect then that the domesticated animal population consuming grain and other plant products that people could eat will be greatly suppressed. In the long run most humans are expected to consume only a little bit of meat, grass-fed or garbage-fed, and fish that is plankton-fed, with the remainder being foodstuffs processed from plants. Then their total requirements come to about a million kilogram calories per year apiece. A hectare of good soil or pond can produce about five times this amount. The average auto with an urban address and an efficient engine will consume the output from about a half a hectare if we similarly increase the efficiency of conversion of biomass into liquid fuel.

Thus the frugal personal auto is expected to require as much land surface for its nutrition as two or three frugal humans. This happens to be the size of the average nuclear-family household when population growth has been reduced to equilibrium. Thus graduating to car ownership in the Third World means doubling the demand made upon the land under circumstances where there is already a scarcity.

In sum, people's demand for private transport and for meat simultaneously have energy implications which force societies to look beyond dry land. They will need to consider ways to harvest the surfaces of protected bays and estuaries, and eventually to move out onto the quieter parts of the open oceans. The invasion of offshore areas started two decades ago with the search for petroleum deposits, and the technique for living and working on the water is becoming steadily more sophisticated.

A large share of the ecumenopolitans may not take the trouble to operate their own vehicles. They could depend instead on a personal-communications system that would reserve space in taxis, buses, trains, ferries and aircraft. The fundamentalists, on the other hand, place a high priority upon the availability of a private vehicle, because obligations to family and the personal network require moving about conveniently in the community. People living on small craft today say they would not feel secure in the absence of powered service vessels.

The growing post-industrial proletariat will aspire to graduate from their bicycles to something quicker and showier, such as mopeds, motorcycles and cars. The bond between people and vehicles is frequently so interwoven with other values that, during times of impoverishment, it is one of the last budgeted items to be relinquished. During normal times we should expect to see 15 per cent of human effort in cities dedicated to the maintenance and replacement of automotives, and in some places even more.

Within the cities themselves the vehicles will require for circulation and parking an area equivalent to the floorspace assigned to housing people. The competition between humans and vehicles for living space will be so strong that one expects many special solutions to the automobility problem. Other space-related values, such as for recreation, will be blended into the allocation, depending on the time of day. Asian cities have already begun to question single-purpose land use when they transform parking lots into night-time, open-air restaurants, and ferries into hostels for street people.

Finally we come to the most problematic population of all—the automata. As well as the phenomenon can be measured, the automata seem to be multiplying in number at the rate of 30-40 per cent per year. They are also shrinking in size, so there is rarely an impression of crowding. The species extinction rate is extraordinary, since it takes only two to five years before an item is displaced by a more competitive breed. The current best example of an automaton is a published piece of software, but as many hand-tooled special programs exist as there are customers using a mainframe. Robots, which comprise another category of automata, are less of a problem, because they are automata doing physical work. They take on jobs that people prefer to avoid, because the work is dirty, dangerous, or too demanding, or because an extraordinary degree of error prevention is involved.

About the turn of the century, at the present rate of development of capabilities, a new stage will be reached—computers will then be able to respond adequately to simple spoken commands. The transition may well be troublesome, because an automaton is less accepting of ambiguity than a horse or a dog. We also expect it to be virtually error proof. Our science fiction has glossed over the variety of difficulties inherent in the man-automaton relationships, just as artificial intelligence enthusiasts underestimated the problems of natural language translation and production.

Nevertheless, a human generation hence, we should expect that the population of automata will exceed the populations of humans in our cities, and keep on expanding, perhaps at a reduced rate. Very likely a new breed of people will evolve to specialise in intimate relationships with the more competent automata. The hackers of today are the pre-speech primitives of that anticipated population. This is an as yet unnamed, and not yet emerged, component of the urbanites. They would be using information systems for both work and fun, but many other new institutions for patterning urban life will also be involved. The automata and the ecumenopolitans are inherently symbiotic, but the new breed will be specialists who are virtually bionic.

These perspectives lead us to a vaguely-perceived image of the social structure of the dominant, most populous agglomerations—the Ecumenopolis and the mainstream in its wake. It is arrived at by extrapolating to a future when the following conditions hold:

- net migration to cities has been reduced to a trickle;
- dependence upon fossil fuels has been reduced to a trickle;
- demographic transitions have been almost universally achieved;
- levels of consumption approach adequacy for all but a few problem communities; and
- a global catastrophe (cosmic collision, nuclear winter, greenhouse effect climatic change, transition to Ice Age, etc.) has not yet occurred.

In Figure 26.1 we see a hierarchy of social status and function. The automata, some of which are robots, at the base. This job-serving slave population will be far more numerous than humans, living in constant risk of displacement. A fair proportion of the humans will also be job-servers living in constant risk of displacement. However, the bulk will be family-servers and community-servers operating within traditional belief systems. Organisation-serving occupations range from microregional, through national, to global scales. The knowledge-manipulating professions appear to be predominantly world-serving. At the top are unique specialists, whose intellectual commitments are to the discovery and dissemination of universal truths.

Figure 26.1 Social Structure Approaching Steady State.

```
U
A
U                                       Universe-serving
AU
UA
BAU                                       World-serving
BAAB
BBAAB
BBAABB                                       Organization-serving
BBBAABBB
BBFAAAAFBB
BBFFAAAAFFBB                                       Nation-serving
BFFFFAAAAAFFFFB
FFFFFFAAAAAAFFFFFF
FFFFFPPAAAAAAPPFFFFF                                       Region-serving
FFFFPPPPAAAAAAAPPPPFFFF
FFFPPPPPPAAAAAAAPPPPPPFFF
FPPPPPPPPPAAAAAAAAPPPPPPPPPF                                       Community-serving
PPPPPPPPPPAAAAAAAAAPPPPPPPPPPP
PPPPPPPPPPAAAAAAAAAAAAAPPPPPPPPPPPP                                       Family-serving
PPPPPAAAAAAAAAAAAAAAAAAAAAAAAAAAAAAPPPPP
PPAAAAAAAAAAAAAAAAAAAAAAAAAAAAAAAAAAAAAPP
AAAAAAAAAAAAAAAAAAAAAAAAAAAAAAAAAAAAAAAAAAA                                       Job-serving
AAAAAAAAAAAAAAAAAAAAAAAAAAAAAAAAAAAAAAAAAAAAAA
```

U = Unique Specialist   B = Bourgeois Ecumenopolitan

F = Fundamentalist   P = Proletarian

A = Automata & Robots

It is interesting to note that all human levels will be closely associated with automata, so that the latter may be assigned a social status based upon the work that they do. This surprise-free version of the future differs from all the scenarios reported in the periodical *Futures* by Dennis Livingston, its reviewer of science fiction.

FURTHER EXPECTATIONS

When looking beyond post-industrial, one must seek a limiting factor that inhibits explosive growth in organisations, automata and accumulating knowledge. Such a factor does not seem to exist in communications channel capacity or in the structure of matter. I expect that the ultimate constraint in man-automation relations—which includes the incentives for producing more automata—will come in the perception and ordering of visual images. As yet we do not have an equivalent of an alphabet for efficient retrieval of imaged knowledge. A few scientific, artistic and technical explorations have touched on this problem, but I have not yet seen a satisfactory formulation.

In the course of the search for a robust, sustainable settlement system we should expect that life in the cities will become infinitely richer and more diverse for all who are interested. It may well become too rich for a significant fraction, and instances of burn-out will be experienced. At present such people retreat to the countryside, causing a reversal in the decline of many small towns and villages. In the future they may also retreat to aquatic communities.

REFERENCES

Bell, D. (1974) 'Controversy—Is There a Post-Industrial Society?', *Society*, 11, 10-2.

Doxiadis, C. A. and Papaioannou, J. (1974) *Ecumenopolis—the Necessary City of the Future*, Norton, New York.

Fakiolas, T. (1985) 'The USSR in the International Division of Labor up to the End of the 20th Century', *Futures*, 17, 331-47.

Hall, P. (1984) *The World Cities*, Wiedenfeld and Nicolson, London (3rd edn).

Hawley, A. H. (1981) *Urban Society: An Ecological Approach*, Wiley, New York (2nd edn).

McNulty, W. K. (1985) 'UK Social Change Through a Wide-Angle Lens', *Futures*, 17, 375-84.

Meier, R. L. (1969) 'The Social Impact of Nuplex', *Bulletin of the Atomic Scientists*, 25 March, 16-21.

——— (1974) *Planning for an Urban World: Design of Resource-Conserving Cities*, MIT Press, Cambridge, Mass.

——— (1984) 'Energy and Habitat: Designing a Sustainable Ecosystem', *Futures*, 16, 351-71.

Oak Ridge National Laboratories (1968) *Nuclear Energy Centers: Industrial and Agro-industrial Complexes*, ORNL-4290, Oak Ridge, Tennessee.
Pipes, R. (1984) *Survival is Not Enough: Soviet Realities and America's Future*, Simon and Schuster, New York.
Webber, M. M. (1968) 'Planning in an Environment of Change: Beyond the Industrial Age', *Town Planning Review*, 39, 177-95, 277-95.

Commentary on Table 26.1

Various procedures for projection into the future suggest that 80-90 per cent of the world population will reside in urban settlements—the largest of them approaching 1 million population—when flows to and from urban areas become roughly equal. This condition is anticipated in the latter part of the twenty-first century.

Participation in the Ecumenopolis is expected to include almost all of the cities mentioned above to varying degrees. It already exists in the developed metropolises, but others, such as those in East China, Delhi and Bombay in India, and Sao Paulo and Mexico City in Latin America, are pushing ahead ambitiously to join them.

A disparate set of other metropolises follow in their wake. They are less able to mobilise resources, but they draw on the experience of their predecessors by maintaining open societies. Finally we can identify a small number of problematic urban settlements which may be overwhelmed by Malthusian pressures while dropping further into the rear.

Turbulence arising from unsuccessful attempts to create advanced societies based upon the Koran is expected to postpone development. Most Islamic cities are expected to go through episodes similar to those experienced by Beirut and Teheran in the 1980s.

The evolution of large socialist cities lags behind the mainstream, but is forced to move in parallel in order to compete economically and ideologically. A recent systematic projection of trends shows these cities slowing to a near halt by the end of the century, so they will fall even further behind the pacemakers unless a successful overhaul is achieved (Fakiolas, 1985). The Chinese strand is progressing more rapidly at present, but is far from catching up with the prior achievements of Soviet cities, while the Vietnamese seem to be stuck in the muck, far, far behind. Finally one finds a residuum of cities for which progress seems blocked, and it is not easy to propose how the hurdles will be overcome:

Chapter 27

URBAN ISSUES IN AN ADVANCED INDUSTRIAL SOCIETY

G. Gappert

As we approach the year 2000, more and more of our planning time will be spent on prognostications about life in the twenty-first century. The usefulness of these speculations will be related to how well they help us deal with our sense of uncertainty about the future. In the 1960s the key words were certainty and planning while in the 1970s they were uncertainty and management. But in the 1980s the key words appear to be surprises and adaptation.

Planners, especially urban planners, do not like uncertainty and they hate surprises. Therefore, the psychological adjustment of urban planners to the uncertain and inchoate realities of the next decade and beyond must be an item of special concern. In this chapter I attempt to describe a pattern of socio-economic development in order to provide at least a tentative framework for understanding the recent past and the immediate future. Cities do not change quickly but evolve into their future out of an historical context. The suspicion is that in an advanced industrial society, cities (and urban development) will become effectively urbane.

THE POST-SOMETHING DILEMMA

Many futurists and other social observers believe that Western society is post-something but disagree on post-what. For the last twenty years or so, Daniel Bell (1983) has befuddled us with his notion of a post-industrial society. This metaphor has become extremely pervasive as a framework for analysing the conditions and problems of contemporary society. Others feel that the future is technologically determined and inevitable. Alvin Toffler (1981) proposes a society of electronic cottages and Marilyn Ferguson (1980) reveals the 'Aquarian Conspiracy.' To these observers the apparent social disarray is just the prelude to a new stage of higher consciousness and cultural achievement. Still other observers believe that the future, especially the American future, is out of control. No one is in charge, there is no conspiracy and even the emerging economic anarchy seems poorly organised. All these concerns, images and metaphors represent attempts to bring some kind of conceptual unity to the myriad bits of the future emerging around us in contemporary society.

The framework of a post-affluent society developed in the mid-1970s (Gappert, 1979) was one attempt to create a model which could be used to assess the myriad trends and events associated with the fundamental dislocation of the American economy caused by OPEC seizing the international resource initiative which accelerated the start of a new international economic order. The assumption of the post-affluent framework is that the economic events of 1971-74 represented a significant turning point and developmental shift in our society. These events—price controls, the oil boycott, the dramatic increase in energy costs—triggered the start of a post-affluent transition. This transition was preceded by the beginnings of a post-industrial shift out of traditional manufacturing employment. The significant beginning of this post-industrial shift can be symbolised by the Soviet launching of Sputnik in 1957. The American response to that event led to the *National Defense Education Act*, and to the acceleration of space and defence programs out of which both new technologies and new sunbelt cities have developed.

The post-industrial framework, however, has lost conceptual vigour as the American economy has evolved towards a further stage of industrial development. The usefulness of the post-affluent framework is also declining as a reconstruction of the American economy commences. But we are still in the midst of a transitional phase. The processes of what Schumpeter once referred to as 'creative destruction' are still rampant. But a restructured economy is emerging.

A NEW PROSPERITY PARADIGM

Research from several sources (Knight, 1979; Bradshaw, 1978; Hamrin, 1980; Hall and Markusen, 1985) can be used to identify an emerging but elusive picture of the economic future of the United States. This is the idea of an advanced industrial society. There is more than a faint hint that we might be experiencing the start of another long wave of economic development based on the new information, communications, medical and biological technologies.

Before he died, the controversial futurist, Herman Kahn, had begun to herald a forthcoming economic boom. Heilbroner and other more seasoned economic observers, while sanguine about the future of capitalism, have begun to identify the massive institutional efforts that will be required to achieve a new, lasting and stable economic prosperity. These efforts are symbolised on the Right by the attempt to secure a Balanced Budget Constitutional amendment and on the Left by discussions about credit allocations, wage-price control mechanisms and some kind of new Reconstruction Finance Corporation. In the Centre is the new legislation introduced by Gary Hart and others for an American Defense Education Bill.

The emergence of a new and more dynamic national economic prosperity has been obscured by the normal short-sighted economic debates having to do with quarterly corporate earnings, the annual federal budget and the monthly behavior of the Dow-Jones Index, interest rates, the rate of inflation and unemployment statistics.

These short-sighted concerns have deflected attention from other significant national economic developments which include the tremendous expansion of employment in America as the national labourforce expanded with the influx of the baby-boom generation and women of all ages, the substantial shortfall in technically-trained American labour which has been met by a substantial influx of educated immigrants or young student immigrants flooding our technical and engineering educational institutions, and the expansion of a global manufacturing economic system which will ensure that large numbers of unskilled and semi-skilled manufacturing jobs will be perhaps forever lost to the American worker.

In *Post-Affluent America* (Gappert, 1979) it was noted that the US labourforce was expanding at a rate of 68,000 jobs a week. Unfortunately, the labourforce was expanding even faster, at a rate of 72,000 workers a week, creating a job shortfall of 4000 jobs. Therefore, we have had a steady increase in the unemployment rate. It might also be remarked that in mid 1982, total employment in the so-called service industries surpassed for the first time total employment in manufacturing, mining and construction (*New York Times*: 6 July 1982)—and service activities over all sectors grow to 72 per cent of the total workforce (Feketekuty and Hauser, 1985). It is important to note, however, that much of that service sector activity provides technical support (legal, financial, engineering, etc.) to manufacturing enterprises. The evidence then is not yet conclusive that a new wave of American prosperity will in fact develop or will be strong enough to provide 'a rising tide that will float all boats.' In this particular context we do not care to debate the probability of its occurrence. Instead we wish to outline the elements of a new social-economic paradigm which might be associated with the emergence of an advanced industrial society if it in fact does develop as the elusive economic recovery of the mid 1980s leads us through a major transformation of our socio-economic technostructure and the achievement of a new prosperity.

THE ADVANCED INDUSTRIAL ALTERNATIVE

A model of an advanced industrial society can be developed as an alternative framework for the interpretation and projection of emerging trends and realities in American society. As shown in Figure 27.1, there are at least three different ways to perceive the social-economic future. The first perspective, System X, assumes that issues of economic productivity and efficiency will lead to both fiercer competition and greater disappointment unless the vitality of the national economy dramatically improves. In this perspective we can expect more outbreaks of social hostility and underclass turmoil.

Figure 27.1 Perspective on Societal Futures.

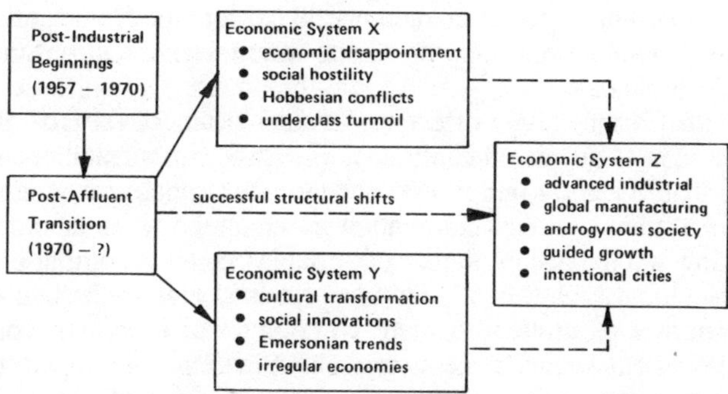

The second perspective, System Y, assumes that although the growth rate may decline, the underlying strength of the national economy is undiminished even though individual gains take longer to achieve. In this perspective individual and household pursuit of purely economic objectives are postponed, deflected or sublimated by activities associated with quality-of-life concerns. Cities and suburbs both contribute to the growth of self-reliance and a self-directing transformation.

Substantial elements of both System X and System Y exist today as part of a constrained economy. The third perspective, that of System Z, is more elusive and harder to establish. The key assumption of this perspective is that a new kind of technological-industrial progress is possible, and probable. It assumes that a second Industrial Revolution is at hand based upon technologies that are knowledge intensive, resource conserving and environmentally benign. Several specialists have made contributions to this perspective. Hamrin (1980) lays out in a comprehensive fashion the elements of an economy centred on information activity. He claims that 'The adjective "information" is more precise than "service" or "post-industrial" for describing the fundamental dynamic element that will shape the economy and society of the future.'

A critical element in Hamrin's analysis is that the ultimate limits to growth in an advanced industrial society 'will basically reflect not resource scarcities but a steadily rising preference for noneconomic endeavors and satisfactions.' Developing what he calls a 'transindustrial world view', Hamrin also proposes a new paradigm for economic growth in which productivity and technological change are important factors along with capital, labour (human resources) and land (natural resources and energy). The guiding norm or ethic within his model of an advanced industrial society is 'selective growth', whereby institutions need to make more holistic decisions in the context of long-range policy frameworks. Although he is pessimistic about the prospects for an immediate adoption of such a new paradigm, his use of a return-on-resources decision-making model for corporations confirms the need to explore different forms of innovative management more appropriate to the changing realities in an advanced industrial society.

Hirschhorn (forthcoming) offers yet another perspective. His reference point is the idea of advanced capitalism as distinct from traditional forms. He claims that 'There is increasing evidence that capitalism is entering a new historical phase of disaccumulation in which the wage labour system is contracting and the accumulation of variable capital is coming slowly to a halt.' Coincident with this, a new work system, new productive sources and new sources of productivity are emerging which are not consistent with traditional categories of economic classification. The problem which Hirschhorn then goes on to explore is the ways in which older social forms and institutional structures may stifle the developmental processes emerging in an advanced form of capitalistic or industrial society. Hirschhorn attempts to develop a model of the advanced industrial-labour process based on seven related hypotheses. A key issue for him is the distribution of intelligence and information in work systems and the ways in which advanced technologies diffuse and change the management functions.

Knight (1979), analysing employment expansion in several older metropolitan regions, has also developed a perspective on industrial transition and transformation. Knight's model is based on the assumption that the large industrial corporations have reached a stage where their technology-transfer functions have expanded their multinational status in a global economy which has grown by a factor of three in the last decade while world trade has increased by a factor of seven. This has both generated new streams of real income into older regions and the export of traditional production jobs to other states and nations. Two constructs are essential to Knight's analysis. The first focuses on the changing nature of industrial knowledge to include such elements as research and development, environmental constraints, international marketing and transfer of technologies, etc. The second is the changing nature of social and cultural development as industrial communities either decline or evolve into a higher stage of development.

Other efforts to formulate the social and economic consequences of a new wave of technological development can be found in Ouchi (1981), Mensch (1982) and Masuda (1981). All suggest a new economic system based on knowledge-intensive technologies.

In the absence of any policy which advocates, or even just articulates, the nature of such a different socio-economic system organised around new technological opportunities, the question is: Will such a system emerge anyhow? One answer is: Let's wait and see. Another is: It's happening already. But in either case even the mere possibility of an advanced industrial economy and society offers a more dramatic set of opportunities to education than the grim sense of decline associated with the post-industrial construct. If one accepts the realities of the current post-affluent condition, but still appreciates the dynamic qualities of American society, its economic strengths and its technological opportunities, then it becomes possible to concentrate on the identification and elaboration of new roles and opportunities in the new order of things. But the transition is not yet concluded and perhaps the transformation has not yet begun. Therefore it is still difficult to forecast the exact nature of our social and economic development between now and the twenty-first century. But Theory Z offers a significant starting point.

THE URBAN CONSEQUENCES OF A THEORY Z PERSPECTIVE

At one level Theory Z is what Naisbitt (1982) refers to as high tech- high touch: '. . . Whenever new technology is introduced into society, there must be counterbalancing human response, that is, high touch, or the technology is rejected.' Naisbitt goes on to assert that 'We must learn to balance the material wonders of technology with the spiritual demands of human nature.' He also uses this particular formulation to conclude that the high-tech isolation of the so-called electronic cottage will send people back to the office after they try working at home.

Ouchi (1981), in his formulation, has suggested that 'A Theory Z culture has a distinct set of such values, among them long-term employment, trust and close personal relationships . . . Of all its values, commitment of a Z culture to its people—its workers—is the most important.' Ouchi's development of a Theory Z framework grew out of his work with a number of corporations, based primarily in California (a System Z State?) where Japanese management practices had been adapted to the American context. The resulting synthesis was dubbed Type Z in an intentional reference to the Theory X and Theory Y management styles which were originally characterised by Douglas McGregor in his seminal study *The Human Side of Enterprise*.

McGregor argued that much which is important about a manager can be understood by knowing that manager's underlying assumptions about human nature and society. McGregor felt that these assumptions were primarily of two kinds, which he labelled Theory X and Theory Y assumptions. A Theory Y manager assumes that people are fundamentally hard-working, responsible, and need only to be supported and encouraged. With the advent of the so-called knowledge worker, the simple Theory X and Theory Y distinctions are no longer adequate and a new Theory Z is also motivated by the need to understand 'how the structure of society and the management of organization can be co-ordinated.' There is a perception that the growing dichotomy between workstyle and lifestyle creates stress and social dysfunction (alcoholism, drug misuse, spouse abuse, workplace vandalism, white-collar crime, etc.).

Theory Z, a Z culture, System Z, are all emerging because of a growing cultural perception that a more humanistic integration of new forms of workplace needs with new levels of personal and psychological fulfilment will be required in the twenty-first century. But, in contrast with the System Y perspective, a Theory Z perspective is more cognizant of economic concerns and global competitive realities.

The Elements of System Z

A System Z perspective—which suggests the possibilities and the prospects—of a new and higher stage of industrial development are represented in Figure 27.2. These elements of an advanced industrial society—of a new kind of social economic system—suggest that growth and development could be directed or guided towards some definite social and economic goals at different levels of our society over some period of time. These elements can also be shown to flow out of the elements of the post-affluent transition as outlined earlier. Some of the manifestations of social and economic change likely to occur in the next two decades are suggested in Figure 27.2. For the next decade or so, a young adult (under 40) population will occupy our cities, with growing demands for recurrent education and training. A new national economic expansion driven by information-intensive technologies and the extension of global business services should be underway by the early 1990s. New biotechnologies will dramatically increase the American capacity to export high-tech agriculture products and new medical technologies will increase the prospects for extending the span of life. The marketplace reorientation to conservation, self-reliance, amenities and quality of life issues will be further along.

Theory Z planning will displace strategic planning as the dominant

Figure 27.2 The Dynamic Emergence of a SYSTEM Z.

Post Affluent Transition		Advanced Industrial Condition
1. Dominance of Baby Boom population	→	1. Higher productivity from the Young Adult learning force
2. Post-Affluent consciousness	→	2. Acceptance of new knowledge-intensive resource conserving technologies
3. Recognition of the transcendental nature of many needs and wants	→	3. Reorientation of consumer behaviour towards frugality and self-reliance
4. Tinkering with household and workplace reforms	→	4. Dramatic reorganization of workplace with Theory Z concerns and Theory Z planning
5. Emergence of new synergistic lifestyles to reflect androgynous values	→	5. Widespread application of new multi-dimensional personality theories which appreciate the needs of the 'self'
6. Discontent over income distribution	→	6. New mechanisms of redistribution including the expansion of the 'underground' economy
7. Confusion in large organizations corporations and governments	→	7. New forms of System Z policy and management

mode as the mid-career crunch and the mid-life crisis of the baby boom population assume epidemic proportions in the early 1990s. At the same time the expanding needs of the self will increase the utilisation of new personality theories such as that provided by Ogilvy (1979) in *Many Dimensional Man*. Furthermore, the neighborhood garage sale, flea market and clothing exchange will become significant components of the recycling and underground economies. Also significantly, new political leaders will emerge with new policy innovations as the baby-boom population increase their political participation in the elections of the late 1980s and early 1990s. But other changes are also in sight.

Cities will be more polynucleated, with the development of more multiple-use megastructures, and medium-density planned housing unit developments. Large tracts of sub-standard housing are likely to be replaced by new towns-in-town while older, solid housing stock will continue to experience gentrification. Urban farming, wood lots and fish ponds will not be unusual. Residential, work and leisure activities will be more integrated in medium and low-density developments.

Individual housing units will find their design influenced by the conversion of family rooms into work-study centres and by the use of passive solar techniques. The distinction between work and leisure will begin to break down for several kinds of workers. At the same time the growth of involuntary leisure-time among the elderly, the young and the semi-skilled will require new social inventions for its constructive utilisation. The more efficient and effective use of urban space, both internal and external, private and public, will follow some of the innovative design arrangements in European cities. The post-modernist, neo-romantic school of urban architecture is likely to flourish as the members of several generations of urban citizens become more urbane consumers.

Indeed, the post-modernist school of architecture specifically and post-modernism in general reflect interesting manifestations of how North American culture is responding to the post-industrial, post-affluent conditions of our times. The Alcan building in downtown Montreal is a striking Canadian example of this new type of urban design that combines economic frugality with social effectiveness, a sense of play and pleasure. The Benjamin Franklin museum in Philadelphia is an interesting combination of both a post-modernist respect for the archaeological value of the site and the striking use of a high-tech style. It is likely that as the advanced industry society of the twenty-first century unfolds, post-modernism will flourish and may even lead to a neo-romantic style or a more exaggerated style of advanced romanticism. Environments which are dominated by, and dependent upon, advanced technologies will cry out for a deliberate humanistic and natural quality.

And so will our workplaces, our households and certainly our cities. It may well be that post-modernist architecture, and not high-tech *per se*, is going to be what emerges as strikingly different about cities in the twenty-first century. But some other issues and questions also need to be addressed. For instance, it might be that the electronic townhouse rather than the electronic cottage better represents a manifestation of the integration of workplace space into the American single-family home.

In an open competition in 1984, the Minneapolis College of Art and Design asked for house designs that would be more appropriate for the contemporary American family, with emphasis on three ideas: the dramatic number of households not made up of the the nuclear family; the rising cost of housing; and the desire or need to work at home. The instructions called for six attached units with accommodations for a car and a home office or work space in each unit, on a specific site in a middle-ring, working-class neighborhood in Minneapolis.

The modified townhouses that won first place were proposed by Troy West, an architect, and Jacqueline Leavih, an urban planner. Behind individual front gardens in their plan is a row of six small, one-storey office structures, each of which links through a corridor kitchen to the three-storey main part of the house which is behind a second private courtyard garden adjacent to the kitchen corridor. This main house opens into 'a community area that is back street, driveway, carport, basketball court, yard and entry—an active neighborhood area' (*New York Times*, Joseph

Giovannini, 13 June 1985). These electronic townhouses combine home-like elements such as nooks and bays and fireplaces and trellises with an appreciation that active adults will have little time for housework and little need for large entertainment spaces. Other design solutions that focus on the workplace-household problem (or is it an opportunity?) are likely to emerge as many cities or urban regions continue to accommodate residential intensification and other medium-density, in-fill solutions to the high energy and travel costs associated with the spread-city manifestation.

Another emerging urban design phenomenon that is likely to persist and extend itself in the twenty-first century is the use of multi-block, multi-building pedestrian walkways that include the street level arcade, the second-storey skywalks and the downtown underground mall (such as Place Ville Marie in Montreal). These new enclosures that explode the scale and quality of interior space for pedestrians, consumers and white-collar workers present a new centralising focus in cities that offsets the decentralising tendency of the electronic townhouse and electronic cottage.

Increasing leisure and use of telecommunications will facilitate increasing low-density developments in which residential, work, leisure activities and perhaps gardening, local food production, solar energy utilisation and other local life-support activities are integrated, and increased emphasis on lifestyle and quality will ensure an increasing range and diversity of these developments within, at the periphery and beyond the urban area. The new affluence (for some) created by new technology will further add to this diversity and to the range of spatial development activities including global networks and virtual (global) cities.

CONCLUDING NOTE

The cities of an advanced industrial society—the future metropoli—will be primarily engaged in indirect, and partially abstract, transactional activities, and may be hungry for collective rites to offset social fluidity, economic transience and electronic isolation. They may also be oriented to both material and non-material standards and satisfactions which are integrated at a community and regional scale. In a Western industrial society, the city has rarely been an acceptable symbol of a collective consumption ethic or of an integrative cultural style which links both pauper and prince. Instead the industrial city has been more of a symbol of the individual struggle for survival or of private success or the competition for control of public resources.

But if we are beginning to develop into some kind of knowledge-driven economy, what is the role of the great urban and metropolitan centres in a Theory Z society? What is our commitment to an urbane civilisation which can support networks of social relationships in an atmosphere of pluralistic tolerance, humane subsistence and, dare we say it, technological progress? Let us agree to attempt to create and design living and working environments in which the human spirit will soar and the quest for renewed wisdom will flourish.

REFERENCES

Bell, D. (1983) *The Coming of Post-Industrial Society*, Basic Books, New York.
Bradshaw, T. K. (1978) *California As a Post-Industrial Society*, Institute of Governmental Studies, University of California, Berkeley.
Ferguson, M. (1980) *The Aquarian Conspiracy, Personal and Social Transformation in the 1980s*, J. P. Tarcher, Los Angeles.
Feketekuty, G. and Hauser, K. (1985) 'Information Technology and Trade in Services', *Economic Impact*, 4, 22-8.
Gappert, G. (1979) *Post Affluent America*, Franklin Watts, New York.
Hall, P. and Markusen, A. (eds) (1985) *Silicon Landscapes*, Allen and Unwin, Boston.
Hamrin, R. (1980) *Managing Growth in the 1980s*, Praeger, New York.
Hirschhorn, L. (forthcoming) *Beyond Mechanization: Flexibility and the Theory of Post-Industrial Technology*, MIT Press, Cambridge, Mass.
Johnson, W. G. (1979) *Muddling Toward Frugality*, Shambhalla, Boulder.
Knight, R. V. (1979) *The Region's Economy: Transition to What?*, Cleveland State University, Ohio.
Lifton, R. J. (1975) *The Life of the Self*, Simon and Schuster, New York.
McGregor, D. (1960) *The Human Side of Enterprise*, McGraw-Hill, New York.
Masuda, Y. (1981) *The Information Society as Post Industrial Society*, World Future Society, Washington, DC.
Mensch, G. (1982) *Stalemate in Technology*, Ballinger, Cambridge, Mass.
Naisbitt, J. (1982) *Megatrends*, Warner Books, New York.
Ogilvy, J. (1979) *Many Dimensional Man, Decentralizing Self, Society and the Sacred*, Harper Colophon, New York.
Ouchi, W. G. (1981) *Theory Z*, Avon, New York.
Toffler, A. (1981) *The Third Wave*, Bantam Books, New York.

Chapter 28

THE TRANSITION TO AN INFORMATION SOCIETY

J. Brotchie, P. Hall and P. Newton

The industrialised world is entering a new era of development: a transition from an industrial economy to a post-industrial or information economy. One feature of this new era is greater interdependence between communication, transport and production. A key factor in this trend of interdependence has been the rapid development of information technology including the microcomputer and telecommunications networks and their use in combination. This has resulted in an enormous increase in the rate of transmission and processing of information. This increase in speed has been compared with that in other related industries. The motor vehicle represented an increase in speed of transport of approximately ten-fold (and an increase of similar order in vehicle size and capacity) in comparison with horse drawn transport. Steam, and later electric power, produced an increase of perhaps 100-fold in the speed of processing goods. The impacts of these changes on our cities and industries and on our lives have been substantial. Electronic information technology on the other hand represents at least a million-fold increase in the speed of processing and transmission of information (and a large increase also in capacity and scope) compared with that previously available, and the effects are already proving to be profound.

This technological change is permeating the production of goods and services, increasing productivity, reducing costs, facilitating diversity and opening up new and global markets for these products, particularly for new, information-based service industries (information-based industries already employ a substantial part of the workforce in the developed world and are continuing to expand; Lamberton, 1985). This new technology is, in fact, beginning to affect all levels of human endeavour, producing both quantitative and qualitative change. Some of these changes are now considered.

INTEGRATION OF COMMUNICATIONS, TRANSPORT AND PRODUCTION

The integration of information technology (communications and processing), transport and production is resulting in increasing interdependence between industries. Industries in a region are being linked into integrated industrial systems. Information technology allows as never before the separation of information from the process of production and from the movement of goods and the use of that information for optimal control of production and distribution. This facilitates the concepts of flexible manufacturing systems, small production runs, economies of scope rather than economies of scale, and even production on demand. It also allows reduction of inventories and storage facilities, and the concept of just-in-time. For example, windscreen wipers and other components made in one city can be organised to arrive just-in-time for assembly in cars in another city, eliminating costly inventory and storage, but requiring reliable high-quality communication, transport and production systems. In the tertiary sector, design and planning processes, management functions, transaction and delivery of services may be similarly co-ordinated.

INFORMATION AND KNOWLEDGE AS FACTORS OF PRODUCTION

Taking a long-term view, the information-technology revolution will produce changes to our patterns of working and living just as profound as did the Industrial Revolution and the Agricultural Revolution before it. The Industrial Revolution was an extension of human muscle, the information revolution is an extension of our minds and senses; its effect is certain to be far-reaching. The Industrial Revolution represented a shift from land to materials and energy as the prime factors of production. The information-technology revolution represents a further shift to information and knowledge as prime factors in production. This shift is reflected in more efficient and innovative design of and introduction of new technologies into vehicles, goods, services, and facilities, with consequent reductions in material and energy use per unit of performance. It allows increasing use of micro-electronics in the various processes involved in new service industries, and in existing secondary industries—for ordering, selecting materials, controlling production, routing-distribution, and overall management, resource allocation, planning and design.

This transition is only just beginning. It is being facilitated by the continually increasing cost-effectiveness of information technology, and the knowledge being fed into its utilisation. This knowledge, up to now, has been concerned with automation of physical tasks and simple, predefined mental (information-processing) tasks. We are at the threshold of a new phase in the information revolution, however, in which other intellectual tasks utilising expert knowledge and simple reasoning and learning are also beginning to be automated. Artificial intelligence will facilitate a qualitative as well as quantitative increase in goods and services produced and provide comparative advantage to the industries and countries which make the first and most effective use of these fifth generation computing tools (Fiegenbaum and McCorduck, 1984). Japan, the US and some European countries are competing in the race to take commercial advantage of this second phase with their fifth-generation computing projects (see Gordon and Kimball, this volume).

INTELLIGENT MACHINES AND SPATIAL DEVELOPMENT

Intelligent machines have been under development for more than two decades (see also Deken and Meier, this volume). They have tended to retain that status partly because of an initial underestimation of the complexity of the problem and partly because of the continual redefinition of what constitutes intelligence—as being those processes still beyond electronic imitation. However, computers have already come a long way towards incorporating certain aspects of intelligence. They can observe and recognise patterns, optimise the organisation of a system for given goals and simulate the reasoning of experts in specialised problem areas such as fault diagnosis, analysis, and engineering design.

The application of artificial intelligence to the design and planning process is also being studied by simulating the more creative processes of human thought. One advanced discovery machine involves the automated association of ideas from different areas to provide new innovations in spatial organisation of human activities or in planning policies (the ideator; Dickey, 1985). Another explores the relationships between inputs and outputs of a system—such as that between the stimuli for urban development and the activities resulting—by evaluating levels of support for potential hypotheses, using simple reasoning and heuristic search (e.g. the Boltzman Machine, Hinton, 1985; Hastings and Waner, 1986). Related optimisation techniques (Sharpe and Marksjo, 1985; Brotchie *et al.*, 1986) also facilitate the efficient arrangement of an urban system. Still more expert advisory systems are being developed for computer-aided planning and design (Marksjo, 1986); for example, to electronically monitor, diagnose faults and advise corrective measures for urban infrastructure systems such as water supply.

These models of intelligence are presently crude and slow but continual and discrete improvements may be expected. The expected consequences of their further development are greater quality and diversity of design and construction of living, working and leisure facilities in human settlements and greater freedom of location with increasing automation of their various service systems, and increasing quality of communications among them, including the increased ability to transmit presence, as outlined by Deken in this volume.

Thus this second phase of the information revolution will provide a range of increasingly intelligent machines and services, from household appliances to motor vehicles, computer workstations, health, education and leisure-oriented systems, and construction, extraction, production equipment, to planning and design machines. The consequences will be increased quality and diversity of spatial development and lifestyle in a wide range of desirable environments for those with a capacity (financial and intellectual) to take advantage of this revolution but, during this development stage at least, also an increased gap between this group and those to whom, for one reason or another, the benefits of this technology have not spread.

Such machines will eventually provide a new population, as observed by Meier (this volume), capable of inhabiting those parts of the earth and beyond unsuited to a satisfying human existence—the sea-bed, the inland deserts, atomic-energy plants, mines and sewers, and other less attractive or hazardous environments, including outer space. They can potentially extend human activity, physical and mental, to these locations, and beyond normal human working hours and life spans. They would also possess the capabilities to construct production plants in these environments. They will increase the quantity, quality and diversity of products and processes—goods and services—and provide comparative advantage to those nations and organisations that possess them and develop them furthest and fastest (Fiegenbaum and McCorduck, 1984). Because they facilitate qualitative improvement, including reliability, and better matching of human needs without necessarily increasing cost, the competitive advantage they provide would be decisive.

INCREASING INFORMALITY

The Industrial Revolution represented a peak in formality of organisation of work and living patterns in both space and time. Further technological developments such as the motor car, distributed electricity, the telephone and the computer have resulted in increasing informality of these working and living patterns. This trend is continuing with formal employment systems already partly replaced by a range of employment and self-employment or subcontracting arrangements including part-time employment, job sharing, flexitime, telecommuting and informal-sector activities (see also Pym, this volume). With the move to cleaner, information-based industries, there is less need for formal zonings and working hours which confine this industrial development. A further and related trend is that of personalised goods and services or do-it-yourself, with the private car replacing public transport, self-service increasing in retailing activities such as supermarkets, petrol stations and banks, and do-it-yourself in home appliances, and in home maintenance and decorating. Further trends include the increasing shift from manufacturing employment to service employment, counterbalanced in part by the substitution of manufactured goods for service activities (e.g. the washing machine, TV, video, home computer).

Increasing female participation in the workforce—itself a result in part of the shift from a muscle-based to an intelligence-based economy—increasing ageing of the population, increasing leisure time and leisure activities, increasing importance of amenity and increasing reduction of institutional care in favour of home care, are all changing our demands for goods and services, transport, and infrastructure, and for housing types and locations (AIRG-CSIRO, 1980). The increasing trend towards informality of employment, automation and self-service is also spreading production of and demands for goods and services, including transport, more evenly throughout the day (and night).

Electronic communications technology is itself a medium conducive to both increasing volume and informality of communication: the telephone, the facsimile machine, the telex and television all facilitate informality. The video recorder and computer add flexibility of timing also.

GLOBAL NETWORKS, GLOBAL, REGIONAL, NATIONAL NODES

Electronic networks are spreading globally, incorporating satellite systems, fibre optics, package switching, microprocessors and various terminal facilities. These global networks are enabling the operations of transnational companies, allowing information to be centralised for key decisions in head offices in New York, London and Tokyo and transmitted to branch plants around the world for further decision making and for dispersed production and distribution (a one-to-many communication network, see Figure 28.1). Thus they facilitate the process of concentration of capital into increasingly large transnational organisations.

These networks are also providing new global channels and markets for the export of services and for the development of new information-based service industries. They enable dispersal of various transactions and interactions (a many-to-many network). They also facilitate the TV broadcast of international events from one or multiple centres to all parts of the world.

Figure 28.1 Dispersal of Locations of Production Versus Dispersal of Interactions (Trips or Communications) with Dispersed Consumers. Trends for Corporate Decision Making and for Production in Information-based and Energy-based Industries at Urban, Regional and Global Levels.

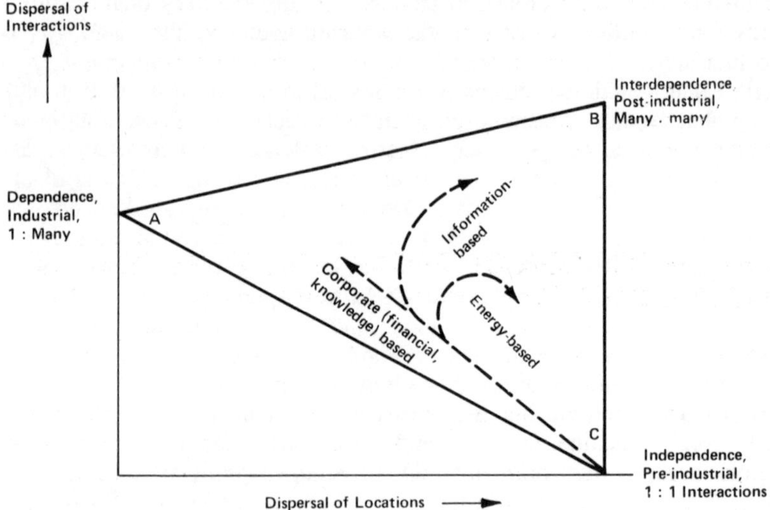

These factors are tending to promote the largest and best intraconnected and interconnected centres such as New York, London and Tokyo—distributed at roughly equal intervals around the northern hemisphere—as global information centres providing 24-hour service in stock and currency trading; San Francisco, Hong Kong, Singapore and Sydney tend to be developing as world-scale regional information centres around the Pacific Rim, the world's fastest-growing economy, with cities such as Sydney and Melbourne as national centres (Langdale, this volume). These centres are being reinforced for global and regional transactions involving both electronic and face-to-face communication and transactions based on trust. Other back-office functions such as credit-card operations, insurance, other services and retail activities are moving to suburban sub-centres within these major information-exchanging metropolitan areas, while production is tending to move to peripheral regions or beyond, over recent years to newly industrialising countries. Older industrial cities, unable to develop a niche in the new global system of information exchange, may be the chief losers.

TIME CONTINUOUS GLOBAL MARKETS

The Chicago Mercantile Exchange and the Singapore Stock Exchange now have an electronic hookup that will permit global trading around the clock (Feketekuty and Hauser, 1985). Market centres for other information-based transactions such as securities, finance or commodities are expected to be similarly linked to facilitate 24-hour trading (sequentially through centres in different time zones). Linking these regional centres to form a single global market produces virtual global centres, and may be the first stage of more distributed global marketing arrangements. New York, London and Tokyo seem to be emerging as the top-level centres in this global exchange, partly because of their geographical position, though strictly speaking, American west coast centres would more logically perform that role.

CONCENTRATION OF DECISIONS, DISPERSAL OF PRODUCTION: THE TRANSNATIONALS

Information technology in the form of both telecommunications and the computer is facilitating the global organisation of production of goods and services. Transnational organisations with headquarters in centres such as New York or London, Tokyo or Denver, and other offices in various regions around the world, are able to organise production on a global basis. A car may be styled in Italy, engineered in the USA and assembled in Spain from parts produced in several other countries. Non-routine functions, such as R & D and key decision-making, seek locations with skilled personnel, high amenity and high quality international transport and communications, whereas routine production seeks locations with low factor costs (Malecki, 1985). As automation displaces blue-collar labour as a key factor of production, low factor costs will tend to shift manufacturing and goods handling back to the more developed and more stable areas of the world, widening the global gap between the advanced industrial societies and those which are beginning to industrialise. The resultant international division of labour will then increasingly depend on national trade barriers and regulations, as witness Japan's move into manufacturing in Europe and the United States.

THE TRANSNATIONAL ENTERPRISE vs LOCAL PLANNING

New technology is finding its first, widest and most effective use in the private sector as large transnational enterprises develop or adapt to exploit it. Many of the key decisions affecting local development are made by these transnationals: where to place a regional office, a research laboratory, an engine plant or a computer production plant. Local authorities and even national governments have little power to initiate these decision processes; the powers they do have are generally devoted to competition with other authorities vying to capture development within their own constituency. Inducements typically include provision of low-cost land, services, infrastructure—in technology parks or enterprise zones —with communications and transport access, and waiving of any conflicting planning regulations. Co-ordinated community planning at a national, State or local level is difficult under these circumstances. It is also a time when more concern is given to economic growth than to its equitable distribution.

SOCIETAL TRANSITIONS

It has been argued that we are in transition from an industrial to a post-industrial society. The changes to expect might be indicated by those of the previous transition—from an agricultural to a post-agricultural society. The features of this shift were:

1. production of agricultural goods actually increased over the longer term,
2. the number of producers (workers) declined significantly (in many societies to less than 6 per cent of total workforce),
3. the productivity of workers increased substantially,
4. the per capita consumption of these goods eventually increased but their costs and the proportion of income spent on them decreased,
5. new sectors employed the remainder of the workforce,
6. these new sectors provided the remainder of the goods and services consumed,
7. they also contributed technology which helped improve the productivity of the older sector(s),
8. they provided both manufactured goods for final consumption (consumer goods) and goods that aided in the production of other goods (producer goods).

In the transition to a post-industrial society similar changes are again occurring:

1. production of manufactured goods is generally increasing—in number and diversity,
2. number of producers (workers) is decreasing with automation and

may reduce to 10 per cent or less of the workforce over time, perhaps by the end of the century in advanced industrial countries,

3. productivity per worker is increasing greatly,
4. consumption of manufactured goods per capita is increasing but their costs and their proportion of total consumption expenditure are decreasing,
5. the service sector is expanding (and high-tech industry is developing) to employ the remainder of the working population in the formal economy—if not the remaining potential workforce,
6. the rest of consumption expenditure is on post-industrial products, e.g. services—tertiary, quaternary, quinary—and high-tech manufactured products,
7. these new products contribute to the productivity and the quality of product of the older sectors. They are basic to the newer sectors,
8. services thus form an increasing part of consumption (consumer services) but also contribute to production (producer services).

These two transitions are contrasted further in Table 28.1.

The similarity of behaviour of these two revolutions may result from the fact that the same economic and human behaviour laws apply to each. The forces driving this change on the supply side are industrial-commercial competition for markets and profits and the search for new products and lower factor costs. On the demand side it is the satisfaction of human wants and needs. Previous revolutions met basic human needs—largely physiological, security and some social. The present, information revolution, can potentially meet further, higher-level human needs—other aspects of security, social esteem and self-realisation. Previous revolutions were based on land, materials, energy, manual labour and skill as factors of production. The present revolution is information-knowledge based.

LOSS OF THE STABLE STATE

Each technological revolution represents not only a loss of the stable state, but also a transition towards another stable state based on the new technology and its changed factors of production, inducing further innovation and development, allowing the further satisfaction of some human needs and the emergence of others. During the transition, greater income differentials may be expected as those possessing and developing the new technology receive a greater share of the income than those merely consuming its products, that is, there is a polarisation between the technological haves and have nots—the new producers and consumers—at the international, inter-regional, firm and individual levels. This differential may be expected to eventually reduce as the technology diffuses, its availability

Table 28.1: Industrial Transitions

Transition from Agricultural to Industrial Society	
In the old (agricultural) sector:	*In the new (industrial) sector:*
Employment went down	Employment went up through industrial growth
Production went up—eventually	Production went up (greatly—through mechanisation)
Productivity went up (greatly)	Productivity went up (greatly)
Costs went down—eventually	Costs went down through mechanisation and scale economies
Market share went down	Market share went up
New technology both caused and assisted this change	New technology was basic to this change

Transition from Industrial to Information Society	
In the old (industrial) sector:	*In the new (information-tertiary) sector:*
Employment is down through automation	Employment is up through sector growth
Production is up both in number and diversity of goods produced	Production is also up both in quantity and range of services provided
Productivity is up	Productivity is up
Costs of goods are down through automation and productivity increases	Costs are down through information technology and productivity increases
Market share is down	Market share is up through the range and quantity of new services introduced
Information technology is facilitating productivity, diversity and quality increases, and new products	Information technology is basic to many of the new services being developed

increases and its cost decreases, and a new stable state is approached (although at this stage there are contrary trends; see Batty, Gordon and Kimball, and Newton and O'Connor; this volume). In the meantime, however, the rapid rates of change of technology, income, power and consequent individual human status are sources of turmoil and conflict, leading to real losses in security and esteem for sections of the population (see Encel, this volume).

Automation of production has resulted in deskilling and loss of blue-collar jobs. Increasing service activities based on information technology have tended to increase white-collar and particularly pink-collar jobs as more and more women are employed in the service and information industries. However, even if the total number of jobs does not reduce as a consequence, a problem results from a conjunction of factors: an increasing potential workforce (due to previous baby booms), deskilling of blue-collar jobs and the difficulty of matching redundant blue-collar workers to new service jobs. Together, these mean that structural unemployment or underemployment may continue for some time into the future, perhaps, indeed, through the lifetimes of the previous adult generation, thereby increasing socio-economic inequality within society, between those with white or pink collars on the one side and those with blue collars or no collars on the other. Socio-political responses of income redistribution may reduce these differentials and conflicts but could also reduce the rate of attainment of the stable state. The fact that information and knowledge are cumulative resources will allow the transition to proceed and indeed accelerate.

The unknown factor concerns the future of work. During the first Industrial Revolution, too, it appeared that many displaced farm workers and hand craftsmen would never be absorbed by the industrial economy, yet they—or their children—were. It seemed inconceivable that within a few decades agriculture's share of the workforce would shrink from some 70 per cent to less than 10 per cent, yet it did. Today, also, it seems that many low-skilled workers may never find work in the new information economy. Yet there are service jobs—in personal and community service, in entertainment, in tourism, in recreational and leisure services—that they could do and do well. Currently, the economy cannot absorb them. It may be that this will require an expansion of public or at least community services (although expansion of public services is predicated upon increase in national industrial productive capacity); it may be that new private sector entrepreneurship will create opportunities in these sectors, ministering to the needs and the incomes of the information workers.

SPATIAL IMPACTS

A framework for exploring the broad spatial impacts of technological change is presented in Figure 28.1. The horizontal axis gives the extent of dispersal of locations, for example, of production (consumption is assumed to be uniformly distributed). The vertical axis gives the dispersal of interactions, for example, between producers and consumers or average interaction distance or trip length. A point in this space represents the extent of dispersal of locations (e.g. of industry) and interactions (e.g. average trip length). The triangle ABC represents the practical limits of this dispersal. Point A represents one centralised location (e.g. of production). Line BC indicates completely dispersed locations (of production). Lines AB and AC represent the practical limits of dispersal of the corresponding interactions. AC represents the most efficient limit, AB the least efficient, or random distribution of interactions.

Point C represents dispersed production and localised, one-to-one interactions (e.g. cottage industries, pre-industrial settlement, or the informal economy). Point A represents centralised production and radial distribution to surrounding consumers or one-to-many interactions, typified by the industrial city or industrial centre with its radial transport (and communication) network, or regional, national or global information centres of corporate headquarters and support services. Point B represents dispersed production and consumption but with distance no deterrent to spatial interaction, allowing many-to-many interactions; for example, electronic cottages on an electronic network, one popular image of the post-industrial city.

Thus point C represents pre-industrial activity, no established networks and independence (of producers and consumers). Point A represents the industrial city, or information node, established radial networks and dependence (of consumers on producers). Point B represents post-industrial society, electronics, grid networks and interdependence between producers (and consumers). Point B represents the ultimate dispersal of production of information services—telecommuting, teleshopping, a society based on the dispersed production and consumption of electronic information services, the virtual city or global village or part thereof. The scale may be urban, regional or global.

Point B, however, does not represent the whole of post-industrial society. As noted earlier the transition from an industrial society does not represent a decrease in production of goods, but rather an overall increase (witness the shift from the cut-throat razor to the safety razor to the plastic disposable razor, or the growth of packaging of all kinds). Thus production of goods will continue but not necessarily in the largest cities and certainly not in their inner areas, but at their peripheries or beyond.

Some activities, such as headquarters' functions of large national and transnational companies continue to centralise, as some markets, (e.g. financial, securities, commodities, etc.) based on electronic communications, also continue to centralise (towards A), reinforcing selected global, regional and national cities. The linking of regional centres to form virtual global markets for transactions is a move towards dispersal. Other activities are also free to disperse. Even some activities that are centralised within a firm (e.g. car-rental bookings) may be located outside the central business district (CBD) in smaller centres where reserve capacity in communications is available but rents are low.

Thus many activities, when viewed at an overall metropolitan level or even at a national or global level, are moving towards the point B, with some becoming or remaining centralised near point A where agglomeration economies, face-to-face communications and transactions of trust are involved, or where a new technology is developed and not yet diffused. Thus at the global or regional level, movement from the point A towards the line BC represents extent of diffusion of new innovations and hence also extent of polarisation between producers and consumers of the new technology.

Activities with largely electronic interactions are moving towards the line AB. Production of goods with high knowledge inputs and low material and energy inputs (e.g. micro-electronics, optics, etc.) also occurs near the line AB. Where large material inputs are required for production with large energy inputs for distribution then energy efficient distribution systems are required representing a movement towards the line AC. The location of production along the line AC (i.e. degree of dispersal or concentration) depends on the relative economies of scale of production vs economies of distribution.

Distance above the line AC depends also on level of diversity or differentiation of product between competing firms. The line AC represents no diversity; the line AB indicates extremely high diversity (and extremely low transport costs). Information technology is facilitating diversity of product in both goods and services and hence a movement towards the line AB. It is also facilitating reductions in weight and increased performance of goods, lower costs, more efficient vehicles and distribution systems, hence a further move towards the line AB and particularly towards the point B. Increasing energy costs, on the other hand, can cause a movement towards the line AC and particularly the point C for those industries in which energy is a major factor in distribution. Thus the general movement is towards the line AB and the point B where knowledge and information are the prime factors of production, and towards the point C where high material and energy inputs in production and distribution are involved. Low technology home services and the informal sector are also dispersed towards the point C. The trend towards transnational corporations with corporate headquarters in global cities is a move towards the point A for corporate decision making (Figure 28.1).

Thus the triangle ABC may be used to indicate spatial impacts of technology change on corporate decision making, production and distribution at the metropolitan, national, regional and global levels. At each of these levels, the centralisation of corporate decision making is indicated by a move towards the point A, and the development and diffusion of the new information-based industries is represented by a movement towards the line AB and generally towards the point B. Energy-based industries, however, particularly where transport costs are high and product differentiation is low, will tend to approach or remain nearer the line AC and, where scale economies are small, near the point C. The informal sector is also at point C. However the shift towards increased knowledge and reduced materials as factors of production is a move towards the point B. Hence the trend overall is towards increased dispersal of locations and interactions for production and consumption of goods and services as new technologies diffuse, but increasing centralisation of corporate decision making as transnational organisations emerge.

THE GEOGRAPHY OF THE INFORMATION SOCIETY: THE WORLD OF 2025

We can thus see the main outlines of technological and economic change that will occur, some with reasonable certainty, some less so, in the coming decades. Nineteen eighty-five marked the fortieth anniversary of the end of World War II. The intervening period saw the start of the transition from the manufacturing to the information economy. It corresponded almost precisely to the upswing—and the start of the downswing—phases of the fourth Kondratieff long wave. If Kondratieff's concept continues to govern technological and economic development, the coming forty years will correspond to the depression phase of the fourth Kondratieff and the recovery phase of the fifth.

We would argue that the events of the depression phase are determined by technological developments that have already taken place. There will be increased manufacturing production and productivity, with increasing diversity of goods produced, yet accompanied by reduced factor costs through automation, further deskilling and job loss, and thus by further large increases in unemployment, perhaps exceeding those of the 1930s. It is no longer fanciful to foresee a vast increase in the production of manufactured goods accompanied by a reduction in manufacturing's share of the workforce in the typical advanced industrial country, to 10 per cent or less by the twentieth century's end.

The same period will show corresponding changes in the information-handling sector, again resulting from technological developments already made or now being made. They include increased production of services and increased productivity in their provision, with reduced factor costs, increased diversity and specialisation of these services and increased market sizes as electronic networks spread. There will thus be increased employment in these services and an increased proportion of consumer expenditure on them.

Underlying these manifestations, of course, are deeper trends: a shift from land, materials, energy and labour to knowledge, information and intellect as key factors of production; increased interdependence between manufacturing, commerce and consumption as the new information channels, as well as traditional transportation systems; and their integration into an integrated system, with flexible manufacturing just-in-time and on-demand; and the emergence of a new global hierarchy of information storage and distribution, interlinked to provide continuous operation of financial, securities and commodities markets.

The consequences for the nature of work, and consequently for social structure, are already profound and will prove even more so as these changes work through to their logical conclusion. Knowledge and intelligence will become an increasing element in all work. Manufacturing and service jobs will be increasingly similar in character, as the actual work of production and goods-handling is automated. Traditional distinctions between male and female jobs will erode. So will traditional boundaries between education, work and leisure, as all become more informal in character.

Equally significant will be the shifts in location of economic activity. The general trend will be towards dispersal, including extension of informal work into the home. Work, residential, shopping and leisure facilities will become increasingly integrated on the same site, returning us in part to pre-industrial patterns of living and working.

Most will benefit—not only materially, but psychically—from these trends. Yet for a substantial minority, the transition represented by the fourth Kondratieff depression years will be a painful one. Automation will lead to continued deskilling, and redundancy, in much blue-collar work and some white-collar and pink-collar work, especially at the low-education, low-skill end of the scale. Thus there will be mass unemployment, accompanied, paradoxically, by skill shortages in the new information sectors. New, often highly-specialised, services will be developed and will become available in the home, with user-friendly interchanges that will allow them to be self-operated. Thus they may provide the basis for a new household economy, but—a further paradox—the people least likely to exploit the new opportunities will be those made redundant in the traditional economy.

These paradoxes are likely to persist throughout the closing years of the twentieth century and into the first decade of the next. The really difficult question is what will follow. If the previous 200 years of economic history provide a guide, there will be yet another technological revolution, or set of revolutions, that will create new industries, in turn regenerating the global economy. It is likely, from all that has been said, that for the most part these will be no longer manufacturing industries—as was the case in all previous economic upswings—but service industries. Or, more accurately, they will consist of manufactured hardware and software brought together to provide information services, which in turn will provide the basis for a great range of activities both in production and consumption.

The most probable outcome, we believe, is a vast further increase in availability of very low-cost information. This information is likely to be of an increasingly sophisticated nature, consisting essentially of artificial intelligence: aided by the rapid development of biotechnology, the early twenty-first century is likely at last to see a true extension of the human brain, a bio-computer (Masuda, 1985). Artificial intelligence will be able to offer real supplementation to the human brain, so overcoming some of the problems of inadequate education and inadequate skill that now characterise society. At the same time, the range of computer-aided design and manufacture is likely to increase, while its cost will decline to very low levels. So it is likely that people will be able to design their own custom-made products—from cars to computers to homes—and to have them manufactured, under their own supervision in local workshops. Again, this offers the potential prospect of a vast range of new economic activities, conducted on a small scale and with dispersed locational patterns.

The central question for the economy of 2025 is thus likely to be whether, and how, the job-creating capacities of the new technologies can overcome their capacities for job destruction. In the late 1980s, that balance seems negative. So it seemed in the 1880s, and in the 1930s. The true verdict is not yet in.

REFERENCES

AIRG-CSIRO (1980) *Housing 2000*, Australian Industrial Research Group and CSIRO Division of Building Research, Melbourne.

Brotchie, J. F., Newton, P. W., Hall, P. and Nijkamp, P. (eds) (1985) *The Future of Urban Form: The Impact of New Technology*, Croom Helm, London, and Nichols, New York.

Brotchie, J. F., Georgeff, M., Sharpe, R. and Marksjo, B. S. (1986) 'Introducing Intelligence and Knowledge into CAD', Proceedings, Conference 'Applications of Artificial Intelligence to Engineering Problems', Southampton, April.

Dickey, J. W. (1985) 'Ideation as the Ultimate Technology, Processes and Spatial Impacts', Paper presented at Second International Workshop on Technological Change and Spatial Impacts, Melbourne, August.

Fiegenbaum, E. A. and McCorduck, P. (1984) *The Fifth Generation*, Pan Books, New York.

Feketekuty, G. and Hauser, K. (1985) 'Information Technology and Trade in Services, *Economic Impact*, 4, 22-8.

Hastings, H. M. and Waner, S. (1986) 'Biologically Motivated Machine Intelligence', *SIGART Newsletter*, January, No. 95, 29-31.

Hinton, G. E. (1985) 'Learning in Parallel Networks', *Byte*, April, 265-3.

Lamberton, D. McL. (1985) 'Information Age Economics', in National Information Technology Committee, *Impact of Information Technology 1984*, Canberra Publishing and Printing Co., Canberra.

Malecki, E. (1985) 'North America' in J. Brotchie, P. Newton, P. Hall and P. Nijkamp (eds), *The Future of Urban Form: The Impact of New Technology*, Croom Helm, London, and Nichols, New York.

Marksjo, B. S. and Roy, G. G. (1986) 'Consistency Checking of a Planning Database', Paper presented at the TIMS XXVII Conference, Gold Coast, Queensland, July.

Masuda, Y. (1985) 'Hypothesis on the Genesis of *Homo Intelligens*', *Futures*, 17, 479-94.

Sharpe, R. and Marksjo, B. S. (1985) 'New Information Technology for Planning and Building the Future', Paper presented at the Second International Workshop on Technological Change and Spatial Impacts, Melbourne, August.

EDITORS AND CONTRIBUTORS

John F. Brotchie Chief Research Scientist, Commonwealth Scientific and Industrial Research Organization, Melbourne.

Peter Hall Professor of Geography, University of Reading, and Professor of City and Regional Planning, University of California, Berkeley.

Peter W. Newton Principal Research Scientist, Commonwealth Scientific and Industrial Research Organization, Melbourne.

Michael Batty Professor of Town Planning, University of Wales, Institute of Science and Technology, Cardiff.

Joseph Deken Director, Information Science Program, National Science Foundation, Washington, DC.

Aant Elzinga Professor, Department of Theory of Science, University of Gothenburg, Goteborg.

Sol Encel Professor of Sociology, University of New South Wales, Sydney.

Peter M. J. Fisher Manager, Major Technology Initiatives, Department of Industry, Technology and Resources, Melbourne.

Rolf H. Funck Professor of Economics and Director, Institute of Economic Policy and Economic Research, University of Karlsruhe, Karlsruhe.

Gary Gappert Director, Institute for Futures Studies and Research, University of Akron, Akron.

Ernst Giese Professor of Economic Geography, University of Giessen, Giessen.

Richard Gordon Associate Professor of Politics; Director, Silicon Valley Research Group, University of California, Santa Cruz.

Britton Harris UPS Professor Emeritus of Transportation Planning and Public Policy, University of Pennsylvania, Philadelphia.

Bruce Hutchinson Professor of Civil Engineering, University of Waterloo, Waterloo.

Duncan Ironmonger Deputy Director, Institute of Applied Economic and Social Research, University of Melbourne, Melbourne.

Linda M. Kimball Research Associate, Silicon Valley Research Group, University of California, Santa Cruz.

Jan S. Kowalski Institute of Economic Policy and Economic Research, University of Karlsruhe, Karlsruhe.

John V. Langdale Senior Lecturer, School of Earth Sciences, Macquarie University, Sydney.

Stuart Macdonald Reader, Information Research Unit, Department of Economics, University of Queensland, St Lucia.

Richard L. Meier Professor of Environmental Design, Departments of City and Regional Planning, Architecture and Landscape Architecture, University of California, Berkeley.

Arnoud Mouwen Public Transport Department, City of Amsterdam.

Mitchell L. Moss Associate Professor, Graduate School of Public Administration, New York University, New York.

Peter Nijkamp Professor of Economics, Free University, Amsterdam.

Kevin O'Connor Senior Lecturer, Department of Geography, Monash University, Melbourne.

Denis Pym Professor, London Business School, London.

J. David Roessner Associate Professor, School of Social Sciences, Georgia Institute of Technology, Atlanta.

Jarlath Ronayne Professor of History and Philosophy of Science and Pro-Vice-Chancellor, University of New South Wales, Sydney.

Glen H. Searle Town Planner, Department of Environment and Planning, Sydney.

Michael Taylor Research Fellow, Research School of Pacific Studies, Australian National University, Canberra.

Tsunekazu Toda Lecturer, Department of Transportation Engineering, Kyoto University, Kyoto.

Ken Tucker Dean, David Syme Business School, Chisholm Institute of Technology, Melbourne.

Michael Wegener Senior Research Fellow, Institute of Spatial Planning, University of Dortmund, Dortmund.

INDEX